AF556757

The Plastically Crystalline State

The Plastically Crystalline State

(Orientationally-Disordered Crystals)

Edited by

John N. Sherwood
Personal Professor in Chemistry,
University of Strathclyde,
Glasgow

A Wiley–Interscience Publication

JOHN WILEY & SONS
Chichester · New York · Brisbane · Toronto

Library of Congress Cataloging in Publication Data:

Main entry under title:

The Plastically crystalline state.

'A Wiley–Interscience publication.'
Includes bibliographies.
1. Plastic crystals. I. Sherwood, John Neil.

QD921.P53 548 78–16086
ISBN 0 471 99715 3

Photoset in Malta by Interprint Limited
and printed in Great Britain by The Pitman Press, Bath

Contributing Authors

R. T. BAILEY	*Senior Lecturer in Physical Chemistry, University of Strathclyde, Glasgow G1 1XL, Scotland.*
N. BODEN	*Lecturer in Physical Chemistry, University of Leeds, Leeds LS2 9JT, England.*
W. J. DUNNING	*Reader in Physical Chemistry, University of Bristol, Bristol BS8 1TS, England.*
R. FOURET	*Professor of Physics, Universite des Science et Techniques de Lille, Villeneuve d'Ascq 59650, France.*
A. J. HYDE	*Reader in Physical Chemistry, University of Strathclyde, Glasgow G1 1XL, Scotland.*
A. HULLER	*Institut für Feskorperforschung der Kernforschungsanlage, Julich Gmbh, 5170 Julich, West Germany.*
A. J. LEADBETTER	*Professor of Physical Chemistry, University of Exeter, Exeter EX4 4QD, England.*
R. E. LECHNER	*Institut Laue-Langevin 156X, Centre de Tri, F38042 Grenoble–Cedex, France.*
R. A. PETHRICK	*Senior Lecturer in Physical Chemistry, University of Strathclyde, Glasgow G1 1XL, Scotland.*
W. PRESS	*Institut fur Feskorperforschung der Kernforschungsanlage, Julich Gmbh, 5170 Julich, West Germany.*
J. N. SHERWOOD	*Personal Professor in Chemistry, University of Strathclyde, Glasgow G1 1XL, Scotland.*

Contents

6 Rayleigh and Brillouin Scattering from Plastic and Related Crystals

7 Infrared and Raman Studies of Molecular Motion in Plastic Crystals

Preface

Although theoretical predictions of the existence of such a phase were made in the early nineteen thirties, the detailed study of the *Plastically Crystalline State* commenced in 1938 following Timmermans recognition and classification of a group of organic materials with the general physical characteristics of the proposed mesophase. His classification defined these materials as the high temperature cubic phases of solids comprising molecules of globular shape. He noted that, for the most part, they were formed from 'normal' crystals by one or more phase transformations, each accompanied by a relatively high entropy change. These cubic phases finally melted with a low (for organic solids) entropy of fusion: $<2.5R$. The crystals were noted to be highly plastic, in extreme cases flowing at an observable rate under the influence of gravity. It is as a consequence of this last property that these solids have been commonly referred to as *Plastic Crystals*.

Experimental studies performed following Timmerman's classification led to some widening of this definition but it still remains a satisfactory basic description of the general properties of the *Plastically Crystalline State*. Initially, these experiments were directed towards the examination of the basic structural, thermodynamic and dynamic properties of this mesophase. This work, first reviewed at a conference in Oxford, U.K. in 1959 [published in full in *J. Phys. Chem. Solids*, **18** (1960)], showed these materials to be dynamically disordered crystalline solids. Further work over the twenty years since that meeting has led to the more detailed definition of the nature of these materials. Basic thermodynamic and structural studies have been widely extended. An evaluation has been made of the nature and behaviour of lattice defects in these materials. Closer examination of the diffuse scattering which interfered with early diffraction experiments has led to the definition of the orientational disorder in some solids. Simultaneously, the development of pulsed nmr techniques, supplemented by infrared and Raman spectroscopy and by Rayleigh and Brillouin light scattering has yielded much information on the dynamics of the reorientational processes. The recent addition of neutron scattering spectroscopy to these, currently more routine, examinations promises to yield additional information on the dynamic and geometrical aspects of these processes.

Some of these techniques are well developed in their application, some have yet to make their full impact. Others, such as dielectric and acoustic relaxation studies are in a stage of redevelopment or inception. Even at the present time however the application of these experimental techniques to these materials coupled with theoretical studies has resulted in a significant improvement in our understanding of the nature of this state of matter. Consequently it seemed an appropriate time to review this information. The techniques which have been used are wide ranging and it would be an impossible task for one person to summarize adequately the contribution of each technique to the present state of knowledge. It seemed more appropriate that a series of authors, each well known in his particular field, should contribute a chapter outlining the application of their particular technique and the results derived therefrom. Such an approach leads inevitably to overlap and repetition between chapters. It does however have the advantage that it can result in the clarification of the relationship between techniques. We hope that the result is satisfactory.

Although the *Plastically Crystalline State* is the central theme of this book, it is only by comparison of the properties and behaviour of this phase of matter with those of other orientationally disordered crystals that a true understanding of its nature can be obtained. Consequently, none of the contributors has been constrained to concentrate on the main theme alone. Evidence from related fields and on related materials has been included where appropriate. We hope that this book will serve to stimulate further research in this and related areas.

Glossary of Symbols

A	specific activity, Lennard–Jones coupling parameter, elastic anisotropy factor
A_C, A_T	specific activity of ^{14}C, ^{3}H labelled species
A_e	amplitude of x-ray scattering
$A_e(\mathbf{x})$	scattering amplitude of free electron (Thomson scattering)
$A_o(\mathbf{Q})$	elastic structure factor
a	jump distance, lattice parameter, unit cell dimension
a_o	lattice parameter
a^0	lattice parameter of perfect crystal
$\mathbf{a}$	molecular position vector
α	polarizability, acoustic absorption coefficient
B	rotational constant, body-centred cubic
$\mathbf{B}$	static magnetic field
B_k	Debye–Waller factor
b	effective molar volume, unit cell dimension
$\mathbf{b}$	Burgers vector
β	compressibility
C'	molar heat capacity associated with internal degrees of freedom
C_p	heat capacity at constant pressure of real system

$C_p{}^o$	heat capacity at constant pressure of an ideal crystal
C_v	heat capacity at constant volume
$C(t)$	time correlation function, auto-correlation function
c	concentration, unit cell dimension
c_{ij}	elastic constants
c_v	concentration of vacancies
D	number of orientations, self-diffusion coefficient
D_C, D_T	diffusion coefficient of ^{14}C, ^3H labelled species
D_m	self-diffusion coefficient at the melting point
D_o	pre-exponential factor
D_p	pipe (dislocation, grain-boundary) self-diffusion coefficient
D_R, D_r	rotational diffusion coefficient
D_{obs}	observed diffusion coefficient
D_{NMR}	diffusion coefficient measured by NMR
d	dipole separation
ΔC_p	excess heat capacity ($C_p - C_p{}^o$)
ΔG	free energy change
ΔG_m	free energy at the melting point
ΔH	enthalpy change
ΔH_f	enthalpy of fusion
ΔH_E^*	activation enthalpy for dielectric relaxation
ΔK	kinetic energy distribution factor
ΔS_f	entropy of fusion
ΔS_t	activation enthalpy for dielectric relaxation
ΔV	change in volume on melting, activation volume
$\Delta \nu$	frequency shift
δ	pipe (dislocation, grain boundary) cross-section
E, E_A, E_a	activation energy

E_c	activation energy for creep
E_d	activation energy for self-diffusion
E_f, E_i	energy eigenvalues
E_m	energy for vacancy migration
E_o	incident energy, electronic excitation energy
E_v	energy for vacancy formation
$E^c(o)$	energy of dissociation
E^{α}_{β}	isotope mass factor
ε	strain
ε_o	permittivity
ε^*	complex permittivity
ε'	real part of the complex permittivity
ε''	imaginary part of the complex permittivity
η	viscosity
$\eta(T)$	order parameter
F	face-centred cubic, free energy
F'	structure factor
$F_1(t)$	instantaneous structure factor
$F_o(t)$	correlation function for a free linear molecule
$F(\omega)$	frequency domain spectrum
f	frequency (Hz), correlation factor
f_k	atomic scattering factor
G_d	free energy for diffusion
G_m	free energy for vacancy motion
G_s	self-correlation function
$G(s)$	conditional probability
$G(t)$	time domain spectrum

G_v	free energy for vacancy formation
G'	Debye internal friction constant
g	Kirkwood factor
$g(r)$	positional correlation function
$\mathcal{G}$	damping coefficient internal friction, operator angle
$\mathcal{G}'$	Debye internal friction coefficient
Γ	full width at half height of distribution curve or peak, mean jump frequency
γ	geometrical factor in diffusion, gyromagnetic ratio
H	enthalpy, hamiltonian, magnetic field, hexagonal
$\mathcal{H}$	interaction hamiltonian
$\mathcal{H}_H$	Hartree hamiltonian
h	Planck's constant
$\hbar$	$h/2\pi$
I	spin quantum number
I_o	scattered x-ray intensity
I_l	multipole moment of order l
I_S	internuclear scattering factor
$I(\omega)$	intensity of infrared absorption, frequency dependent intensity
$I(\mathbf{x})$	intensity of scattered radiation
J	interaction energy, rotational quantum number
$J(\omega)$	spectral density function
K	transition probability, bulk modulus
$K_{l\mu}(\Omega)$	cubic harmonic
k	rate constant, Boltzmann constant
k_B	Boltzmann constant
$\mathbf{K}, \mathbf{k}$	wave vector

L, L_o	length of crystal, (perfect crystal)
L_s	latent heat of sublimation
LPR	Landau–Placek ratio
$\mathscr{L}$	collisional operator
Λ	thermal conductance
λ	wavelength
M	monoclinic, molecular weight
M_n	moment
M_2	second moment
m	mass
m_{ol}	effective mass of internal motion
$\mathbf{m}$	nuclear magnetization
μ	shear modulus, dipole moment
N	number of jumps, number of cells
N_1, N_2	number of orientations
n	number of radioactive atoms, refractive index, number of spins
n_v	number of vacancies
ν	frequency
ν_o	incident frequency
ν_{ol}	frequency of internal vibration
O	orthorhombic
Ω	molar volume, polar coordinates
Ω_1,Ω_2	Euler angles
ω	rotational frequency, angular frequency, jump frequency, probability term
ω_1,ω_2	configurations
ω_l	orientation of molecule on site l

ω_R	rotational frequency (angular)
$\omega_{OR}, \omega_{VIB}$	orientational and vibrational line widths
P	pressure
P_{ij}	Pockel's coefficient
P_L	orientational probability
$\mathbf{P}$	momentum coordinate
$\mathbf{\Psi}$	non-bonded intermolecular potentials
$\mathbf{\Psi}_{ij}$	non-bonded inter atom potential
$\mathbf{\Psi}^{\circ}$	vibrational wave function
ϕ	orientation
ϕ_{ab}	potential function between atoms a, b
Q	quantity of radioactivity
$\mathbf{Q}$	scattering vector
$\mathbf{q}$	space coordinate
R	intermolecular distance
$\mathbf{R}$	equilibrium positional vector
$\overline{R^2(t)}, \overline{r^2(t)}$	mean square diffusion distance
$\mathbf{R}_l$	cell vector
r_{ij}	internuclear distance
r_{lk}	equilibrium position of atom *k* in molecule *l* in a unit cell
$\mathbf{r}$	positional vector
ρ	bulk density, depolarization factor, position of close approach
ρ_x	density measured by x-rays, electron density
ρ_w	density of states in frequency domain
$\rho(\Omega)$	nuclear density distribution
S	surface area, scattering function
$S(Q,\omega)$	scattering cross-section

S_v	entropy for vacancy formation
S_m	entropy for vacancy migration
$\mathcal{S}$	Streaming operator
$\mathbf{s},\mathbf{s}_o$	unit vectors
σ	stress, cross-section
$\sigma_{inc}, \sigma_{coh}, \sigma_s$	incoherent, coherent, total cross-section
T	triclinic
T	temperature
T_m	melting temperature
T_t	transition temperature
T_λ	lambda-point transition temperature
T_1	spin-lattice relaxation time
T_2	spin-spin relaxation time
$\mathrm{T}_{1\rho}$	relaxation time in the rotating frame
T_{1d}	dipolar relaxation time
t	time
τ	mean jump time, correlation time, relaxation time
τ_D, τ_{IR}	correlation time determined by dielectric relaxation, infrared
τ_J	correlation time for rotational angular momentum
τ_m	correlation time at the melting point
τ_{OR}, τ_L	orientational, librational correlation time
τ_R	rotational correlation time
θ	angle of observation
θ_i	excited state wave function
θ_j	angle between jumps
θ_m	Debye frequency
$\boldsymbol{\theta}$	Librational vector
$\overline{\theta^2}$	mean square amplitude of libration

U_c	core energy of a dislocation
U_{el}	elastic energy of a dislocation
$\mathbf{u}$	displacement vector
$\overline{u}_i$	mean square amplitude of vibration
u_{lk}	displacement of atom k in molecule l from equilibrium
μ_B	Bohr magneton
μ	dipole moment
V, V_m	bulk molar volume, interaction energy, potential energy, coulomb interaction
V	phonon velocity
V_v	vacancy volume
V_x	volume determined by x-rays
$V^{\dagger}$	activation volume
V^{o}	volume of perfect crystal
V_{long}, V_{trans}	longitudinal and transverse phonon velocities
V_v, V_h	vertically or horizontally polarized spectrum from incident vertically polarized light
V_{ST}, V_{FL}	static, fluctuating part of potential
$V(w)$	crystal field
$V(m)$	molecular field
v	specific volume of liquid, velocity of sound
$\mathbf{v}$	velocity
$\mathbf{X}$	scattering vector
x	distance
ξ	correlation length
$Y(\Omega)$	spherical harmonic
Z, z	number of molecules (atoms) per unit cell

1

The Crystal Structure of some Plastic and Related Crystals

W. J. Dunning

1.1 INTRODUCTION

In 1923 when discussing the crystal structure of the face-centred cubic phase of hydrogen chloride, Simon and von Simson[1] introduced the idea that the molecules in a crystal may be rotating around one or more of their axes. They related the onset of such rotation to the increase in the heat capacity observed by Eucken[2] for this substance at the solid–solid transition temperature. This idea was developed by Pauling[3] and used by other workers in the 1930's to account for the similar behaviour of other substances.

Timmermans[4] noticed that certain organic compounds, when in the solid phase which is stable just below the melting point, had relatively high vapour pressures and melting points, a very low entropy of fusion ($\Delta S_f < 20$ J mol^{-1} K^{-1}) and were very soft and plastic. He correlated these characteristics with the molecular shape which he described as 'globular'. That is, the molecules were either of tetrahedral symmetry or were effectively spherical or ellipsoidal. These '*cristaux plastiques*' are usually of cubic symmetry and, at some transition temperature below the melting point, they transform to crystals of lower symmetry. Timmermans supposed that in the plastic–crystalline phase the molecules are undergoing rotational motion. Following examination of the high temperature cubic phases of CBr_4, C_2Cl_6 and camphor, Finbak[5] suggested that the rotation of the molecules cannot be entirely free, since steric considerations showed that the sphere of rotation of any one molecule would penetrate those of its neighbours. Three years later Zernike[6] pointed out that the phenomena observed in plastic crystals may be equally understood as a transition from an orientationally ordered state to an orientationally disordered state.

As a result of the orientational disorder and the large amplitudes of libration of the molecules, x-ray diffraction photographs of plastic crystals usually show considerable background scattering. Above this only a very small number of reflections of diminished intensity are discernible. Thus only a limited amount of information is available for determining the structure. It is often possible to fill the gap by using supplementary information about the other non-plastic phases of the compound. Although the plastic–crystalline phase itself has a special intrinsic interest, its relationship with the other solid states, stable and metastable, should be as complete as possible. Such an extended view will demand detailed theoretical analysis of the energetics and statistical mechanics of the phases; an illustration of this is the theory of James and Keenan[7] which predicted the structures and properties of the solid phases of CD_4. In the same vein, a detailed understanding is required of the various factors that decide that some members of a series of similarly shaped globular molecules form plastic crystals and some do not; such problems are of particular interest to chemists. Alongside these considerations, the question may arise as to whether all the relevant solid phases of a substance have been discovered. Here it must be remembered that kinetics as well as thermodynamics must be taken into account.

1.2 KINETICS OF PHASE TRANSFORMATION

When it is known that a compound can exist in the plastic–crystalline state an understanding of the structure and properties of that state needs a concomitant study of the other phases which may occur. Ideally, as the first requirement the thermodynamic properties of all the phases should be known. The phase diagram is a convenient and simple way of presenting such information. Though such diagrams are considered to be elementary physical chemistry, in practice the experimental delineation of an unambiguous phase diagram, exhibiting the ranges of stability and interrelations of the phases, may be fraught with difficulties; recent work on carbon tetrachloride exemplifies this.[8] Often the work has to be carried out at inconveniently low temperatures or at high temperatures where there is a possibility of thermal decomposition.

On cooling the liquid the phase which crystallizes may be metastable relative to another crystalline phase. Such behaviour would be in accord with Ostwald's 'Law of Stages' which states that a supersaturated or undercooled state does not spontaneously transform directly into that phase which, under the conditions ruling, is the most stable of the possible states but into that phase which is next more stable than itself. There are exceptions to the 'law' and the explanation probably lies in the kinetics of the transformation.[9] The deciding factors would be the relative rates of nucleation and of crystal growth for the stable and the next more stable forms; both these factors may be sensitive to the presence of impurities, homogeneous and heterogeneous. The solid–solid transformation of the resulting unstable modification into the stable form may be very slow. For example, Westrum and McCullough[10] mention that *n*-propylbenzene freezes to a metastable phase II which showed no evidence of transformation to the stable phase II after 8 hours near its melting point. In this case *n*-propylbenzene is monotropic, i.e. at all temperatures below its melting point phase II is metastable with respect to phase I.

In discussing the thermodynamic properties of the tetramethyl and tetraethyl compounds of Group IV elements, Staveley *et al.*[11] say 'it is unwise to conclude that a substance of this kind cannot exist in more than one form simply because the same sample in the same apparatus always crystallizes in the same form even in as many as twenty experiments'. Thus a statement that a certain compound exists in only one crystalline form is always subject to some reservation. It cannot be assumed that techniques such as calorimetry or differential thermal analysis, involving as they do a time element, discover all the forms of a condensed system. It will also be clear that the temperature of a system does not unequivocally characterize the phase of the system; it is not unlikely that some of the discrepancies occasionally found between the results of different x-ray crystallographers may be ascribed to the formation of different or inhomogeneous phases under conditions which are apparently the same. When molten tetraethyltin is cooled below its melting point, two or three forms are produced simultaneously[11] and Perdok and Terpstra[12] found that single crystal $C(SCH_3)_4$ a few degrees below a transition point 296.4 K contained randomly oriented dispersed grains of the undercooled higher-temperature phase.

Undercooling is common with plastic crystals, Thomas, Staveley, and Cullis[13] found that the plastic–crystalline phase I of CBr_4 showed no signs of transformation to phase II after four hours when undercooled by 0.72 K but when undercooled by 0.85 K the change was complete after about twelve hours. The triclinic phase II of hexachloroethane, C_2Cl_6, showed no change to the orthorhombic phase III after being undercooled by 1.52 K for five days.

Superheating can also occur; phase II of CBr_4 was superheated[13] by 0.08 to 0.18 K and it took 24 hours for the change to phase I to reach 85% completion. Hexachlorethane changes very slowly[14] from phase III to phase II when superheated by 0.26 K. In both cases, CBr_4 and C_2Cl_6, the phase transformation takes place more easily from the lower-temperature to the higher-temperature form than *vice-versa*. The work of Wiebenga[14] suggests the same conclusions for the other solid-state transition between phase II and the body-centred cubic phase I of C_2Cl_6. These observations would be consistent with the supposition that the rate of transformation contains a factor of the form $\exp(-E_A/kT)$ where E_A is the activation energy for the transport of a molecule from the less stable to the more stable phase.[15]

The 'drive' or 'driving force' for the transformation is the difference in the chemical potentials of the stable and unstable phases. In the case of a change from the high-temperature form, e.g. phase I, to a low-temperature form, e.g. phase II, the driving force increases the greater the undercooling and this factor alone would be expected to increase the rate of transformation. Should there also be a transport factor $\exp(-E_A/kT)$ in the rate, this would introduce a retardation increasing as the undercooling increases. Thus as the temperature decreases below the transition temperature, the rate of transformation may increase to a maximum and then decrease. When this behaviour occurs it may be exploited. If a plastic–crystalline phase is rapidly undercooled to a low temperature where its rate of transformation is small, then the amplitudes of the thermal motions would be considerably reduced and the x-ray diffraction patterns would be sharper. Such a reduction in molecular motion has been noted for several plastic crystals by Suga and Seki[16] who, by quench cooling, have prepared a glassy state in which the molecules still occupy the phase I lattice but no longer undergo rotational motions.

Solid state transformations may be reconstructive or displacive. This does not exclude the possibility that a phase change may be of a different type under different conditions such as different undercooling or superheating. When considering the possible mechanism it is of some importance to establish whether or not there is a relationship between the old lattice and the new lattice.

In determining the crystal structures of adamantane, Nordman and Schmitkons[17] mounted single crystals of the cubic phase I in their x-ray diffractometer and obtained its diffraction pattern. On cooling with cold nitrogen the low-temperature tetragonal phase II gave a new diffraction pattern which showed that the direction of the tetragonal *C* axis always coincided with one or more of the three *a* directions of the original cubic phase I crystal. The low-temperature structure is formed through the tilt of the molecules by 9° about the *C* axis.

According to Finbak and Hassel[18] it would seem that the simplest mechanism by which the cubic phase I of CBr_4 might transform into the monoclinic phase II would involve only small changes in the orientation of the molecules, accompanied by slight distortion and shrinkage of the lattice. If this occurred a very close orientational relationship between the lattice directions of the mother and daughter crystals ought to be observed. From x-ray Laue patterns and the morphology of the crystals, Mnyukh and Petropavlov[19] have sought and failed to find any evidence for close relationships and conclude that the mechanism of phase transformation does not require either a crystallographically coherent interface with the mother phase or even any structural correlation at the interface. A similar conclusion was reached by Hartshorne and Roberts[20] in the case of the transformation of monoclinic β-sulphur to rhombic α-sulphur.

In grain boundary migration, the boundaries separate grains of the same phase. The driving force may then be due to the free surface energy at the boundaries, to different degrees of strain etc. McCrone[21] studied grain growth in thin polycrystalline films using optical microscopy. He found that boundary migration for TNT, DDT, and vitamin K is definitely dependent on the orientation of the crystals; while for metals, octachloropropane, camphor, pinene hydrochloride, ice, etc. the orientations of the lattice were unimportant. Related to these interesting observations are those of Hartshorne and Swift,[13] who attempted to establish the transition temperature for CBr_4 by finding the temperature at which the boundary between the two phases neither advanced nor retreated. Temperatures could be found at which the boundary remained stationary over most of its length but in some places it advanced and in others it retreated. Since the chemical potentials of the two phases should be equal at the transition temperature the situation is closely similar to that for grain boundary migration.

1.3 ENERGETICS OF MOLECULAR CRYSTALS

Structure determination is of basic importance for the discussion of molecular behaviour in solids. It is a basis from which to develop a detailed understanding of lattice binding energies and of molecular and lattice dynamics. The mean positions of atoms in an infinite crystal are represented by the indefinite extension in three dimensions of a translation lattice or other space group. The configurational potential energy Ψ of a molecular crystal may be defined as the potential energy of the atoms at rest in their represented mean positions. The first assumption that can be made is that Ψ can be separated into two independent parts, an intramolecular part due to the chemical bonds within the molecule and an intermolecular part due to interaction between the molecules.[22]

The simplest procedure for an approximate calculation of the intermolecular (non-bonded) potential energy Ψ^{nb} is to add the energies of interaction of all pairs of non-bonded atoms, i.e. those which do not belong to the same molecule

$$\Psi^{\mathrm{nb}} = \frac{1}{2} \sum_{i \neq j}^{\mathrm{nb}} \Psi_{ij} \tag{1.3.1}$$

When only van der Waals attractive and overlap repulsive forces are present between such non-bonded pairs, the potential functions ψ_{ij} frequently used are either of the Buckingham[23] form

$$\psi_{ij} = A_{ij} \exp(-B_{ij} r_{ij}) - u_{ij} r^{-6} \tag{1.3.2}$$

or of the Lennard–Jones[24] form

$$\psi_{ij} = \lambda_{ij} r_{ij}^{-12} - u_{ij} r_{ij}^{-6} \tag{1.3.3}$$

Here r_{ij} is the internuclear distance between an atom of the kind i and an atom of the kind j while A_{ij}, $B_{ij} \lambda_{ij}$ and u_{ij} are constants independent of r_{ij} but depend on whether the atoms i and j are of the same kind or of different kinds, u_{ij} may be calculated from theory[25].

When a crystal at 0 K is dissociated into molecules at infinite separation, the energy of dissociation, $E^{c}(0)$, will be equal (with the sign changed) to Ψ^{nb} plus changes in the zero-point energy.[26] This assumes that changes in the intramolecular (bonded) potential energies of the molecules are negligible. The changes in the zero point energies of the internal degrees of freedom of the molecules are likewise ignored. Taken into account however are those changes assumed to be ascribable to the motions of the centres of mass of the molecules and to the small librational motions of the molecules, each considered as a rigid body, around their centres of mass in the crystal. Kitaigorodsky *et al.*[27] have suggested a method of estimating this intermolecular contribution to the zero-point energy. For each coordinate q of the centre of mass and of the orientation, the force constant k_{Fq} for harmonic motion in that coordinate is given by the value of $\delta^2\Psi/\delta_q^2$ at the minimum of Ψ. From the k_{Fq}'s, the frequencies of vibration in the coordinates can be obtained using the mass and moments of intertia of the molecule; hence the contributions to the zero-point energy result. Then

$$E^{c}(0) = -\Psi - \frac{h}{2}\sum_{q} \nu_q \tag{1.3.4}$$

This zero-point energy is small[26] or negligible[28] in most molecular crystals. The same principle can be used to calculate the Debye temperature for vibrations and librations as a function of the unit cell volume per molecule. At present only very rough values are obtainable.

Should the chemical bonds in the molecule be polar, the electrostatic interaction between partial monopole charges centred on the appropriate atoms is evaluated. Similarly, when there are hydrogen bonds, Lewis acid-base or other specific interactions between molecules, these must be included.

The geometry of the molecules in the crystal is given by the coordinates of atoms in the unit cell which are obtained from x-ray diffraction and supplemented if necessary, especially in the case of hydrogen atoms, by neutron diffraction. The calculated lattice potential energy should be a minimum at the experimental values of the lattice constants and of the three Eulerian angles describing the orientation of the molecule within the cell. To satisfy these constraints, the parameters A_{ij}, B_{ij} or λ_{ij} together with any parameters associated with other forces are varied to obtain the best fit. It is necessary to avoid having more adjustable parameters than the number of experimental constraints. This requirement may be satisfied by choosing initially a simple homologous series of chemical compounds to derive the parameters for a few atom-pair interactions. Using these values as standards, the examination of other suitable crystalline compounds will systematically add to the standards. Since many compounds exist in polymorphic forms, the number of lattice constants available for deriving or checking parameters is increased.

The assumption is made that these 'standard' parameters are transferable[29] from one kind of molecule to another and from one lattice to another. Equivalently, the potential energy of a non-bonded pair of atoms is assumed to be unaffected by their environments in their molecules and in the crystal and to depend only on their distance apart. That this is unlikely to be exactly true is indicated by the case of the rare gas solids which gives some idea of the influence that crystal structure may have on the crystal energy and of the order of magnitude.

Helium under compression at low temperatures crystallizes in the hexagonal close packed (h.c.p.) structure while the stable forms of neon, argon, krypton, and xenon are face-centred cubic (f.c.c.). Crystals of Ne and Ar in the h.c.p. form have been observed experimentally and found to be metastable. Investigation of stacking faults show that the lattice potential energies of the stable f.c.c. forms are lower than those of the h.c.p. forms by rather more than 0.1 per cent. However calculations[30] using two body potential functions give lattice potential energies for the heavier rare gas crystals which are higher by about 10^{-2} per cent for f.c.c. than for h.c.p., indicating that h.c.p. should be the stable form.

Niebel and Venables[31] explain the discrepancy by showing that the crystal fields, which are different for h.c.p. and f.c.c., split the energy levels of the atoms differently with the consequence that the van der Waals attractive potentials are not quite the same in the two modifications. Calculation of the difference in the lattice potential energies of the two forms show that f.c.c. is the more stable by rather more than 0.1 per cent in accordance with observation.

Mirsky[32] has tested the transferability concept. The potential energy functions between pairs of non-bonded atoms were derived from the structural sublimation and elasticity data of hydrocarbons. Transferred to adamantane, $C_{10}H_{16}$, they were used to calculate the properties of the low-temperature tetragonal and the high-temperature cubic plastic crystalline forms. The lattice energy of the tetragonal form was calculated to be -81.5 kJ mol^{-1} at 0 K. The dimensions of the unit cell were found to be $a = 0.645$ nm and $c = 0.863$ nm at 0 K and, although precise thermal expansion data are not available, appear to be in very good agreement with experimental values. For the cubic phases three models were taken in which the molecules were ordered or orientationally disordered or freely rotating. The rotating phase was least favoured energetically and the orientationally disordered most. The calculated lattice energy of the ordered and disordered phases were found to be in satisfactory agreement with a rather rough value derived from the experimental enthalpy of sublimation. A value of 4.2 kJ mol^{-1} calculated for the energy of the phase transition from the tetragonal to the orientationally disordered phase was the closest to the experimental value[33] of 3.37 kJ mol^{-1}. Mirsky considered that the orientationally disordered model corresponds to the true structure of the cubic phase.

One object of these lattice energy calculations is to correlate the properties of the plastic crystalline phase with those of the other phases of the same compound. Another object is to correlate the phase diagrams of different compounds which have the same geometrical structure, for example the series CH_4, SiH_4, GeH_4, SnH_4 or the series CF_4, CCl_4, CBr_4, CI_4; the molecules of these are all tetrahedral. Pitzer[34] and Guggenheim[35] have specified some of the conditions limiting the applicability of the principal of corresponding states to the thermodynamic properties of compounds in the gaseous, liquid, and solid states. These ideas have been developed by Trappeniers[36] with the aim of correlating the properties of molecules which have a similar geometry. Each species of molecule consists of n atoms of the same sort symmetrically distributed about the centre of the molecule at a distance x from it. The magnitude of x depends on the molecular species. Instead of assuming that the molecules are quasispherical and that the potential energy for a pair of molecules is of the spherically symmetrical form $A\phi(R/R_0)$, where R is the distance between the centres of the molecules, the interaction energies of all pairs of atoms which do not belong to the same molecule are summed. When these and other underlying assumptions are satisfied, such similar molecules will obey a principle of corresponding states. Encouraging results were obtained from a right application of the principle to the phase diagrams of carbon tetrachloride and carbon tetrabromide. The principle of corresponding states has been found to be applicable to a number of the properties of rare gas solids.[37]

It is not to be expected that any but the most favourable series of similar molecules will conform closely with the principle, but at first even qualitative predictions would be useful in correlating phase diagrams. Also the nature of the deviations would be revealed for more detailed study. No doubt the covalent–ionic characters of the bonds between the molecules would have to be taken into consideration. The partial monopole charges to be allotted to the atoms can be calculated by the molecular orbital method.[38] When such charges are present, Coulombic terms are introduced into the expression for the intermolecular potential energy function resulting in a form incompatible with the principle of corresponding states.

When hydrogen bonds are present between molecules further terms are added to the potential energy function.[29] Hydrogen bonds are known to have a marked effect on the solid state transition temperatures. For example tetrafluoromethyl methane,[39] $C(CH_2F)_4$ at 249.7 K undergoes a solid state transformation to a plastic crystal which melts at 367.4 K, while for pentaerythritol, $C(CH_2OH)_4$, the respective temperatures[40] are 457 K and 539 K. In the tetragonal phase of $C(CH_2OH)_4$, which is stable below 457 K, the molecules are linked by O—H—O hydrogen bonds in the planes perpendicular to the $\bar{4}$ axis in such a way that a layer structure with pronounced basal cleavage results.[41] Though indicative, such a comparison is only valid when the two compounds have the same crystal structures and closely similar molecular volumes in their respective solid phases. In many cases the effect of the hydrogen bonds predominate over that of the van der Waals interaction and the molecular packing is altered accordingly.

Molecules may interact with each other by sharing, to a greater or less degree, electron pairs. In reaction kinetics a reagent which readily donates electrons is termed *nucleophilic*, an *electrophilic* reagent is one which readily accepts electrons from a nucleophilic reagent. When equilibria rather than kinetics are being considered the term used for an electron donor is *Lewis base* and for an electron acceptor *Lewis acid.*[42] When such interactions between adjacent molecules in the crystalline state occur, the symmetries of the molecular orbitals involved need to be harmonized with those of the site and the crystal.

In crystalline $(CH_3)_3GeCN$ the cyanide groups are directed towards the germanium atoms in adjacent molecules.[43] The N—Ge distance is found to be 0.357 nm, slightly less than 0.36 nm, the contact distance calculated from the van der Waals radii. In $(CH_3)_3SnCN$ the N—Sn distance is 0.245 nm, considerably less than 0.40 nm the calculated van der Waals distance.[44] This indicates that the intermolecular interactions are much stronger for the tin compound than for the germanium compound, and that tin is a much stronger acceptor than germanium for electrons donated by the cyanide group. This difference is reflected in the two crystal structures. In both the crystalline compounds the molecules are linked in chains—$[-M-C-N-]_n$ where the CN group forms bridges between the M's (Ge or Sn); in the tin compound the trimethyl tin group is planar. The crystal structure of $(CH_3)_3CCN$ is unknown but it may form a plastic crystalline phase in which the molecules do not undergo isotropic re-orientation but only rotate about the C–CN axis.[45] In crystalline $C(CN)_4$ the cyanide groups are not directed at the central carbon atoms because the acid site is not localized on these central carbons but involves adjacent cyanide carbons as well.[46] This is one of the factors that favours for this compound a trigonal structure which is a distortion of the cubic SiF_4 structure.

In orthorhombic trimethyltin fluoride, $(CH_3)_3SnF$, crystals the $(CH_3)_3Sn$ groups and the F atoms are arranged alternately in a chain-like manner along the *a* axis; the tin atoms are five-coordinate and the Sn–F . . . Sn bridges are non linear.[47] Crystalline silicon tetrafluoride has a body-centred cubic lattice of SiF_4 molecules.[48] Each SiF bond points directly toward the Si atom in the adjacent molecule. Thus each Si atom is linked to its eight nearest neighbouring silicons by unsymmetrical linear fluorine atom bridges, Si–F . . .Si. In four of these bridges the short intramolecular Si–F bonds are tetrahedrally disposed, the four long Si . . .F bonds are centred on and normal to the faces of this tetrahedron. In crystalline zirconium tetrafluoride ZrF_4, the structure is similar but the intramolecular bond length is 0.203 nm and the intermolecular length is 0.218 nm, these are to be compared with the distance of 0.211 nm calculated.[49] for a Zr–F bond; the distinction between intra- and intermolecular bonds has almost vanished.[50] Carbon tetrafluoride CF_4 undergoes a solid state transition[51] at 76.2 K to a plastic crystalline form but the crystal structure of this is thought to be not cubic but monoclinic.[52] This difference in structure may be related to the fact that carbon unlike silicon and germanium is unable to expand its valence shell.

It is now possible[53] by x-ray diffraction methods to determine with high accuracy the electron density distribution within a crystal and, by thermal neutron diffraction experiments to locate the nuclei. Let ρ_X be the electron density at a point in the unit cell as evaluated from the x-ray measurements. A notional electron density ρ_N at the same point may be computed by superimposing the electron clouds appropriate to isolated uncombined atoms, each atom with its nucleus at the location prescribed by the neutron experiments. The difference $\rho_{X\text{-}N} = \rho_X - \rho_N$, gives the change in the electron density at that point due to bond formation, overlap, etc. averaged over the thermal motion in the crystal. Since the electron cloud of an uncombined atom in field-free space is averaged over all orientations, Coppens and Coulson[54] point out that for evaluating ρ_N the electron cloud should be orientated appropriately to the directions of the bonds which that atom will form.

Not all crystals are suitable for such an investigation since the outer valence electrons, which are the ones redistributed, scatter at low Bragg angles and there may not be a sufficiency of x-ray reflections in this region to yield high resolution. Thermal motions smear out the density difference and it may be difficult to compensate for this to the precision required. Nevertheless, the possibility exists that the effect of bridging may be revealed by showing that a redistribution of electrons has occurred, for example, not only in the intramolecular Si–F bonds but also between a fluorine atom of one SiF_4 molecule and the silicon atom of a neighbouring molecule.

1.4 REVIEW OF CRYSTALLOGRAPHIC DATA

The preceding sections review the existing information on the kinetics of phase transformations and on the equilibrium phase relationships between plastically crystalline and normal phases. The studies carried out and their intercomparisons are few. Much experimental and theoretical work remains to be done in this area. To provide a basis for this, the remainder of this chapter is devoted to a review of crystallographic data for plastically crystalline solids.

1.4.1 Hydrogen Fluoride

Hydrogen fluoride melts at 190.09 K with an entropy of fusion of 24.06 J mol^{-1} K^{-1}. X-ray diffraction photographs of powders by Gunther, Holm and Strunz[55] gave the same lines at 151 K as at 91 K. At the latter temperature the lines were indexed for a tetragonal unit cell with $a = 0.545$ nm $c = 0.995$ nm and $Z = 16$. Atoji and Lipscomb[56] in later work decided that at 148 K the unit cell was orthorhombic with a space group *Bmmb* (D_{2h}^{17}); the dimensions of the unit cell were found to be $a = 0.342$ nm, $b = 0.432$ nm, $c = 0.541$ nm with $Z = 4$.

Neutron diffraction studies[57] on solid DF at 4.2 K and 85 K show that the DF molecules form parallel zig-zag chains.

1.4.2 Hydrogen Chloride and Deuterium Chloride

The melting point of hydrogen chloride[51] is 158.94 K. In comparison with HF the entropy of fusion is only 12.50 J mol^{-1} K^{-1} [58] and there is a solid-state transformation at 98.38 K. In phase I Simon and von Simson[1] found that at 105 K the central points of the molecular electron clouds lay on a face-centred cubic lattice with $a = 0.554$ nm. To account for this high symmetry these authors[1] suggested (in 1924) that the polar molecules are rotating around one or two axes. Natta[60] found that a is 0.5435 nm and 0.5466 nm at about 98 K and 103 K respectively. Also using x-ray powder diffraction methods Sandor and Farrow[61] found that a is 0.5482 nm at 118.5 K. For comparison DCl which has a transition temperature of 105.0 K[62] is also face-centred cubic[63] with $a = 0.5475$ nm at 118.5 K. Using thermal neutron diffraction for DCl these authors[64] concluded that the pattern for a face-centred structure resulted from a statistical averaging of the disordered deuterium nuclei.

The structure of phase II of HCl was found by Natta[65] to be face-centred orthorhombic with $a = 0.503$ nm, $b = 0.535$ nm $c = 0.571$ nm and $Z = 4$, at 95 K; Natta gave the space group as F^{222} (V^7) or *Fmmm* (V_h^{23}). Later x-ray examination[61] at 92.4 K also decided that phase II was face-centred orthorhombic with $a = 0.5082$ nm, $b = 0.5410$ nm, $c = 0.5826$ nm. For DCl at 92.4 K the unit cell parameters were found[61] to be 0.5068 nm, 0.5399 nm and 0.5828 nm. Neutron diffraction powder patterns for DCl at 77.4 K showed some new lines, the result of neutron scattering by the deuteron nuclei. The space group is then found to be $Bb2_1m$ (C_{2v}^{12}), the unit cell is orthorhombic (parameters 0.5053 nm, 0.373 nm and 0.5825 nm) but it is not face-centred. In this phase zig-zag chains of DCl, all oriented the same way, are arranged in layers parallel to the (001) plane.

1.4.3 Hydrogen Bromide

Hydrogen bromide melts at 186.28 K and there are solid state transformations at 116.9 K, 113 K and 90 K.[58,59] At a temperature between 115 K and 120 K Natta[60] found the crystals to be face-centred cubic with $a = 0.576$ nm and $Z = 4$. According to his x-ray diffraction patterns the space group at 100 K was either $F222$ (V^7) or *Fmmm* (V_h^{23}) and the unit cell was face-centred orthorhombic with $a = 0.555$ nm, $b = 0.5640$ nm, $c = 0.6062$ nm and $Z = 4$.

1.4.4 Hydrogen Iodide

At 222.36 K hydrogen iodide melts, it undergoes transformations at 125.68 K and 69 K. According to Ruhemann and Simon,[63] x-ray diffraction patterns are consistent with a face-central tetragonal unit cell with an axial ratio c/a = 1.08 and values of a of 0.619 nm, 0.603 nm and 0.594 nm at temperatures of 125 K, 82 K and 21 K respectively. A similar conclusion was reached by Natta[60,65] who obtained an axial ratio of 1.075 with a = 0.610 at 100 K.

Using neutron diffraction to study DI Sandor and Clarke[64] found that phase I is face-centred cubic (Z = 4) with orientational disorder; at any moment the orientations of the molecules at equivalent sites are random. Phase II has the space group *Bbcm* (D_{2h}^{18}) with two-fold disorder whilst the lowest temperature phase III has an ordered monoclinic cell and the space group is $C/2c$ (C_{2h}^{6}).

1.4.5 Hydrogen Sulphide

The calorimetric data obtained by Clusius and Frank[66] give the melting point as 187.2 K with an entropy of fusion ΔS_f = 12.71 J mol^{-1} K^{-1}; three solid phases were found with transition temperatures 126.2 K and 103.6 K. X-ray powder photographs[67] taken at 20.4 K, 90 K, 120 K and 155 K showed 'identical' diagrams. According to these, all three phases are face-centred cubic with a = 0.577 ±0.005 nm the lattice constant changes by less than 1% over the temperature range.

1.4.6 Methane, Tetradeuteromethane

Methane melts at 90.675 K[68,69,70] and has an entropy of fusion[70] of 10.37 V mol^{-1}; it has one solid–solid transition[68] at 20.49 K and there are indications of another[68,71] at about 8 K. X-ray diffraction, just above and just below 20.5 K showed that the lattice is face-centred cubic[72] with a = 0.590 nm, Z = 4. Measurements on polycrystalline CH_4 at many temperatures between 4.2 K and 75 K by Greer and Meyer[73] revealed a small sigmoid increase in the lattice constant at 20.4 K and a small bump at 65 K. Below 10 K inconsistencies in the lattice constant were observed from experiment to experiment. Interpreting the results obtained from inelastic, incoherent neutron scattering, Janik *et al.*[74] conclude that near the melting point CH_4 molecules undergo hindered rotation.

The melting point of deuteromethane is 89.78 K[68,75] and the entropy of melting[68] is 10.11 J mol^{-1} K^{-1}; it has two λ-type transition points[68,75] at 22.4 K and 27.2 K. Bol'shutkin *et al.*[76] have found that the high-temperature phase is face-centred cubic with a = 0.587 nm at 40 K and Z = 4. Press *et al.*,[77] using coherent neutron scattering from powder samples, derived the space group as *Fm*3*m* with four molecules in the unit cell and a = 0.596 nm at 77 K; there was evidence that the lattice is partially ordered at 77 K and completely disordered at 35 K.

Phase II of CD_4 was found by Bol'shutkin *et al.*[76] to be also face-centred cubic at 25 K with a unit cell parameter of 0.584 nm at 25 K, while Press *et al.*[77] obtained a = 1.164 nm at 24.3 K with Z = 32; the space group found was $Fm3c$; some ordering is present, six out of eight molecules order with a local symmetry of $\bar{4}2m$ and the remaining two are orientationally disordered with almost-free rotation.

The structure of the lowest temperature phase III is described by Bol'shutkin *et al.*[76] as face-centred tetragonal with a = 1.148 nm, c = 1.166 nm and z = 32 at 6 K. The reflections observed by Press[77] gave no clear answer for the structure of phase III and he suggested that it might be cubic primitive with a = 1.161 nm and z = 32 at 17.5 K. Arzi and Sandor[78] also using neutron diffraction techniques find that at 4.2 K the cell is tetragonal with a = 0.8183 nm, c = 1.1671 nm and sixteen molecules in the primitive cell; the sublattice of carbon atoms is body-centred tetragonal. That the structure of phase III is non-cubic is supported by the optical measurements of Ballik *et al.*[79] who found it to be birefringent. Assuming that it is tetragonal the observed values of the birefringence may be used to calculate the lattice parameters, these were found to be a = 0.577 nm and c = 0.584 nm.

Greer and Meyer[73] using x-rays have studied CD_4 also over the temperature range 4.2 K to 80 K and found an anomalous increase in the lattice constant near 20 K and a smaller increase is noticeable near 27 K. Below 10 K there is again some inconsistency in the values of the parameter.

A statistical mechanical theory of the phase structures in CD_4 has been published by James and Keenan,[7] the treatment is classical and cannot be applied to CH_4 since quantum effects are neglected. In the intermolecular potential energy function interactions involving electrostatic dipoles and quadrupoles vanish as a result of the symmetry and only octopole-octopole interactions survive if those of higher-poles are neglected. The theory predicts that in phase I the molecules are orientationally disordered while in phase II one molecule in four rotates freely within a shell of librating nearest neighbours. Phase III is predicted to be tetragonal with all molecules librating about equivalent equilibrium orientations. After selecting a value for the octopole moment which allows the theoretical value of the upper I–II transition temperature to be matched with 27.1 K (the value observed experimentally for CD_4), the predicted lower transition temperature is 24.4 K to be compared with the observed 22.2 K.

The investigations of Press and Kollmar[80] using incoherent neutron scattering tend to confirm that phases I and II of CH_4 are isostructural with the corresponding phases of CD_4. During calorimetric measurements[68,71] on CH_4 it was found that the time required to reach equilibrium was long, it varied from several hours at 5 K to 20 minutes at 11 K and similar times were observed during measurements of the optical birefringence.[79] The slow changes may result from slow transformations of the same nature as those that occur in the deuterated methanes when their I–II transition points are traversed. Alternatively there is the possibility that they originate from the slow interconversion between the different nuclear-spin species of the methane.[81]

An interesting approach is that of Nakamura and Miyagi.[82] Taking quantum effects into account they used the principle of corresponding states and were able to predict values for the transition temperatures in the series CH_4, CH_3D, CH_2D_2, CHD_3 and CD_4 in good agreement with those observed.

The solid state phase diagrams for the two-component systems CH_4/Ar and CD_4/Ar have been determined by Greer and Meyer.[73] These show an upper consolute temperature on the co-existence curve for methane-rich and argon-rich face-centred cubic crystalline phases.

1.4.7 Silane, Trideuterosilane, Germane, and Tetradeuterogermane

Silane melts at 88.5 K and $\Delta S_f = 7.52$ J mol^{-1} K^{-1}. There is a λ-type heat capacity anomaly[83,84] at 63.75 K and a smaller one at about K. The high-temperature phase is weakly birefringent and phase II strongly birefringent.[83] The infrared and Raman spectra of the solids have been redetermined by Fournier *et al.*[85] who find that the spectra of the liquid and of phase I are closely similar, this suggests that orientational disorder prevails in phase I. From the multiplicity of the components in the Raman spectrum of phase II, it appears that there are at least two molecules in the primitive unit cell. The properties[84] of $SiHD_3$ are close to those of silane; it melts at 86.8 K with $\Delta S_f = 7.19$ J mol^{-1} K^{-1} and has a solid state transition point at 66.05 K.

The melting point of germane is 107.26 K and the entropy of fusion is small, 7.77 J mol^{-1} K^{-1}. In their calorimetric study Clusius and Faber[86] found two transitions at 76.5 K and 73.2 K at both of which undercooling and superheating phenomena occurred; there is a third λ-type anomaly at 62.9 K. Phase I was weakly birefringent and the lower-temperature phases strongly birefringent. Passing through the temperature 62.9 K had no discernible effect on the birefringence, however The *et al.*[87] found marked changes in the infrared and Raman spectra; there were also changes at the higher transition temperatures.

Extending these spectroscopic investigations to GeD_4, Wilde[88] observed transitions at 68.5 K and 77 K. Thus CD_4, SiD_4 and GeD_4 all exist in three crystalline modifications, but Wilde considers that the point group symmetries of the molecules at their sites in the various crystals indicate that the corresponding phases of these compounds are not isostructural.

1.4.8 Carbon Tetrafluoride

Bol'shutkin *et al.*[89] have recently determined the crystal structure of CF_4 (T_m 89.47 K, $T_{t\,I\text{-}II} = 76.23$ K) in a low temperature phase at 10 K and find it has a monoclinic lattice with $a = 0.8435 \pm 0.0005$ nm; $b = 0.4320 \pm 0.0002$ nm., $c = 0.8369 \pm 0.0005$ nm, $\beta = 119.40° \pm 0.05$; space group $P2_1/c$ $Z = 4$.

The authors also measured the temperature dependence of these lattice parameters at intervals from 10 K to 75 K and find that at 75 K just below the transition point $a = 0.8597$ nm, $b = 0.4433$ nm both dimensions increasing monotonically from the values at 10 K while the value of c is 0.8381 nm at 75 K having gone through a maximum near 63 K. The angle β decreases to 118.73° at 75 K.

Greer and Meyer[90] has previously indexed the diffraction pattern of the low temperature phase at 40 K on the basis of a monoclinic unit cell. The high temperature form (I or β) they found could not be indexed but it was established that it was not cubic but probably also monoclinic.

1.4.9 Silicon Tetrafluoride

In the solid form SiF_4 does not undergo an orientational disordering transition below the melting point.[91] From the results of powder photographs Natta[92] obtained the space group $I\bar{4}3m$ (T_d^3) for SiF_4 crystals; a body-centred cubic cell with a parameter $a = 0.541$ nm. These results were confirmed by Atoji and Lipscomb[93] from a study of single crystals at 126 K. Since the atoms could be located in the lattice, 2 Si at 0, 0, 0 and $\frac{1}{2}, \frac{1}{2}, \frac{1}{2}$ the 8 F at x, x, x etc. in the eightfold positions of $I\bar{4}3m$, the molecules are not disordered.

Atoji and Lipscomb calculated the crystal energy as -17.56 kJ mol^{-1} in comparison with -25.92 kJ mol^{-1} obtained experimentally by Patnode and Papish.[94] The crystal energy has also been calculated more recently by Shinoda[95] using Lennard-Jones potential functions and octopole–octopole interaction terms between neighbouring molecules. A value of -31.01 kJ mol^{-1} was found, to be compared with a value of -28.17 kJ mol^{-1} derived from calorimetric data of Pace and Moser.[91] The question of fluorine atom bridges between the silicon atoms was not considered.

1.4.10 Carbon Tetrachloride

The difficulties found in establishing the phase relationships of this compound are not uncommon and are due to the often sluggish kinetics of phase changes. Up till 1966 it was considered that solid CCl_4 melted at 250.4 K and underwent a phase transformation at 225.4 K[96,97] from a high-temperature crystalline phase, deemed to be optically isotropic, to a low-temperature, optically anisotropic phase. Post[96] found that the intensities of the x-ray reflections from the isotropic phase decreased very rapidly as the Bragg angle increased and that heavy diffuse scattering was present. Hence only poor diffraction data was obtainable, but Post found that at 235 K the unit cell is face-centred cubic with $a = 0.834 \pm 0.003$ nm and four molecules in the unit cell.

In a later x-ray investigation Rudman and Post[97] discovered that the cubic phase (now designated Ia) transformed into a hitherto unrecognized rhombohedral modification Ib. They reported that when either modification is cooled below 225.4 K a monoclinic form II results. A sensitive method of measuring birefringence had been developed by Ballik, Gannon and Morrison[98] and using this Koga and Morrison[8] showed that the birefringence of the cubic phase Ia is extremely small, That of the rhombohedral phase Ib is larger, while that of the monoclinic phase II was too large to measure by this method. Kotake, Namakura and Chihara[99] found that the melting point of the cubic phase Ia was 3.86 K lower than that of phase Ib and Silver and Rudman[100] confirmed this. Using thermal analysis and x-ray diffraction methods Badiali, Bruneax-Poulle and Defrain[101] have clarified the situation and find that the cubic f.c.c. form melts at 245.4 K and is always metastable, standing in monotropic relationship to Ib. The rhombohedral phase Ib is stable between 225.4 K and 250.4 K. The transformations relating to the three solid phases are illustrated in the following diagram.[101] The liquid remains in the undercooled state down to temperatures of about $T_1 = 239$ K. Phase Ia can be preserved down to $T_2 = 221$ K when it transforms into phase II. Phase Ib can be undercooled down to about $T_3 = 200$ K.

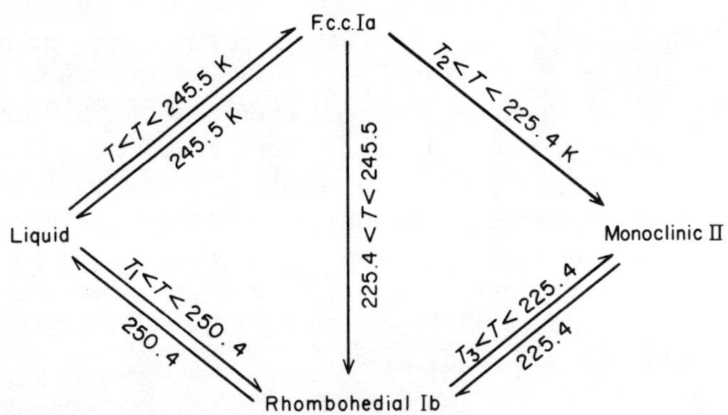

The crystal structure of the rhombohedral Ib has been determined by Rudman and Post.[47] There are 21 molecules in the unit cell which has the dimension $a = 1.49$ nm with $\alpha = 90.0°$. The same authors find that the monoclinic phase II has 32 molecules in the unit cell, the dimensions of which are $a = 2.03$ nm, $b = 1.16$ nm, $c = 1.49$ nm with $\beta = 111°$. The monoclinic unit cell corresponds to an assembly of slightly distorted cells related to the face-centred cubic cell of phase II; this should be compared with the structure of phase II of CBr_4.

The pressure–temperature diagram of CCl_4 has been studied by Bridgman[102] who found a high temperature polymorph, phase III at 29 Nm^{-2} (20 kbar). This solid III was reported to crystallize in the orthorhombic crystal system and to have a space group $C222_1$ (D_2^5). Recently Piermarini and Braun[103] have found that it is monoclinic with four molecules in the unit cell and $a = 0.9079$ nm, $b = 0.5764$ nm, $c = 0.920$ nm, $\beta = 104.29°$. This phase is isostructural with $SnBr_4$.[104]

At still higher pressures phase III is transformed to phase IV; the former is highly birefringent while the latter is not birefringent so the change of phase can be recognized when it takes place. At room temperature the rate of transformation becomes extremely slow.

1.4.11 Titanium Tetrachloride

The x-ray diagrams of single crystals of $TiCl_4$ were found by Brand and Sackmann[105] to be closely similar to those obtained for $SnBr_4$ crystals and the distribution of reflection intensity ratios was almost the same for the two compounds. After further analysis it was concluded that $TiCl_4$ at 241 K had the same crystal structure as $SnBr_4$. The space group for $TiCl_4$ is $P2_1/c$ (C_{2h}^5) with 4 molecules in the unit cell of dimensions: $a = 0.970$ nm, $b = 0.648$ nm, $c = 0.975$ nm, $b = 102.6°$.

1.4.12 Tin Tetrachloride

Brand and Sackmann[105] determined the structure of $SnCl_4$ and as in the case of $TiCl_4$ found it to have (at 234 K ($T_m = 240$ K) the same structure as $SnBr_4$. The space group of $TiCl_4$ is $P2_1/c$ (C_{2h}^5) with 4 molecules in the unit cell of dimensions $a = 0.985$ nm, $b = 0.675$ nm, $c = 0.998$ nm, $\beta = 102.25°$.

1.4.13 Carbon Tetrabromide

Numerous investigations by Finbak and Hassel[18,106] were carried out on powders and single crystals of CBr_4-1, the phase stable above the transition point at 319.86 K. Even in the most favourable cases only very weak reflections were obtained from which a face-centred cubic cell with a about 0.87 nm and $Z = 4$. Mark[107] had previously obtained a lattice constant of 0.567 nm Å with $Z = 1$ and a space group $P\bar{4}3$ m (T_d^1) but withdrew these evaluations in a later paper.[108]

According to Finbak and Hassel[18,106] CBr_4-II, the phase stable below the transition point, is monoclinic; the unit cell probably contains 32 molecules and has the dimensions $a = b = 1.226$ nm, $c = 2.412$ nm, $\beta = 125.05°$. The axial ratios agree with the crystallographic measurements of Ziengiebl.[109] These monoclinic crystals are, as mentioned previously by Groth,[110] pseudo-cubic. The monoclinic cell may be regarded as derived from eight of the pseudo-cubic cells. Assuming that the pseudo cubic cells have a dimension $a_{pseudo} = 0.86$ nm (c.f. $a = 0.87$ nm for the cell of phase I), the dimensions calculated for the pseudo-cubic-monoclinic cell are $a_{calc.} = 2.106$ nm, $b_{calc.} = 1.216$ nm, $c_{calc.} =$

2.432 nm, $\beta_{calc.}$ = 125.25°: figures which are very close to those determined experimentally for phase II. The actual monoclinic cell of phase II may be regarded notionally as being derived from the unit cell of phase I by a slight reduction in the dimensions of the cubic unit cell, followed by a slight angular distortion involving eight originally cubic cells.

More recent studies[111,112] of both phases of this material confirm that $Z = 4$ for phase I and define a as 0.882 nm. On the same basis as above the authors calculate the dimensions of the pseudo-cubic-monoclinic cell as $a_{calc.} = 2.160$ nm, $b_{calc.} = 1.247$ nm, $c_{calc.} = 2.160$ nm and $\beta = 109.47°$. These values are in much better agreement with the newly measured values:[111] $a = 2.143$ nm, $b =$ 1.212 nm, $c = 2.102$ nm and $\beta = 110.88°$ than are those of Finbak and Hassel.

1.4.14 Silicon Tetrabromide

By rapidly cooling $SiBr_4$ Pohland[113] obtained an anisotropic birefringent form II which melted at 275.6 K. On heating this solid phase from 243 K a transformation, said to be irreversible took place to form a cubic isotropic modification which melted at 278.6 K. These observations were confirmed by Kennard and McCusker[114] and by Sackmann, Demus, and Pankow;[115] the latter produced the isotropic phase by slowly cooling the melt.

1.4.15 Titanium Tetrabromide

Three modifications have been observed.[116] On cooling the melt, a cubic form I (melting point 312.3 K) usually appears. In the course of a few days at room temperature phase I changes to a birefringent phase II (T_m 312.8 K). When the melt is quenched a weakly birefringent form III is produced which changes to phase I slowly at 268 K and rapidly at 293 K. If this system is again cooled before the transformation III → I is complete, then at 253 K a slow regrowth of III is observed over hours. Unless a residue of nuclei of phase III is left, phase I will not transform back into III. Biltz and Jeep[116] had found previously that, in the presence of Br_2, there was a solid-state transformation of $TiBr_4$ at 258 K. It appears that in the transformation I → III, the rate of nucleation of phase III from phase I is low and that Br_2 increases this rate.

$TiBr_4$, crystallizes with the same structure[117] as SnI_4, body-centred cubic with a lattice parameter $a = 1.125$ nm.

1.4.16 Tin Tetrabromide

Biltz and Jeep[116] found that $SnBr_4$ in the presence of Br_2 underwent a solid state transformation at 267 K, but Sackman, Demus and Pankow[115] did not succeed in verifying this transition. In their experiments a birefringent form was always obtained. An x-ray determination of the structure by Brand and

Sackmann[118] established the unit cell as monoclinic with the dimensions at 293 K: $a = 1.059 \pm 0.003$ nm, $b = 0.710 \pm 0.002$ nm, $c = 1.066 \pm 0.003$ nm, $\beta = 103.6 \pm 0.17$; $Z = 4$, space group $P2_1/c$.

$SnBr_4$ is isomorphous with $SnCl_4$, $TiCl_4$ and CCl_4-III.

1.4.17 Carbon Tetraiodide

Carbon tetraiodide at room temperature had been reported as being cubic by Mark[119] but Finbak and Hassel[18,106] found that as a result of Lamellar twinning, many crystals simulated optical isotropy. Others, small single crystals, showed strong birefringence. Because of the strong similarity of habit and optical properties, Finbak and Hassel consider that CI_4-II and CBr_4-II are isomorphous.

They then interpreted their x-ray data for CI_4-II on the assumption that CI_4-II was monoclinic and pseudo-cubic like CBr_4-II. The dimension of the pseudo-cubic cell of the monoclinic crystals was evaluated as 0.914 nm from which they derived for the pseudo-cubic monoclinic cell.

$c_{calc} = 2.239$ nm, $b_{calc} = 1.293$ nm, $c_{calc} = 2.585$ nm, $\beta = 125.25°$. Since such calculated dimensions for CBr_4-II agreed closely with the measured dimensions, they considered that this would also be the case for CI_4-II. With these assumptions the dimensions we have for monoclinic CI_4-II, $a = 2.24$ nm, $b = 1.29$ nm, $c = 2.58$ nm and $\beta = 125°$, $Z = 32$.

Shortly before reaching its melting point (444 K with decomposition) CI_4 becomes optically isotropic; by analogy with CBr_4 and CCl_4 it is presumed that this indicates that there may be a high temperature cubic phase.

1.4.18 Silicon Tetraiodide

Using sublimed SiI_4 for powder diagrams, single octahedral and flat triangular crystals (from CS_2 as solvent) for rotation and Laue photographs, Hassel and Kringstad[120] determined its structure. They found it to be face-centred cubic with a space group Pa3 (T_h^6). The unit cell contains 8 molecules which are geometrically associated in pairs and has the dimension $a = 1.199$ nm.

The structure of SiI_4 is very closely similar to those of GeI_4 and SnI_4.

1.4.19 Titanium Tetraiodide

TiI_4 crystallizes as octahedra and melts at 423 K. Hassel and Kringstad[121] reported that it has the same structure as SnI_4, a body-centred cubic cell with the space group Pa3 (T_h^6) and $a = 1.200$ nm and $Z = 8$.

Later investigation by Rolsten and Sisler,[122] using powder diffraction patterns taken at room temperature, 398 K and 423 K showed that TiI_4 was cubic at 398 K with a lattice dimension of $a = 1.221$ nm and $Z = 8$. They found that the room temperature phase was of lower symmetry than the high temperature modification. The transformation temperature lies between 373 K and 398 K.

1.4.20 Germanium Tetraiodide

Jaeger, Terpstra, and Westenbrink[123] investigated the structure of GeI_4, and found (as in the cases of SiI_4 and SnI_4) that the unit cell is face-centred cubic with space group Pa3 (T_h^6). The unit cell contains 8 molecules and its lattice dimension is 1.189 nm.

1.4.21 Tin Tetraiodide

Early work on this crystal was carried out by Mark and Weissenberg[124] and by Ott.[125] The later results by Dickinson[126] showed that the structure is face-centred cubic with space group Pa3 (T_h^6). There are eight molecules in the unit cell which has a dimension of 1.223 nm (compare SiI_4 and GeI_4). Meller and Gankuchen[127] confirmed these results, refining the value of a to 1.226 nm.

1.4.22 Neopentane (2,2-Dimethylpropane)

Neopentane melts at 254 K and undergoes a solid state transition at 140 K[128] Wahl[129] observed that neopentane is birefringent at 90 K and becomes isotropic above the transition point. Mark and Noethling[130] obtained powder photographs at liquid air temperature (~90 K); they considered that the neopentane had crystallized in the cubic system at this temperature and that the unit cell had a dimension of 1.126 nm, with $Z = 8$. Later work by Mones and Post[131] gave powder diagrams of phase I above the transition temperature but only three or four very broad lines were observed. However single crystals obtained at 240 ± 5 K, just below the melting point, appeared to be isotropic and oscillation diagrams, indexed on the basis of a face-centred cubic unit cell, yielded 0.878 ± 0.005 nm as the dimension of the cell with $Z = 4$.

Single crystals of phase I invariably fragmented when cooled below the transition-point. Mones and Post suggested tentatively that phase II might be tetragonal with $a = 1.12$ nm and $c = 1.15$ nm. Recently Rudman and Post[132] have studied phase II again at 118 K and have indexed the diffraction patterns on the basis of an hexagonal unit cell for which $a = 1.43$ nm, $c = 0.884$ nm, and $Z = 4$.

1.4.23 Tetramethyltin, Tetramethyllead

Staveley[133] states that the solid forms of these compounds have no transitions and melt with entropies of fusion $\Delta S_f = 43.2$ and 44.4 J mol^{-1}K^{-1} respectively.

1.4.24 *t*-Butyl chloride (2-Chloro-2-methyl propane)

t-Butyl chloride melts at about 248 K.[100,134,135] Phase I is formed from the melt[136] at about 245 K. The crystal structure of this phase was first determined by Schwartz, Post and Fankuchen.[137] Single crystals were found to be optically isotropic and oscillation diagrams showed the crystal structure to be face-centred cubic; the dimension of the unit cell was given as $a = 0.840$ nm. A later investigation[136] at 230 K refined the unit cell parameter to 0.862 nm with $Z = 4$.

At 209.8 K[100] phase I changes to phase II, the structure of which is tetragonal, space group $P4/nmm$, $a = 0.708$ nm, $c = 0.614$ nm with 2 molecules in the cell. The bond between the central carbon and chlorine atoms lies along the C axis; since this axis has fourfold rotational symmetry the three methyl groups must be rotating or orientated randomly around this axes.

At above 178 K phase II transforms to phase III, the reverse transformation occurs at about 182 K.

1.4.25 *t*-Butyl bromide (2-Bromo-2-methylpropane)

The solid state transition II → I takes place at 231.6 K[138] and another III → II at 208.6 K. Optical examination by Schwartz, Post, and Fankuchen[137] showed that phase I is optically isotropic and phase II anisotropic. Single crystals of the isotropic phase were subjected to x-ray investigation and the structure of phase I was found to be face-centred cubic with $a = 0.878$ nm and $Z = 4$.

1.4.26 Chloroform

Chloroform melts at 210.5 K, when studying the powder diagrams Fourme and Renaud[139] found no evidence of a phase transition between 98 K and 209 K. Single crystals were grown in a capillary tube and orientated in the camera using Laue photographs. Chloroform was found to be orthorhombic at 185 K with a space group *Pnma* (D_{2h}^{16}). The unit cell has the dimensions $a = 0.7485 \pm 0.0005$ nm, $b = 0.9497 \pm 0.001$ nm, $c = 0.5841 \pm 0.001$ nm and there are 4 molecules in the unit cell.

1.4.27 Iodoform

Khostsyanova, Kitaigorodskii, and Struchov[140] derived the space group $C6_3(C_6^6)$ for iodoform, the dimensions of the unit cell are, $a = 0.683 \pm 0.002$ nm, $c = 0.753 \pm 0.002$ nm and the cell contains 2 molecules.

1.4.28 Methylchloroform (1,1,1-Trichloroethane)

Rudman and Post[136] discovered that methylchloroform formed a cubic phase from the melt at a temperature of 235 K, just below the melting point

240 K. X-ray diffraction data on this phase (I) showed that its structure was face-centred cubic with a = 0.839 nm and 4 molecules in the unit cell.

The transition temperature for II → I has been found by other workers[141] to be about 224 K while Silver and Rudman[142] found that the change I → II took place at about 227 K. The structure of the lower temperature phase was studied at two temperatures, 213 and 128 K and at both temperatures was found to be orthorhombic with space group *Pnma*. There are 4 molecules in the unit cell. The dimensions of the cell at 213 K are a = 0.1.153 nm, b = 0.80 nm, and c = 0.588 nm. Being more precise these new figures at 215 K supercede those published earlier.[136]

The figures obtained at 128 K are a = 1.146 nm, b = 0.789 nm and c = 0.581 nm not very different from those obtained at 213 K and, as mentioned, with the same space group. It was decided that the form of methylchloroform at this lower temperature was still phase II and the previous assessment[136] that there existed a monoclinic modification III was withdrawn.

1.4.29 Methylene iodide

Although no crystallographic data is available for this substance Marzocchi, Schettino and Califano[143] have analysed the infra-red spectra of the two crystalline forms at temperatures ranging from 77 K to 276 K. Phase I which is metastable (T_m = 278.7 K) is probably orthorhombic with space group D_{2h}^{5} or D_{2h}^{13}. Phase II which is stable (T_n = 279.2 K) is probably monoclinic with space group C_{2h}^{5}. The transformation[144] of I to II was observed with a polarizing microscope.

1.4.30 Isopropylidene chloride (2,2-Dichloropropane)

Using x-ray diffraction Rudman[145] found that a high temperature phase Ia was formed from the undercooled melt. This phase has a face-centred cubic unit cell of dimension a = 0.845 nm and with 4 molecules per unit cell.

Another phase, Ib, is formed from the melt at temperatures between 227 K[100] and 233 K:[146] this phase melts at about 240 K. At 228 K its crystal structure has been determined[136] as having a rhombohedral unit cell with a = 1.468 nm, α = 90.0° with 21 molecules in the unit cell.

Phase Ib transforms to a lower temperature phase II at a temperature of about 186 K[100,146] which has an orthorhombic unit cell of dimensions a = 0.967 nm, b = 0.609 nm, c = 0.940 nm and Z = 4.

1.4.31 Neohexane (2,2-Dimethylbutane)

This molecule is not so compact as neopentane, $C(CH_3)_4$, since one of the methyl groups has been replaced by an ethyl group. In the solid state it has two transition points, the first at 127 K and the second at 140 K[147]. Turkevich and Smyth[148] observed the lower transition with a polarizing microscope and reported that neohexane was anisotropic below and becomes isotropic above this temperature. This combined with the large entropy of transition and a small entropy of fusion suggests that molecular re-orientation may begin at the lower transition.

X-ray diffraction[149] of the high temperature form was difficult. Only four reflections could be observed 111, 200, 220, and 222 from which it was deduced that the structure was face-centred cubic with four molecules in the unit cell and $a = 0.890$ nm.

1.4.32 Tetrakis(methylmercapto)methane (Tetramethylorthothiocarbonate)

Between 288 K and its melting point at 338 K there are three different crystalline modifications of $C(SCH_3)_4$; all three have a high degree of plasticity. Below 296.3 K the crystal structure is tetragonal, the space group is $P42_1C(D^4_{2d})$ and the dimensions of the unit cell are $a = 0.8536$ nm $c = 0.6948$ nm at 291 K with $Z = 2$. The x-ray diagrams[150] are normal and a sufficient number of reflections with their intensities were obtained for calculating reasonable electron density contours and for deducing the dimensions and symmetry of the molecule. In the crystal structure there are layers of molecules stacked in the *c* direction, the molecules in one layer are rotated slightly about the *c* axis and those in the next layer are rotated in the opposite direction (c.f. adamantane). The methyl groups are arranged in a pattern resembling a body-centred lattice, the interstices of which are alternatively empty or filled with a CS_4 group.

On heating, a solid state transformation takes place at 296.3 K to another but more strongly birefringent tetragonal phase. The space group[12,15] is $I4/mmm$ (D^{17}_{4h}) with $Z = 2$, $a = 0.817$ nm and $c = 0.796$ nm. In this crystal the CS_4 tetrahedra are disordered, being randomly distributed between two orientations related by a 90° rotation about the *c* axis. The methyl groups are packed in layers paralled to (001) and in this modification each methyl can take either of two equivalent positions with respect to the CS_4 tetrahedron. The diffracting power of this phase is rather weak and decreases with increasing angle of reflection.

Another phase transformation occurs at 318.6 K to a completely optically isotropic cubic modification. The four lines on the powder diagram agree with those for a body-centred cubic lattice of space group $Im3m$ (O^9_h); the dimensions of the unit cell containing two molecules is $a = 0.815$ nm. In this phase the high symmetry of the lattice can be reconciled with the point symmetry of the molecule by an orientational disordering of the molecules similar to that described for phase II but now taking place around the three fourfold axes.

1.4.33 Tetrakis(methylmercapto)silane

This compound has a cubic phase in equilibrium with the liquid at 304 K.[152] A large freezing point depression constant was found indicating that the entropy of fusion is small. $Si(SCH_3)_4$ and $C(SCH_3)_4$ molecules are geometrically similar and together with $C(CH_2CH_3)_4$, $C(CH_2hAl)_4$ and $C(CH_2OH)_4$ form an interesting group of molecules.

1.4.34 Hexafluoroethane

The melting point of C_2F_6 is 173.1 K and in the solid state there is a transition at 103.98 K.[153]

According to Lewis and Pace[154] the structure of the low temperature solid at 83 K is monoclinic, the space group is C_{2h}^2 and the unit cell contains 2 molecules.

1.4.35 Succinonitrile (1,2 Dicyanoethane)

The plastic crystalline phase of this compound is body-centred cubic[155] with $a = 0.637$ nm and $Z = 2$. The melting point is at 327.4 K and the transition temperature[156] is at 233.2.

1.4.36 Dimethyl acetylene

Dimethyl acetylene melts at 241 K and there is a solid-state transition at 154 K. The structures of both phases I and II have been determined by x-ray analysis.[157] Both phases are tetragonal, the unit cell parameters of the high temperature form at 223 K are $a = 0.542 \pm 0.002$ nm, $c = 0.689 \pm 0.002$ nm with $Z = 2$, and those of the low-temperature form are $a = 0.536 \pm 0.0006$ nm, $c = 1.367 \pm 0.003$ nm at 138 K.

Below its melting point acetylene, C_2H_2, crystallizes in a plastic phase in which molecules undergo reorientation and self diffusion, reorientations of the whole molecule occur also in the low-temperature form.[158] In contrast, with dimethylacetylene, any type of overall molecular rotation about an axis perpendicular to the molecular axis does not take place. On the other hand the methyl groups are mobile in both phases. In phase I the methyl groups are undergoing slightly hindered rotation about the molecule axis and in phase II the rotation is more severely hindered.

1.4.37 Cyclopropane

Cyclopropane melts at 146.6 K. Bates[159] has analysed the polarized infra-red spectrum of a single crystal at 85 K and after making certain assumptions proposed that the space group of the orthorhombic crystal is $Pnm2_1$ (C_{2v}^7) and that there are two molecules in the primitive unit cell. No evidence was found for a solid state phase change at or above 85 K.

Bond density maxima are shifted away from the bond axis in cyclopropane[160] as in other strained three-membered rings such as ethylene oxide[161] and ethyleneimine.[162]

1.4.38 Hexachlorocyclopropane

For crystals of this compound Tanako *et al.*[163] derive a space group $P2_1/a$ (C_{2h}^5). The monoclinic unit cell has the dimensions $a = 1.260$ nm, $b = 0.613$ nm, $c = 1.094$ nm and $\beta = 112.0°$.

1.4.39 Cyclobutane

Cyclobutane melts at 182 K and has a solid state transition temperature[164] at 145 K. In powder photographs[165] of the high temperature phase I, only a single line was observable. Single crystal rotation patterns showed only reflections of the forms (110) and (200) and from these a body-centred cubic lattice was deduced with the unit cell dimension $a = 0.606 \pm 0.003$ nm, $Z = 2$. To achieve cubic symmetry the molecules are orientationally disordered.

1.4.40 Octachlorocyclobutane

Margulis[166] has refined the earlier analysis of the x-ray data of Owen and Hoard[167] and has found the space group of the crystals to be $P2_1/m(C_{2h}^2)$. The monoclinic unit cell has the dimensions $a = 0.800$ nm, $b = 1.064$ nm, $c = 0.628$ and $\beta = 107.45°$ and $Z = 2$.

1.4.41 Cyclopentane

Cyclopentane has a very similar heat capacity curve[168] to that reported for neohexane.[169] There are two solid state first order transitions, one at 122 K and the other at 138 K; the melting point is at 179.7 K. Post, Schwartz, and Fankuchen[149] found that single crystals of cyclopentane are uniaxial above 138 K, on cooling below this temperature they changed into a polycrystalline modification of seemingly lower symmetry.

The intensities of x-ray reflections diminish rapidly as the Bragg angle increased, but they could be indexed on the basis of an hexagonal unit cell with $a = 0.583$ nm, $c = 0.933$ nm.

The space group was uncertain but either $C6/mmc$, (D_{6h}^4), $C6mc(C_6^4)$ or $C\overline{6}2c$ (D_{3h}^4). There are 2 molecules in the unit cell.

The structure is very close to hexagonal close packing, which has only one set of parallel close packed planes compared with face-centred cubic close packing which has four equivalent sets of such planes. There is lower possibility of gliding occurring in hexagonal close-packed crystals.

1.4.42 Cyclohexane

Cyclohexane melts at 279.8 K and undergoes a second order solid state transition at 186.1 K.[170] Hassel and Sommerfeldt[171] established that the crystals are optically isotropic above the transition temperature. They investigated the structure of the high temperature form at two temperatures, 233 K and 268 K and found it to be face-centred cubic with $a = 0.876$ nm, Z = 4. The space group was not definitely established but the authors consider that it is likely to be $P23$ (T′). The investigation of this phase by Oda[172] using single crystals, found the unit cell to be face-centred with $a = 0.873$ nm and $Z = 4$ in close agreement with Hassel and Sommerfeldt. The space group is either $F\bar{4}3m$ (T_d^2 or $F43$ (O^3) or $Fm3m$ (O_h^5). Besides the Bragg reflections Oda records the presence of diffuse spots and halos. Murti[173] found $a = 0.880$ nm at 268 K.

1.4.43 Cyclopentanone and Cyclohexanone

Cyclopentanone[171] does not form plastic crystals Cyclohexanone[171] has a transition point at 224.8 K and melts at 242 K. Above the transition point the crystals are optically isotropic, cubic and well-formed, below the transition point they are birefringent. At 228 K the crystal structure of the high temperature form is face-centred cubic with a unit cell dimension $a = 0.861$ nm and $Z = 4$.

1.4.44 Chlorocyclohexane and Bromocyclohexane

Chlorocyclohexane[171] has a face-centred cubic unit cell, with $a = 0.905$ nm and $Z = 4$, at 228 K. It melts at 232 K and the transition point is at 224 K.

Bromocyclohexane[171] does not form cubic crystals.

1.4.45 Cyclohexanol

Cyclohexanol[171] melts at 298.3 K. Above its transition point at 244.6 K it crystallizes as cubic crystals; below this temperature it forms doubly refracting non-cubic crystals. X-ray examination at 280 K gave the structure of the high-temperature form as face-centred cubic with $a = 0.833$ nm and $Z = 4$. Oda[172] has studied the x-ray diffraction of single crystals and discusses the diffuse scattering in relation to a molecular rotation and the effect of hydrogen bond formation in correlating molecular orientations.

1.4.46 Cyclooctane

The triple point of cyclooctane[174] is 287.98 K. The structure of phase I is primitive cubic with eight molecules in the unit cell[175] and the dimension $a =$

1.19 ± 0.005 nm at 273 K and $a = 1.182$ nm at 209 K. The suggested space group is $Pm3m(O_h^1)$.

1.4.47 Cubane

Cubane melts at 473 K; at room temperature the crystal structure[176] has the space group $R\overline{3}$. The rhombohedral unit cell contains 1 molecule and has the parameters $a = 0.534 \pm 0.0002$ nm, $\alpha = 72.26 \pm 0.05°$.

1.4.48 Triethylenediamine (1,4 Diazabicylco [2,2,2]-octane)

The shape of the molecules of this compound are roughly oblate spheroidal. The melting point is 433 K and there is a solid state transition temperature at 351 K.[177] Bruesch[178] gives the structure of the high temperature form as face-centred cubic with 4 molecules in the unit cell.

The vapour pressures of the two forms as functions of temperature were carried out by Wada *et al.*[179] who also determined the structure of the low temperature crystalline modification II. They found it to be an hexagonal close packed arrangements with the space group $P6_3/m$ (C_{6h}^2). The dimensions of the unit cell are $a_{hex} = 0.620 \pm 0.001$ nm and $c_{hex} = 0.958 \pm 0.002$ nm with Z = 2. These conclusions are supported by the later work of Weiss[180] who confirmed the space group and found slightly different unit cell dimensions: $a_{hex} = 0.614 \pm 0.0015$ nm, $c_{hex} = 0.946 \pm 0.002$ nm. The last authors discuss the possibility that one half of the molecules might be twisted around the N–N molecular axis relative to the other half, so that each six-membered ring would be in a slightly skewed boat configuration. If the molecules, some with left-handed and others with right-handed twist, were located at sites randomly, a measure of disorder would be present in the low-temperature crystal. Another possibility is that there is disorder in the ABC stacking sequence of the hexagonal close-packed planes: no evidence of this was found. However the crystal structure of II has been determined by Nimmo and Lucas using three dimensional neutron diffraction by a single crystal. They found that only the model with non-twisted molecules resulted in an acceptable structure.

1.4.49 Quinuclidine (1-Azabicyclo [2,2,2]-octane)

The x-ray investigation of chinuclidin was carried out by Nowacki[182] at room temperature. Only a small number of reflections were observable above a strong background. The unit cell is face-centred cubic with $a = 0.8977 \pm 0.0009$ nm and Z = 4. A later determination[183] of the unit cell dimension gave $a = 0.895 \pm 0.001$ nm. Since the molecules of Quinuclidine have trigonal symmetry around the N–C axis it must be assumed that these molecular axes are rotationally disordered statistically parallel to the four space diagonals of the cubic unit cell.

The low temperature form was found to be hexagonal close packed. The hexagonal cell has $a_{hex} = 0.61$ nm and $c_{hex} = 1.00$ nm.

1.4.50 Bicyclo[2,2,2]-octane

Bruesch[183] has determined the structures of the high and low temperature solid phases of this compound. The high temperature form was found to be face-centred cubic with $a = 0.914 \pm 0.0001$ nm and $Z = 4$. The low temperature modification is hexagonal: the unit cell has the parameters $a_{hex} = 0.64$ nm $c_{hex} = 1.51$ nm with $Z = 3$.

1.4.51 Adamantane (Tricyclo [3,3,1,1^{3,7}] Decane

Adamantane has a high melting point, 542 K and there is a solid state phase transition at 208.6 K[184], it crystallizes from methyl alcohol as octahedral crystals. Nowacki[185] could observe only sixteen lines in his powder photographs which he indexes on the assumption of a face-centred cubic unit cell of dimensions 0.9426 nm. He assumed that the space group was the ordered $F\bar{4}3m$ and calculated the positions of the atoms. At the same time Giacomello and Illuminati[186] independently reported the results of their powder investigations and reached essentially the same conclusions as Nowacki but derived a slightly different unit cell parameter $a = 0.954$ nm. Using single crystal data, Nordman and Schmitkons[17,187] redetermined the structure of the room temperature phase. They found that their data satisfied Nowacki's assumed space group very well. However, they favoured the possibility that the space group is $Fm3m$ with the molecules orientationally disordered: the molecules being randomly distributed between two orientations related by a 90° rotation and indicated by

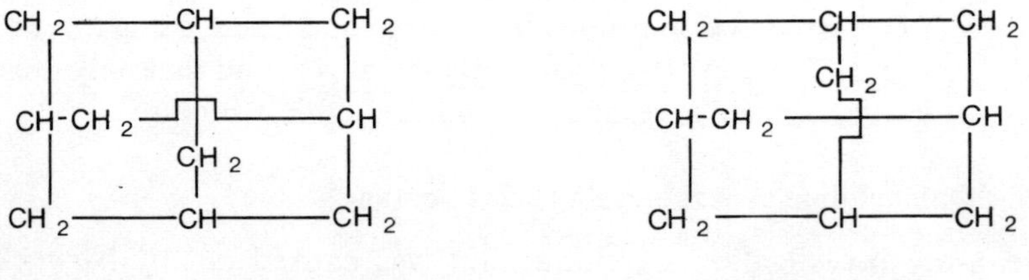

```
CH2 ------- CH ------- CH2          CH2 ------- CH ------- CH2
 |          |           |            |          |           |
 |          |           |            |         CH2          |
CH-CH2 ----+-+-------- CH           CH-CH2 ----+-+-------- CH
 |          |           |            |          |           |
 |         CH2          |            |          |           |
 |          |           |            |          |           |
CH2 ------- CH ------- CH2          CH2 ------- CH ------- CH2
```

(here one of the central methylene groups, CH_2, lies immediately above the other). By considering how the ratio of observed to calculated structure amplitudes depended on the calculated phase angle, they were able to conclude that the structure is disordered and the space group is $Fm3m$ (O_h^5). The dimension of the unit cell is 0.945 nm and $Z = 2$. Lucas[188] confirmed these results.

The low temperature phase II was also studied by Nowacki,[185] by Nordman and Schmitkons[187] and by Lucas.[188] This phase has a tetragonal unit cell, the space group is $P\bar{4}2_1C$ (D^4_{2d}) and $Z = 2$. The tetragonal cell is related to the cubic unit cell of phase I.[17,187] When the molecules in the cubic lattice are tilted through an angle of 9° around one *a* axis, this axis is preserved as the *c* axis of the tetragonal cell. The dimensions of the cell are $a = 0.660$ nm and $c = 0.881$ nm.

Using powder diffraction methods Mirskaya[189] has determined at room temperature the thermal expansion coefficients of the two phases and found $(dV/dT)/V = \alpha = 0.7 \times 10^{-4}K^{-1}$ (phase II) and $\alpha = 4.4 \times 10^{-4}K^{-1}$ (phase I).

1.4.52 Hexamethylenetetramine (1,3,5,7-tetraazatricyclo [3,3,1,1^{3,7}] Decane)

The molecule has a cage-like structure similar to that of adamantane, with four trivalent nitrogens replacing the four methane CH groups. It crystallizes with a body-centred cubic lattice,[190] space group $I\bar{4}3m$ (T^3_d) with $a = 0.702$ nm.

More recent studies[191] have confirmed and extended this data, giving *a* (298 K) = 0.7021 nm, *a* (100 K) = 0.6931 nm. and *a* (34 K) = 0.6910 nm. No evidence of a transition was found in calorimetric investigations[192] up to 500 K and Staveley[193] concludes that it does not form a plastic crystal.

1.4.53 Phosphorous Trioxide (P_4O_6)

The molecules of this compound and of others such as As_4O_6, Sb_4O_6, As_4S_6 have the same cage-like structure as those of adamantane and hexamine.

1.4.54 Congressane

Congressane[19] melts at 510 K, almost as high as adamantane, its density (1.210 g cm^{-3}) as determined by x-ray is unusually high for a hydrocarbon. The crystal structure[195] at room temperature has the space group $Pa3$ (T^6_h): the face-centred cubic cell has the dimensions $a = 1.0109 \pm 0.002$ nm with 4 molecules in a unit cell.

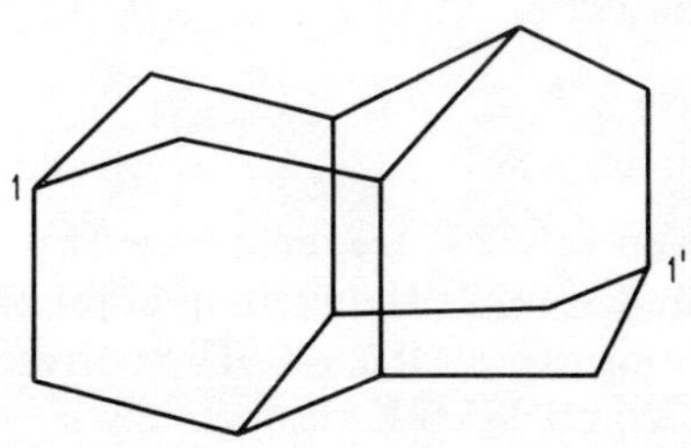

1.4.55 Norbornane (Bicyclo [2,2,1] Heptane), Norborylene (Bicyclo [2,2,1] Hept-2-ene) Norbornadiene (Bicyclo [2,2,1] Hepta-2,5-diene)

This group of materials has been examined by powder diffraction methods by Jackson and Strange.[196] Norbornane forms two plastic phases, a hexagonal close packed phase a = 0.617 nm, c = 1.003 nm (131 K – 306 K) and a face centred cubic phase a = 0.873 from 306 K to the melting point 360 K. The other two materials form hexagonal close packed plastic phases with unit cell parameters norbornadiene, a = 0.589 nm, c = 0.951 nm, T_m = 254 K and norbornylene, a = 0.608 nm, c = 0.981 nm, T_m = 320 K.

1.4.56 *dl*-Camphor

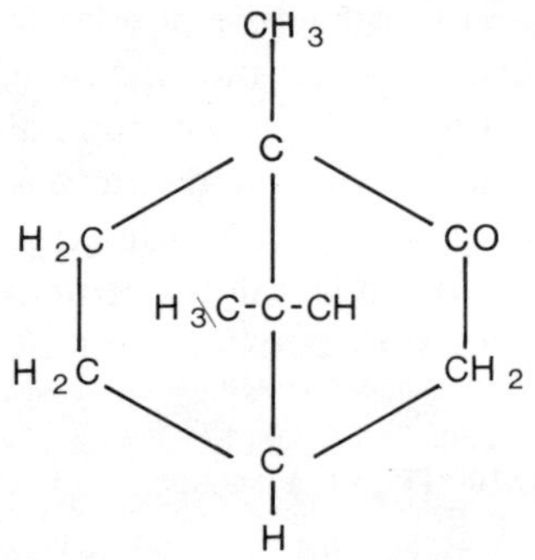

CH_3
C
H_2C CO
H_3C-C-CH_3
H_2C CH_2
C
H

A 'cage' molecule, *dl*-camphor melts at 449 K and has two solid state transitions at 350 K and 203 K.[156] Finbak[155] has studied this compound at 423 K. The x-ray diagrams showed strong background blackening and the results are best represented by a face-centred cubic cell with a = 1.01 nm and Z = 4.

1.4.57 *d*-Camphene and *dl*-Camphene

Aston[156] gives the melting point of *d*-camphene as 313.6 K and the transition temperature as 149–155 K. The corresponding temperatures for *dl*-camphene are 324 K and 153 K. The plastic crystalline phase has a body-centred structure with a = 0.800 nm at 293 K according to Finbak[155] and Wheeler[197] reports a = 0.798 nm at 298 K.

1.4.58 Borneol

X-ray diagrams[155] taken at 463 K are represented by a face-centred cubic cell having a = 1.025 nm and Z = 2. The melting point of *l*-borneol is given[156] as 476.6 K and that of *dl*-borneol as 486.6 K. The corresponding transition temperatures are 344–348 K and 345.6 K respectively.

1.4.59 Bornyl Chloride

The melting point of *dl*-bornyl chloride is 400 K and it has a transition temperature[156] of 163 K. Finbak[155] obtained powder photographs at 385 K showing heavy background scattering. The unit cell is face-centred cubic with $a = 1.039$ nm and four molecules in it.

1.4.60 Pivalic Acid (Trimethylacetic Acid)

The initial study of pivalic acid yielded only three x-ray reflections from which Namba and Oda[198] assigned it a face-centred cubic symmetry, $a = 0.882$ nm, $z = 4$.

A more comprehensive study of the diffuse scattering involved is described in Chapter 3.

1.4.61 Pentaerythritol and Derivatives

As indicated above (Section 1.3) this material transforms[40] from a tetragonal structure to a cubic, plastic crystal at 457 K. Doshi, Furman, and Rudman[199] have examined a series of substituted derivatives: R.C. $(CH_2OH)_3$. All form disordered face-centred cubic phases at high temperatures. For all substituents $-COOH$, $-CH_3$, $-NO_2$ and $-NH_2$ the cell parameter $a \cong 0.89$ nm.

REFERENCES

1. F. Simon and C. von Simson, *Z. Physik*, **21**, 168 (1924).
2. A. Eucken (priv. comm. to Simon and von Simson, referred to above).
3. L. Pauling, *Phys. Rev.*, **36**, 430 (1930); see also S. B. Hendricks, *Nature*, **126**, 167 (1930); *Z. Krist.*, **67**, 106, 475 (1930); **68**, 189 (1930).
4. J. Timmermans, *Bull. soc. chim. Belge*, **44**, 17 (1935); *J. chim. phys.*, **35**, 331, (1938); *J. Phys. and Chem. Solids*, **18**, 1 (1961).
5. C. Finbak, *Arch. Math. Naturvidenskab*, **B42**, No. 1, 71 (1938); B. Post, *Acta Cryst.*, **12**, 239 (1959); W. J. Dunning, *J. Phys. Chem. Solids*, **18**, 21 (1961).
6. F. Zernike, *Ned. Tijdschr. Natuurkunde*, **8**, 66 (1941).
7. H. M. James and T. A. Keenan, *J. Chem. Phys.*, **31**, 12 (1959).
8. Y. Koga and J. A. Morrison, *J. Chem. Phys.*, **62**, 3259 (1975).
9. G. Tamman, *Z. Physik. Chem.*, **25**, 441 (1898); '*States of Aggregation*' van Nostrand, New York (1926) p. 220; P. Othmer, *Z. Anorg. Chem.*, **9**, 209 (1915); I. N. Stranski and D. Totomanow, *Naturwiss*, **20**, 905 (1932); *Z. Physik. Chem.*, **A163**, 399 (1933); M. Volmer, *Kinetic der Phasenbildung*, Steinkopff, Dresden, (1939), Chapter 3: R. Lacmann, *Z. Naturforsch.* **17a**, 812 (1962) L. Dufour and R. Defay, *Thermodynamics of Clouds*, Academic Press, London, (1963), p. 201.
10. E. F. Westrum and J. P. McCullough, in '*Physics and Chemistry of the Organic Solid State*', ed. D. Fox, M. M. Labes and A. Weissberger, Interscience Wiley, London, 1963, p. 47.
11. L. A. K. Staveley, J. B. Warren, H. P. Paget, and D. J. Dowrick, *J. Chem. Soc.*, 1992 (1954).

12. W. G. Perdok and P. Terpstra, *Rec. Trav. Chim. Pays Bas*, **65**, 493(1946).
13. D. G. Thomas, L. A. K. Staveley, and A. F. Cullis, *J. Chem. Soc.*, 1772(1952); see also N. H. Hartshorne and P. McI. Swift, *J. Chem. Soc.*, 3705(1955).
14. E. H. Wiebenga, *Z. Anorg. Chem.*, **225**, 38(1935).
15. R. Becker, *Ann. Physik*, **32**, 128(1938); J. Frenkel, *Phys. Z. Sowjet*, **1**, 489(1932). S. H. Bransom, B. Millard and W. J. Dunning, *Disc. Faraday Soc.*, **5**, 83(1948).
16. H. Suga and S. Seki, *J. Non-Crystalline Solids*, **16**, 171(1974).
17. C. E. Nordman and D. L. Schmitkons, *Acta Cryst.* **18**, 764(1965); see also P. Coppens and T. M. Sabine, *Mol. Cryst.*, **3**, 507(1968).
18. C. Finbak and O. Hassel, *Z. Physik. Chem.*, **B36**, 301(1937).
19. Yu. V. Mnyukh and N. N. Petropavlov, *J. Phys. Chem. Solids* **33**, 2079(1972); **34**, 159(1973).
20. N. H. Hartshorne and M. H. Roberts, *J. Chem. Soc.*, 489(1955).
21. W. C. McCrone, *Disc. Faraday Soc.*, **5**, 158(1949); see also G. Tammann, *Z. Anorg. Chem.*, **182**, 289(1929).
22. A. L. Kitaigorodsky, *J. Chim. Phys.*, **63**, 9(1966).
23. R. A. Buckingham, *Proc. Roy. Soc.*, **A168**, 264(1938).
24. J. E. Lennard-Jones, *Proc. Roy. Soc.*, **A106**, 463(1924).
25. R. A. Scott and H. A. Scheraga, *J. Chem. Phys.*, **45**, 2051(1966).
26. L. L. Shipman, A. W. Burgess, and H. A. Scheraga, *J. Phys. Chem.*, **80**, 52(1975).
27. A. I. Kitaigorodsky, B. D. Korechkov, and A. G. Kulkin, *Fiz. Tverd. Tela SSSR*, **7**, 643(1965).
28. A. I. Kitaigorodsky and K. V. Mirskaya, *Mol. Cryst. Liq. Cryst.*, **6**, 339(1970).
29. A. I. Kitaigorodsky, *Molecular Crystals and Molecules*, Academic Press, New York (1973); A. I. Kitaigorodsky and K. V. Mirskaya, *Mat. Res. Bull.*, **7**, 271(1972).
30. R. Kihara and S. Koba, *J. Phys. Soc. Japan*, **7**, 348(1952).
31. K. F. Niebel and J. A. Venables, *Proc. Roy. Soc.*, **A336**, 365(1974).
32. K. Mirsky, *Acta Cryst.*, **A32**, 199(1976).
33. E. F. Westrum, *J. Phys. Chem. Solids*, **18**, 83 (1961).
34. K. S. Pitzer, *J. Chem. Phys.*, **7**, 583(1939).
35. E. A. Guggenheim, *J. Chem. Phys.*, **13**, 253(1945).
36. N. Trappeniers, *Physica*, **17**, 501(1951). *Ind. Chim. Belge*, **16**, 372(1951); *Acad. Roy. Belge Classe Sci.*, **27**, 5(1952); *Compt. Rend. Reunion Ann. Avec Comm. Thermodyn.*, Union Intern. Phys. (Paris) 1952, *Changements de Phases*, p. 241.
37. A. O., Urvas, D. L. Losee, and R. O. Simmons, *J. Phys. Chem. Solids*, **27**, 2269 (1967); D. L. Losee and R. O. Simmons, *Phys. Rev.*, **172**, 944(1968); J. Naghizadeh and S. A. Rice, *J. Chem. Phys.*, **36**, 2710(1962).
38. F. A. Momany, L. M. Carruthers, R. F. McGuire, and H. A. Scheraga. *J. Phys. Chem.*, **78**, 1595(1974).
39. E. F. Westrum and J. P. McCulloch, in '*Physics and Chemistry of the Organic Solid State*', ed. D. Fox, M. M. Labes and A. Weissberger, Interscience, London 1963, Vol. 1.
40. E. F. Westrum, *Proceedings of the Symposium on Thermodynamics*, Fritzens-Wattens, 1959. '*Thermodynamics*', Butterworths London, 1961, p. 241.
41. F. S, Llewellyn, E. G. Cox, and T. H. Goodwin, *J. Chem. Soc.*, **1937**, 883; I. Nitta and T. Watanabe, *Nature*, **140**, 365(1937); *Sci. Papers Inst. Phys. Chem. Res.*, **34**, 1669(1938); E. W. Hughes, unpublished, see L. Pauling '*Nature of the Chemical Bond*', Cornell Univ. Press, 1939, p. 294.
42. D. P. N. Satchell and R. S. Satchell, *Quart. Rev.*, **25**, 171(1971).
43. E. O. Schlemper and D. Britton, *Inorg. Chem.*, **5**, 511(1966).

44. E. O. Schlemper and D. Britton, *Inorg. Chem.*, **5**, 507 (1966); H. A. Bent, *Chem. Rev.*, **68**, 587 (1968), p. 598.
45. Z. M. El Saffar, P. Schultz, and E. F. Meyer, *J. Chem. Phys*, **56**, 1477 (1972); C. Clemett and M. Davies, *Trans. Faraday Soc.*, **58**, 1705 (1962); E. F. Westrum and A. Ribner, *J. Phys. Chem.*, **70**, 1208 (1966).
46. D. Britton, *Acta Cryst.*, **B30**, 1818 (1974).
47. H. C. Clarke, R. J. O'Brien, and J. Trotter, *J. Chem. Soc.*, 2332 (1964).
48. G. Natta, *Gazz. Chim. Ital.*, **60**, 911 (1930); M. Atoji and W. M. Lipscomb, *Acta Cryst.*, **7**, 597 (1954); R. W. G. Wyckoff, '*Crystal Structures*' 2nd edn. Vol. II, Interscience, New York (1963), D. Britton Ref. 5.
49. H. A. Bent, *J. Chem. Educ.*, **42**, 348 (1965).
50. H. A. Bent, ref. 3, p. 623. *Chem. Rev.*, **68**, 623 (1968)
51. L. A. K. Staveley, *Ann. Rev. Phys. Chem.*, **13**, 351 (1962).
52. S. C. Greer and L. Meyer. *J. Chem. Phys.*, **51**, 4583 (1969).
53. P. Coppens in '*Chemical Crystallography*' International Reviews of Science, Physical Chemistry, Series II, Volume 11, ed. J. M. Robertson, Butterworths, London 1975, p. 21.
54. P. Coppens and C. A. Coulson, *Acta Cryst.*, **23**, 718 (1967); C. A. Coulson in '*Thermal Neutron Diffraction*, ed. B. T. M. Willis, Oxford, Univ. Press, 1970, p. 68.
55. P. G. Gunther, K. Holm, and H. Strunz, *Z. Phys. Chem.*, **B43**, 229 (1939).
56. M. Atoji and W. N. Lipscomb, *Acta Cryst.*, **7**, 173 (1954).
57. E. Sandor, *Disc. Faraday Soc.*, **48**, 117 (1969).
58. A. Eucken and C. Karwat, *Z. Phys. Chem.*, **112**, 467 (1927).
59. W. F. Giauque and R. Wiebe, *J. Amer. Chem. Soc.*, **50**, 2193 (1928); H. L. Johnston and W. F. Giauque, *J. Amer. Chem. Soc.*, **51**, 1441 (1929).
60. G. Natta, *Gazz. Chim. Ital.*, **63**, 425 (1933).
61. E. Sandor and R. F. C. Farrow, *Nature*, **213**, 171 (1967); Nature **215**, 1265 (1967); *Disc. Faraday Soc.*, **48**, 78 (1969).
62. K. Clusius and G. Wolf, *Naturforsch*, **2a**, 495 (1947).
63. B. Ruhemann and F. Simon, *Z. Phys. Chem.*, **B15**, 389 (1951).
64. E. Sandor and J. Clarke, *Acta Cryst*, **A28**, 5188 (1972).
65. G. Natta, *Nature*, **127**, 235 (1931).
66. K. Clusius and A. Frank, *Naturwiss.*, **24**, 62 (1936); K. Clusius, *Gottingen Nachr.*, **GrIII**, 171 (1933); *Z. Elektrochem.*, **39**, 598 (1933); K. Clusius and F. Weigand, *Z. Elektrochem.*, **44**, 676 (1938).
67. E. Justi and H. Nitka, *Phys. Z.*, **37**, 435 (1936); L. Vegard, *Nature*, **126**, 916 (1930); G. Natta, *atti R. Accad. Lincei, Ser.*, 6, **9**, 679 (1930).
68. J. H. Colwell, E. K. Gill, and J. A. Morrison, *J. Chem., Phys.*, **39**, 635 (1963).
69. A. Frank and K. Clusius, *Z. Phys. Chem.*, **B36**, 291 (1937); W. F. Giauque, R. W. Blue, and R. Overstreet, *Phys. Rev.*, **30**, 196 (1933).
70. K. Clusius, *Z. Phys. Chem.*, **B3**, 41 (1929).
71. J. H. Colwell, E. K. Gill, and J. A. Morrison, *J. Chem. Phys.*, **36**, 2223 (1962).
72. J. C. McLennan and W. G. Plummer, *Phil. Mag.*, **7**, 701 (1929); H. H. Mooy, *Comm. Phys. Lab. Univ. Leiden* No. 313d (1931); *Proc. Acad. Sci.*, **34**, 550 (1931); *Nature*, **127**, 707 (1931); A. Schallamach, *Proc. Roy. Soc.*, **A171**, 569 (1939); *Nature*, **143**, 375 (1939); J. Herzey and R. E. Stoner, *J. Chem. Phys.*, **54**, 2284 (1971).
73. S. C. Greer and L. Meyer, *J. Chem. Phys.*, **52**, 468 (1970); *Z. Angewandte Phys.*, **27**, 198 (1969).
74. J. A. Janik, K. Otnes, G. Solt, and G. Kosaly, *Disc. Faraday Soc.*, **48**, 87 (1969).
75. H. A. Kruis, L. Popp, and K. Clusius, *Z. Elektrochem.*, **43**, 664 (1937); K. Clusius and L. Popp, *Z. Phys. Chem.*, **B46**, 63 (1940).

76. D. N. Bol'shutkin, V. M. Gason, A. I. Prokhvatilov, and A. I. Erenburg, *J. Structural Chem.*, **12**, 313 (1971).
77. W. Press, B. Dormer, and G. Will, *Phys. Letters*, **31A**, 253 (1970); W. Press, *J. Chem. Phys.*, **56**, 2597 (1972).
78. E. Arzi and E. Sandor, *Acta Cryst.*, **A28**, *S*188 (1972).
79. E. A. Ballik, D. J. Gannon, and J. A. Morrison, *J. Chem. Phys*, **57**, 1793 (1972); **58**, 5639 (1973).
80. W. Press and A. Kollmar, *Solid State Comm.*, **17**, 405 (1975).
81. M. Bloom and J. A. Morrison, '*Orientational Order and Disorder in the Solid Isotopic Methanes*', Specialist Periodic Reports, The Chemical Society, London, 1972, Chap. 6, pp. 140–159.
82. T. Nakamura and M. Miyagi, *J. Chem. Phys.*, **54**, 5276 (1971).
83. K. Clusius, *Z. Phys. Chem.*, **B23**, 213 (1933).
84. M. L. Klein, J. A. Morrison, and R. D. Weir, *Disc. Faraday. Soc.*, **48**, 93 (1969). W. M. Sears and J. A. Morrison, *J. Chem. Phys.*, **62**, 2736 (1975).
85. R. P Fournier, R. Savoie, N. D. The, R. Belzile, and A. Cabana, *Canad. J. Chem.*, **50**, 35 (1972).
86. K. Clusius and G Faber, *Z. Phys. Chem.*, **B51**, 352 (1942).
87. N. D. The, J.-M. Gagnon, R. Belzile, and A. Cabana, *Canad. J. Chem.*, **52**, 327 (1974).
88. R. E. Wilde, *J. Phys. Chem. Solids*, **38**, 257 (1977).
89. D. N. Bol'shutkin, V. M. Gasan, A. L. Prokhvatilov, and A. I. Erenburg, *Acta.*, **B28**, 3542 (1972).
90. S. C. Greer and L. Meyer, *J. Chem. Phys.*, **51**, 4583 (1969); V. M. Gasan, A. I. Prokhavatilov, and A. I. Erenburg (1970).
91. E. L. Pace and J. S. Moser, *J. Chem. Phys.*, **39**, 154 (1963).
92. G. Natta, *Gazz. Chim. Ital.*, **60**, 911 (1930).
93. M. Atoji and W. N. Lipscomb, *Acta Cryst.*, **7**, 597 (1954).
94. W. I. Patnode and J. Papish, *J. Phys. Chem.*, **34**, 1494 (1930).
95. T. Shinoda, *J. Phys. Soc. Japan*, **38**, 224 (1971).
96. B. Post, *Acta Cryst.*, **12**, 349 (1959); V. Goldschmidt, *Z. Krist.*, **51**, 26 (1912); W. Wahl, *Proc. Roy. Soc.*, **A89**, 327 (1914).
97. R. Rudman and B. Post, *Science*, **154**, 1009 (1966).
98. E. A. Ballik, D. J. Gannon, and J. A. Morrison, *J. Chem. Phys.*, **59**, 5639 (1973).
99. K. Kotake, N. Namakura, and H. Chihara, *Bull. Chem. Soc. Japan*, **40**, 1018 (1967).
100. L. Silver and R. Rudman, *J. Phys. Chem.*, **74**, 3134 (1970).
101. J. P. Badiali, J. Bruneaux-Poulle, and A. Defrain, *J. Chim. Phys.*, **73**, 113 (1976).
102. P. W. Bridgman, *Phys. Rev.*, **3**, 153 (1914).
103. G. J. Piermarini and A. B. Braue, *J. Phys. Chem.*, **58**, 1974 (1973); C. E. Weir, G. J. Piermarini, and S. Block, *J. Chem. Phys.*, 2089 (1969).
104. P. Brand and H. Sackmann, *Z. Anorg. Chem.*, **321**, 262 (1963).
105. P. Brand and H. Sackmann, *Z. Anorg. Chem.*, **321**, 262 (1963).
106. C. Finbak and O. Hassel, *Tidskr. Kjemi Bergves*, **17**, 145 (1937), via *Chem. Zentr.*, **88**, 1077 (1938).
107. H. Mark, *Ber Deuts. Chem. Ges.*, **57**, 1820 (1924).
108. H. Mark, *Z. Physik. Chem.*, **B38**, 209 (1937).
109. Cited in P. von Groth, *Kristallographie*, Vol. I. p., 230(1906).
110. P. von Groth, Kristallographie, Vol. I. p., reference 4(1906).
111. M. More, F. Baert, and J. Lefebvre, *Acta Cryst.*, **B33**, (1977).
112. M. More, J. Lefebvre, and R. Fouret, *Acta Cryst.*, **B34**, (1978).

113. E. Pohland, *Z. Anorg. Chem.*, **201**, 265 (1931).
114. S. M. S. Kennard and P. A. McCusker, *J. Amer. Chem. Soc.*, **70**, 1039(1948).
115. H. Sackmann, D. Demus, and D. Pankow, *Z. Anorg. Chem.*, **318**, 257(1962).
116. W. Blitz and K. Jeep, *Z. Anorg. Chem.*, **162** 32(1927).
117. O. Hassel and H. Kringstad, *Z. Physik. Chem.*, **B15**, 274(1931).
118. P. Brand and H. Sackmann, *Z. Anorg. Chem.*, **321**, 262(1963); *Acta Cryst.* **16** 446(1963).
119. H. Mark, *Ber. Deuts. Chem. Gesell*, **57**, 1820(1924).
120. O. Hassel and H. Kringstadt, *Z. Physik. Chem.*, **B13**, 1(1931).
121. O. Hassel and H. Kringstad, *Z. Physik. Chem.*, **B15**, 274(1931).
122. R. F. Rolston and H. H. Sisler, *J. Amer. Chem. Soc.*, **79**, 5891(1957).
123. F. M. Jaeger, P. Terpstra, and H. G. K. Westenbrink, *Proc. Koningk. Akad., Wetensch. Amsterdam*, **28**, 747(1925).
124. H. Mark and K. Weissenberg, *Z. Physik*, **16**, 1(1923).
125. H. Ott, *Z. Krist.*, **65**, 222(1926).
126. R. G. Dickinson, *J. Amer. Chem. Soc.*, **45**, 958(1923).
127. F. Meller and I. Fankuchen, *Acta. Cryst.*, **8**, 343(1955).
128. J. G. Aston and G. H. Messerly, *J. Amer. Chem. Soc.*, **58**, 2354(1936).
129. W. Wahl, *Proc. Roy. Soc.*, **88**, 356(1913).
130. H. Mark and W. Noethlin, *Z. Krist.*, **65**, 435(1927).
131. A. H. Mones and B. Post, *J. Chem. Phys.*, **20**, 755(1952).
132. R. Rudman and B. Post, *Mol. Cryst.*, **5**, 95(1968).
133. L. A. K. Staveley, *J. Phys and Chem. Solids*, **18**, 46(1961), L. A. K. Stavely, J. B. Warren, H. P. Paget, and D. J. Dowrick, *J. Chem. Soc.*, 1992(1954); L. A. K. Staveley, H. P. Paget, B. H. Goalby, and J. B. Warren, *J. Chem. Soc.*, 2290(1950).
134. L. M. Kushnem, R. W. Crowe, and C. P. Smyth, *J. Amer. Chem. Soc.*, **72**, 1091 (1950).
135. A. Dworkin and M. Guillanin, *J. Chem. Phys.*, **63**, 53(1966).
136. R. Rudman and B. Post, *Mol. Cryst.*, **5**, 95(1968).
137. R. S. Schwartz, B. Post, and I. Fankuchen, *J. Amer. Chem. Soc.*, **73**, 4490(1956).
138. W. O. Baker and C. P. Smyth, *J. Amer. Chem. Soc.*, **61**, 2798(1939).
139. R. Fourme and M. Renaud, *Compt. Rend.*, **B263**, 69(1966).
140. T. L. Khostsyanova, A. I. Kitaigorodskii, and Yu. T. Struchov, *Zh. Fiz. Khim.*, **27**, 647(1953).
141. T. R. Rubin, B. H. Levedahl, and D. M. Yose, *J. Amer. Chem. Soc.*, **66**, 270(1944).
142. L. Silver and R. Rudman, *J. Chem. Phys.*, **57**, 210 (1972).
143. M. P. Marzocchi, V. Schettino, and S. Califano, *J. Chem. Phys.*, **45**, 1400 (1965).
144. See also G. Tammann and R. Hollman, '*Kristallisieren und Schmelzen*' (Barth, Leipzig, 1903) p. 278; W. Wahl, *Z. Phys. Chem.*, **88**, 143(1914).
145. R. Rudman, *Mol. Cryst. Liq. Cryst.*, **6**, 427 (1970).
146. A. Turkwich and C. P. Smyth, *J. Amer. Chem. Soc.*, **62**, 2468(1940).
147. J. E. Kilpatrick and K. S. Pitzer, *J. Amer. Chem. Soc.*, **68**, 1066(1946).
148. A. Turkevich and C. P. Smyth, *J. Amer. Chem. Soc.*, **64**, 737(1942).
149. B. Post, R. S. Schwartz, and I. Fankuchen, *J. Amer. Chem. Soc.*, **73**, 5113(1951).
150. W. G. Perdok and P. Terpstra, *Rec. Trav. Chim. Pays-Bas*, **62**, 687(1943).
151. W. G. Perdok and P. Terpstra, *Rec. Trav. Chim. Pays-Bas*, **65**, 493(1946); W. G. Perdok, *Helv. Chim. Acta*, **30**, 1782(1947).
152. H. J. Backer and W. G. Perdok, *Rec. Trav. Chim., Pays-Bas*, **61**, 533(1943).
153. E. L. Pace and J. G. Aston, *J. Amer. Chem. Soc.*, **70**, 566(1948).
154. A. Lewis and E. L. Pace, *J. Chem. Phys.*, **58**, 3661(1973).

155. C. Finbak, *Arch. Math. Naturvidenskab*, **B42**, no. 1, 71(1938); C. Finbak and H. Viervoll, *Tidskr, Kjemi, Bergvesen*, **2**, 35, (1942); via *Chem. Zentr.*, **88**, 1077 (1938).
156. J. G. Aston in '*Physics and Chemistry of the Organic Solid State*', ed. D. Fox, M. M. Labes, and A. Weissberger, Interscience, London, 1963, p. 543.
157. E. Pignatoro and B. Post, *Acta Cryst.*, **8**, 672(1955); M. G. Miksic, E. Segerman, and B. Post, *Acta Cryst.*, **12**, 390(1959).
158. S. Albert and J. A. Ripmeester, *J. Chem. Phys.*, **57**, 3953(1972).
159. J. B. Bates, *J. Chem. Phys.*, **58**, 4236(1973).
160. A. Hartman and F. L. Hirschfeld, *Acta Cryst.*, **20**, 80(1966).
161. D. A. Mathews and G. D. Stuckey, *J. Amer. Chem. Soc.*, **93**, 5954(1971); *Trans. Amer. Crystallogr. Assn.*, **8**, 113(1972); D. A. Mathews, J. Swanson, and G. D. Stuckey, *J. Amer. Chem. Soc.*, **93**, 5954(1971).
162. T. Ito and T. Sakurai, *Acta Cryst.*, **B29**, 1594(1974).
163. T. Tanako, T. Chiba, Y. Sasada, M. Kakudo, S. Nasakura, and S. Murahahi, *Bull. Chem. Soc. Jap.*, **38**, 157(1965).
164. W. Rathje and A. Gwinn quoted in reference 2.
165. G. F. Carter and D. H. Templeton, *Acta Cryst.*, **6**, 805(1953).
166. T. N. Marfulis, *Acta Cryst.*, **19**, 857(1965).
167. T. B. Owen and J. L. Hoard, *Acta Cryst.*, **4**, 172(1951).
168. J. G. Aston, A. L. Fink, and S. C. Schumann, *J. Amer. Chem. Soc.*, **65**, 341 (1943).
169. J. G. Aston, B. Bolger, R. Trambarula, and H. Segall, *J. Chem. Phys.*, **22**, 460 (1954).
170. J. G. Aston, G. J. Szasz, and A. L. Fink, *J. Amer. Chem. Soc.*, **65**, 1135(1943); E. R. Andrew and R. G. Eades, *Proc. Phys. Soc. (London)*, **A65**, 371 (1952); *Proc. Roy. Soc. (London)*, **A216**, 398(1953).
171. O. Hassel and A. M. Sommefeldt, *Z. Phys. Chem.*, **40B**, 391(1938).
172. T. Oda, *X-Sen*, **5**, 26(1948).
173. G. S. R. K. Murti, *Indian J. Physics*, **32**, 460(1958).
174. H. L. Finke, D. W. Scott, M. E. Gross, J. F. Messerly, and C. Waddington, *J. Amer. Chem. Soc.*, **78**, 5469(1956).
175. D. E. Sands and V. W. Day, *Acta Cryst.*, **19**, 278(1965).
176. E. B. Fleischer, *J. Amer. Chem. Soc.*, **86**, 3889(1964).
177. S. S. Chang and E. F. Westrum, *J. Phys. Chem.*, **64**, 1557(1960); J. C. Trowbridge and E. F. Westrum, *J. Phys. Chem.*, **67**, 2381(1963).
178. P. Bruesch, *Spectrochim. Acta*, **22**, 861(1966); P. Bruesch and H. H. Funthard, *Spectrochim Acta*, **22**, 877(1966).
179. T. Wada, E. Kishida, Y. Tomiie, H. Suga, S. Seki, and I. Nitta, *Bull. Chem. Soc., Japan*, **33**, 1317(1960).
180. G. E. Weiss, A. S. Parkes, R. E. Nixon, and R. E. Hughes, *J. Chem. Phys.*, **41**, 3759 (1964).
181. J. K. Nimmo and B. W. Lucas, *Acta Cryst.*, **B32**, 348(1976).
182. W. Nowacki, *Helv. Chim. Acta*, **29**, 1798(1946).
183. P. Bruesch, *Spectrochim. Acta*, **22**, 861(1966).
184. S. C. Chang and E. F. Westrum, *J. Phys. Chem.*, **64**, 1547(1960).
185. W. Nowacki, *Helv. Chim. Acta*, **28**, 1233(1945).
186. G. Giacomello and G. Illuminati, *Gazz. chim. Ital.*, **75**, 246(1945).
187. C. E. Nordman and D. L. Schmitkons, *Acta Cryst.*, **18**, 764(1965); J. Donohue and S. H. Goodman, *Acta Cryst.*, **22**, 352(1967).
188. B. W. Lucas, *Thesis*, Queen Mary College, University of London, 1965.

189. K. V. Mirskaya, *Sov. Phys. Crystallography*; **8**, 167 (1963).
190. P. A. Schaffer, *J. Amer. Chem. Soc.*, **69**, 1557 (1947).
191. L. N. Becka and D. W. J. Cruickshank, *Proc. Roy. Soc.*, **A273**, 423 (1963); J. A. K. Duckworth, B. T. M. Willis, and G. S. Pawley, *Acta Cryst.*, **A25**, 482 (1969).
192. S. S. Chang and E. F. Westrum, *J. Phys. Chem.*, **64**, 1547 (1960); E. F. Westrum, E. T. Chang, S. S. Chang, and J. C. Trowbridge, *18th Intern. Congr. Pure and Appl. Chem.*, Montreal (1961).
193. L. A. K. Staveley, *Ann. Rev. Phys. Chem.*, **13**, 351 (1961).
194. C. Cupas, P. von, R. Schleyer, and D. J. Trecke, *J. Amer. Chem. Soc.*, **87**, 917 (1965).
195. L. Karle and J. Karle, *J. Amer. Chem. Soc.*, **87**, 918 (1965).
196. R. L. Jackson and J. H. Strange, *Acta Cryst.*, **B28**, 1645 (1972).
197. R. D. Wheeler, priv. comm., J. R. Green, and C. E. Schele, *J. Phys. Chem. Solids*, **28**, 383 (1967).
198. Y. Namba and T. Oda, *Bull. Chem. Soc. Japan*, **25**, 225 (1952).
199. N. Doshi, M. Furman, and R. Rudman, *Acta Cryst.*, **29B**, 143 (1973).

2

Lattice Defects, Self-Diffusion, and the Plasticity of Plastic Crystals

John N. Sherwood

2.1 INTRODUCTION

One of the most intriguing properties which provoked the recognition of this state of matter is that which led to the original name—the very high plasticity. At the time of the observation, a comparison was drawn between these materials and the more brittle organic solids which were regarded as normal. This, of course, is a chemists viewpoint based on the fact that most of the organic solids which he examines are brittle or friable. A metallurgist may well have interpreted normality conversely. The observation led however to the usually accepted description of this class of solids as '*Plastic Crystals*'[1] with the unfortunate implication that other molecular solids are not plastic. In truth the orientationally disorderd solids are an extreme; they are highly plastic. A general overview of the different solid types shows that plasticity is very closely interlinked with a general increase in orientational molecular disorder. The tendency to orientational disorder decreases as one proceeds from highly symmetrical systems to highly non-symmetric systems. This is reflected in a corresponding decrease in the plasticity of the material. The most significant example of this change occurs at the phase transformation from the orientationally-disordered phases to the lower temperature, more-ordered, phase. Here we have the same molecule in phases with, respectively, high and low plasticity.

One consequence of the last observation was that the high temperature orientationally disordered phase came to be regarded as a mesophase of a kind parallel to the well-investigated *liquid crystal* systems. It was suggested that the phase represented a restricted liquid state. This is an attitude which has unfortunately persisted despite the accumulation of evidence to the contrary. Whereas, obviously, both plastic and liquid crystalline phases represent gradual approaches to the liquid state, they have little else in common; the former is a distinct high temperature, solid phase. It was in the light of this belief however that Michels[2] made the first measurements of the relative plasticities of a series of organic crystals. He compressed the solids in a plastometer which consisted of a cylindrical chamber with a 1 mm orifice. The pressure required to extrude the compacted, polycrystalline material through the hole was measured. The results which are summarized in Table 2.1 were obtained at various temperatures and hence are not directly equivalent. However some obvious generalizations result.

The four cubic crystals which fall within the class being considered ($\Delta S_f/R \lessapprox 2.5$, where ΔS_f is the entropy of fusion) were expressed at pressures of about 250 kg cm^{-2}. In comparison, the low temperature forms, II, require 6–12 times more pressure, a range paralleled by the normal solids naphthalene and hexamethylbenzene in the lower half of the table. 1,4-dichlorobenzene and the two long chain molecules, all of which undergo limited re-orientation in the solid state are intermediate, requiring 3–5 times the pressure. Thus, for the first time Michels defines experimentally the distinction between the plastic and normal phases and also shows that there exists a gradation in plasticity which parallels the degree of re-orientational disorder in the crystal phase examined. A further observation, not usually noted in subsequent reviews, is that the extruded samples often showed the outward characteristics of extruded crystalline solids and not extruded viscous fluid samples.

Michels also comments for the first time that these materials should provide the metallurgist with substrates on which metallurgical deformation problems could be studied under easily attainable conditions.

This solid state character is adequately confirmed by the numerous x-ray crystallographic examinations reviewed by Dunning in Chapter 1, many of which are contemporary with Michels' studies. These yield only low scatter-

Table 2.1 Deformation of organic crystals[2]

	T_m/K	$\Delta S_f/R$	Crystalline form I	Crystalline form II	*P* kg cm^{-2} I	II
CBr_4	365.5	1.3	F	M	247	1500
C_2Cl_6	4.60	2.2	B	T	250	3600
d-Camphor	451.5	1.4	F	R	247	454
d-Borneol	581.6	2.1	F	—	240	864
Naphthalene	345	6.7	M	—	2240	—
Hexamethylbenzene	438	5.6	T	—	2800	—
1,4-Dichlorobenzene	325.7	6.8	M	M	740	1150
Dotriacontane	343.5	28	M	—	1200	—
Stearic acid	344	19.5	M	M	1120	—

F—f.c.c., B—b.c.c., O—orthorhombic, M—monoclinic, T–triclinic.
I—high temperature phase, II—lower temperature phase.

ing with rather diffuse halos. The results are however interpretable in terms of a basically crystalline lattice with orientational molecular disorder.

Having defined these materials as being predominantly solid-like and not liquid-like it then becomes more fruitful to follow Michels' advice and to examine the plastic deformation of these materials in this light.

2.2 DEFORMATION OF CRYSTALLINE SOLIDS

The subject of deformation of crystalline solids is a wide ranging one and any detailed consideration is outside the scope of this article.[3] It is well established however that the principal mobile species are lattice defects; point defects, dislocations and sub-grain-boundaries, and it is the migration of these through the lattice which gives rise to the necessary mass-transport which accompanies deformation.

There are two basic gross observations which limit the range of microscopic processes which one needs to consider. These relate to the reasons why these materials were initially referred to as *plastic crystals*:

(i) they flow under gravity: an observation usually made on materials such as cyclohexane, perfluorocyclohexane, and *dl*-camphene near their melting points

(ii) they deform easily under stress without fracture: usually noted for materials deformed by manual bending at temperatures further from the melting point.

Rather more scientifically these observations correspond to temperature and stress ranges in which different deformation mechanisms dominate. They represent experiments under

(a) high temperature, low stress

(b) low temperature, high stress

Although this is rather a sweeping generalization it serves to distinguish between the type of experiments performed to date and which have led to a more precise definition of the plasticity of these materials.

2.2.1 Experiments at High Temperature and Low Stress

At high temperatures and low stresses plastic deformation occurs by creep.[3] Measurement of the rate of deformation in tension usually yields the characteristic creep curve shown in Figure 2.1. On loading, after an initial transitory region—primary creep (A)—in which work hardening and changes in the substructure of the material occur, the deformation rate becomes linear with time: secondary or steady state creep (B). Finally, there is an acceleration to breakdown (C). In compression only the first two regions are observed. For the present materials, region A tends to be short indicating that the defects responsible are highly mobile and that equilibrium is rapidly established.

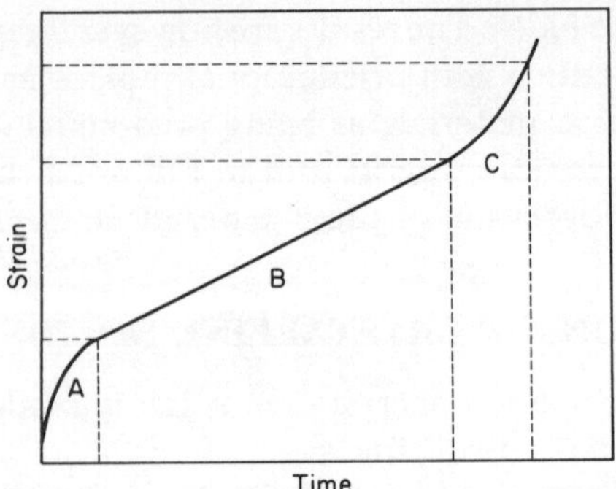

Figure 2.1 Characteristic curve for the high temperature deformation of a crystalline solid

In the secondary, steady-state, creep region, the strain rate $\dot{\varepsilon}$ at constant temperature (T) is empirically related to the stress σ by the expression

$$\dot{\varepsilon}(T) = a \exp(m\sigma) + b\sigma^n \tag{2.2.1}$$

where a, b, m, and n are constants.

At low stresses this expression reduces to the power law

$$\dot{\varepsilon}(T) = b\sigma^n \tag{2.2.2}$$

The temperature dependence of the strain rate can usually be adequately expressed by an Arrhenius relationship of the type

$$\dot{\varepsilon}(T) = d \exp(-E(\sigma)/RT \tag{2.2.3}$$

yielding an overall expression of the form

$$\dot{\varepsilon} = A\sigma^n \exp(-E_c/RT) \tag{2.2.4}$$

At very high temperatures and extremely low stresses flow is believed to occur by the migration of vacancies across the grains in the specimens (Nabarro–Herring Creep).[4] Under these circumstances the rate of deformation will be governed by the rate of migration of vacancies. The flow is Newtonian, thus $n = 1$ and E_c should be equivalent to the activation energy for vacancy diffusion E_d. This type of behaviour has been noted for thin foils of metals close to their melting points.[4] Under the more usual conditions for bulk specimens the mass transport occurs by dislocation climb. In this mechanism lattice vacancies diffuse through the lattice to dislocations where they accumulate to form jogs along the dislocation core (Figure 2.2). The gradual accumulation of vacancies results in the effective climb of the dislocation from the lattice. Again the rate of climb and hence plastic flow is self-diffusion controlled. Theoretical analyses based on this model[5] indicate that the stress exponent $n \cong 4.5$. Experimental

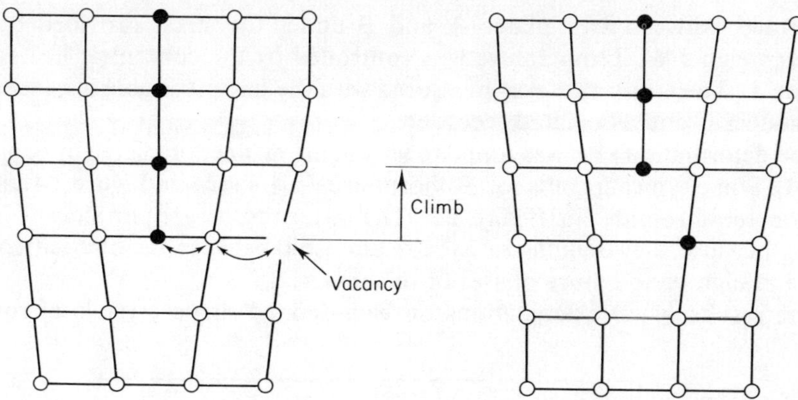

Figure 2.2 The climb of an edge dislocation by the accumulation of lattice vacancies on the dislocation edge

examination of metals yields values in the range $n = 4$–6 and an excellent correlation between E_c and E_d.

Studies of the steady state deformation of the current materials have been made as a function of stress and temperature in a simple parallel plate creep apparatus of the type shown in Figure 2.3. In this, the sample S was

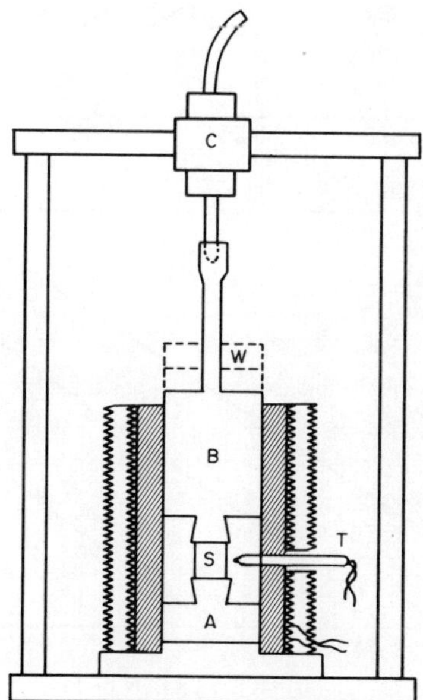

Figure 2.3 A simple compression creep apparatus

deformed between two plates A and B under the stress provided by the added weights W. Temperature was controlled by the concentric heater and sensor T. The strain rate was measured with the linear variable displacement transducer C and associated recorder.

The deformation rate was found to vary in the manner predicted by equation (2.2.4). Some typical results for α-phosphorus[6] are shown in Figure 2.4 and for carbon tetrabromide[7] in Figure 2.5. The latter results are particularly interesting because they demonstrate the relationship between the plasticity of the low and high temperature phases of this material.

The results of all examinations on well-defined single crystals of rotator

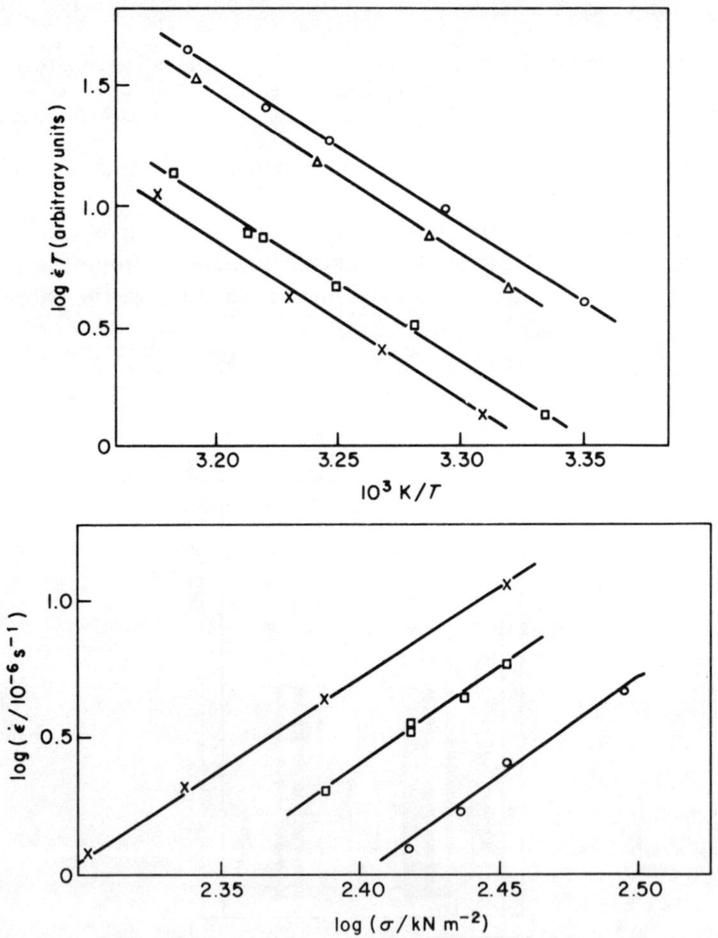

Figure 2.4 The influence of temperature (T) and stress (σ) on the deformation rate ($\dot{\epsilon}$) of α-phosphorus [equation (2.2.4)]

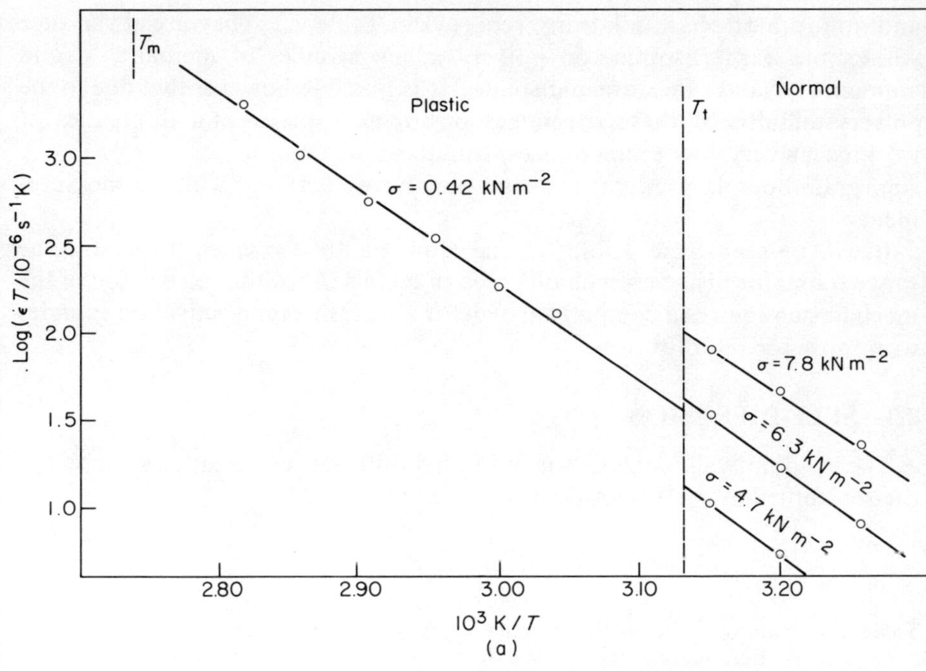

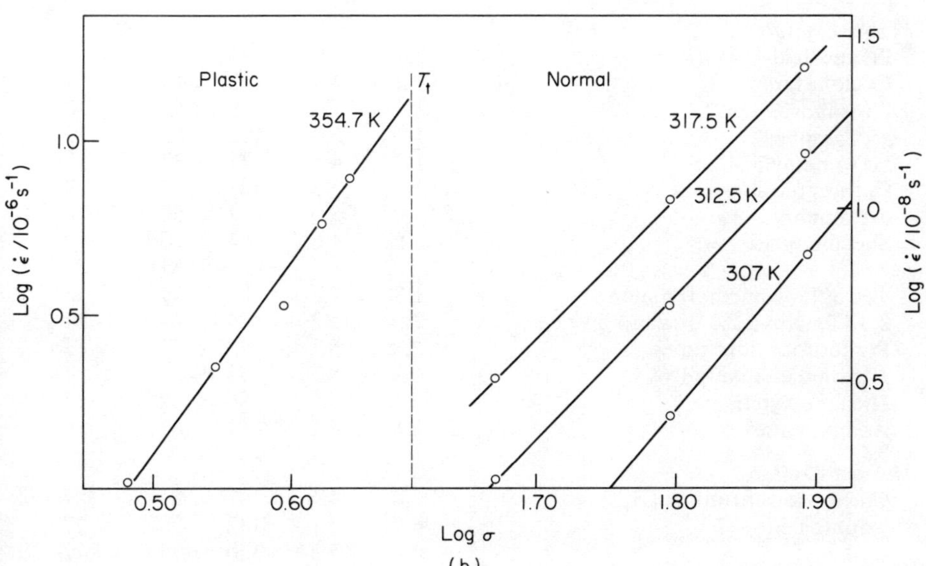

Figure 2.5 The variation of strain rate ($\dot{\epsilon}$) with stress (σ) and temperature (T) for carbon tetrabromide in the nighbourhood of the phase transformation

and non-rotator phase solids are reported in Table 2.2. They are at variance with some earlier studies on polycrystalline samples of methane,[18,19] and ammonia[20–22] and other organic solids.[3] It is possible however that due to the polycrystallinity of these specimens or to the experimental difficulties of working at very low temperatures, simultaneous deformation processes e.g. slip, grain-boundary migration, etc., may have interfered with the measurements.

It will be seen from Table 2.2 that almost all of the values of n lie in the range found for dislocation climb creep in metals. Absolute verification of the mechanism requires a comparison of deformation rates and activation energies with those for self diffusion.

2.3 SELF-DIFFUSION

The random walk analysis for unlimited diffusion yields an expression for the self-diffusion coefficient D of the form

$$D = \overline{R^2(t)}/6t \tag{2.3.1}$$

Table 2.2 Values of the stress exponent (n) and activation energy (E_c/kJ mol^{-1}) for high temperature plastic deformation in organic crystals

	$\Delta S_f/R$	n	E_c	L_s^a	Reference
Plastic Crystals					
Pivalic acid	0.8	4.9	94	59	8
Cyclohexane	1.0	4.9	112	46	9
Phosphorus	1.0	7.0	126	59	6
dl-Camphene	1.2	5.0	102	51	10
Hexamethyldisilane	1.3	4.5	77	37	11
Carbon tetrabromide (I)	1.3	5.2	103	47	7
d-Camphor	1.4	5.0	202	80	10
Succinonitrile	1.5	4.6	57	70	12
				59	24^b
Tetra(fluoromethyl) methane	1.7	4.8	88	39	13
2,3 Dimethyl, 2,3 dicyano butane	2.0	5.2	84	—	14
Perfluorocyclohexane	2.3	5.2	72	36	13
1 Bromoadamantane	—	5.2	133	—	7
Hexamethylethane	2.4	5.6	86	39	15
Adamantane	2.5	4.4	151	63	16
Normal Crystals					
Carbon tetrabromide (II)	—	4.4	107	54	7
Naphthalene	6.7	5.1	142	75	17
Biphenyl	6.4	5.3	130	74	17

$^a L_s$—Latent heat of sublimation.
^{b}by impression creep.

where $\overline{R^2(t)}$ is the mean square distance from the origin moved by the diffusing particle in a period of time t. Transferring the process to a lattice where motion occurs by a series (N) of discrete jumps each of length a then $\overline{R^2(t)} = Na^2$ and

$$D = Na^2/6t = a^2/6\tau = a^2\Gamma/6 \tag{2.3.2}$$

where τ and Γ are the mean jump time and mean jump frequency of the diffusing particle. The experimental evaluation of D requires the measurement of $R^2(t)$ over long periods of time or the assessment of τ. For the present materials, information on the former quantity can be derived from radiotracer self-diffusion studies or from n.m.r. spin–echo studies. The latter can be determined from pulsed n.m.r. relaxation time measurements.

Pulsed n.m.r. measurements, their interpretation and the problems encountered in assessing their relationship to the radiotracer measurements are discussed in full in a later chapter (Chapter 5). In the present section only the performance and reliability of the radiotracer measurements will be considered. These monitor mass transport, and hence correspond most directly to the deformation studies.

2.3.1 Radiotracer Measurements[25]

The basis of the radiotracer measurement is to examine the rate of penetration of a radioactively labelled molecule into the bulk of the specimen under examination. The necessary carbon 14 or tritium labelled radiotracers can in some cases be purchased commercially. In many cases however, labelled forms of the more interesting systems are not commercially available and must be synthesized. The purity of the tracers can be assessed with sufficient confidence to claim that self- and not impurity-diffusion is under examination. On the whole the specific activities obtainable for these materials are somewhat lower than are often achievable in some other solid systems (e.g. for self-diffusion studies in metals or ionic solids). They are high enough however to permit diffusion concentration profiles to be followed over a sufficient range of specific activity and penetration to yield an equivalent precision to that obtainable for these other solids.

Single crystals of high purity provide the best substrates. With these, the contribution of competing processes such as diffusion along dislocations and other linear boundaries in the crystal is minimized. As will be seen below, for the present materials this is the principal limiting factor to the accuracy of the evaluated diffusion coefficient.

With the availability of these starting materials, the rate of radiotracer penetration can be followed by several techniques.

2.3.1(a) Serial Sectioning Experiments

In this method, the radiotracer is applied to one surface (ideally a known crystallographic plane) of the crystal by vacuum deposition or from saturated solution. Preliminary preparation and long term annealing are usually required to reduce the crystal specimen to an equilibrium state. After a suitable isothermal annealing period at a temperature close to the melting point, during which the specimen and deposit are maintained *in vacuo* or in an inert atmosphere, the crystal is cooled to room temperature, the sides removed to avoid contamination due to surface diffusion or evaporation/condensation and the sample sectioned parallel to the initial face. Sectioning is best carried out using a calibrated microtome which allows the removal of the thin sections (2–20 μm) necessary in some experiments. For low melting or highly volatile materials it is convenient to carry out all operations in a deep freeze unit converted to a glove box in order to minimize the loss of the specimen.

Alignment of the crystal with the microtome blade is carried out using an autocollimator or an optical lever. Section thickness can be assessed by weighing or alternatively by spectroscopic methods.

Finally, the radioactivity is measured by liquid scintillation counting. Fuller experimental details for experiments using organic crystals will be found in many of the references quoted in Table 2.3 below.

The integration of Fick's Law for the usual geometry of the diffusion couple, i.e. a thin source of total activity Q diffusing into a semi-infinite solid, yields an expression for the concentration of radioactive tracer (specific activity A) as a function of distance x into the crystal after an annealing period t of the form

$$A = \left[Q/(\pi Dt)^{1/2}\right] \exp(-x^2/4Dt) \qquad (2.3.3)$$

In a perfect, pure, crystal, a plot of log A against x^2 should be linear and D obtainable from the slope of the line. Unfortunately, few crystals are so ideal. Inherent line imperfections induced during growth or by subsequent mechanical handling and impurity precipitation form interconnecting channels within the solid (pipes) along which the tracer can diffuse more rapidly than through the lattice. Where the concentration of such defects is high, this dual penetration process can lead to difficulties in evaluating D. The problem is best exemplified in the present case by considering two extreme face-centred cubic plastic crystals: adamantane (ΔS_f = 2.5 R) and cyclohexane (ΔS_f = 1.0 R).

Adamantane is the least plastic of this range of materials and with careful preparation by growth from solution[26] can be obtained in a high state of perfection with mosaic spreads as low as 50″ and few ($< 10^4$ cm^{-2}) dislocations.[25] The most delicate handling however results in the recrystallization of the specimen with an associated increase in mosaic spread to 0.5°; a figure typical of melt-grown crystals. Figure 2.6(a) shows a typical diffusion profile for this

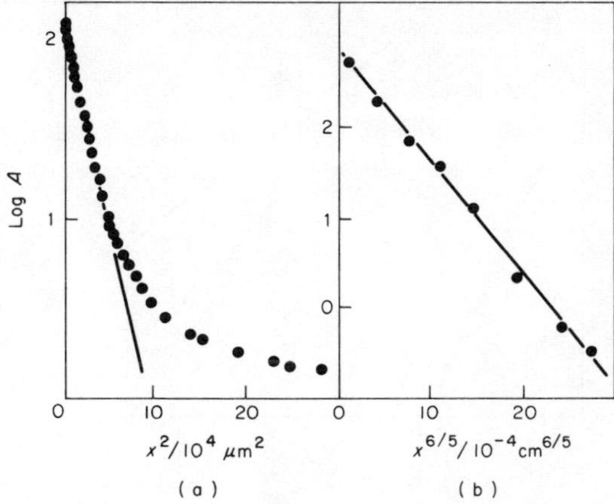

Figure 2.6 Typical plots of (a) log A against x^2 [equation (2.3.3)] and (b) log A against $x^{6/5}$ [equation (2.3.4)] for adamantane[31]

material. Instead of being linear the profile curves sharply to a low level portion at deep penetrations. The initial linear portion is due predominantly to self-diffusion through the lattice, the latter to the collective pipe diffusion process. When the secondary process is of such a low level, it is usually deemed sufficiently accurate to determine D from the initial linear region ($x \leqq 3\sqrt{Dt}$) where lattice self-diffusion is presumed to dominate. It is obvious that the reliability of the result is open to question since the presence of a contribution from the rapid process could lead to a line of lower slope than that which would result from lattice self-diffusion alone and hence to an overestimate of D. The degree to which this occurs can be assessed by determining the pipe self-diffusion coefficients at lower temperatures where penetration into the lattice from the surface is negligible. Lattice diffusion will still have some influence since there will always be some simultaneous diffusion from the pipes into the surrounding lattice. Mathematical analysis of this problem yields an expression for the resulting penetration profile of the form[27–29]

$$D_p\delta = -\left[\frac{d(\ln A)}{d(x)^{6/5}}\right] \cdot \left[\frac{4D}{t}\right]^{1/2} (0.78)^{5/3} \qquad (2.3.4)$$

where δ is the width of the pipe.

Plotting the data in the form log A against $x^{6/5}$ should yield a straight line (Figure 2.6(b)), D_p can be evaluated from the slope of this line using values of D extrapolated from high temperatures. The relative values of the lattice and pipe self-diffusion coefficients are shown in Figure 2.7. The temperature dependence of each can best be expressed as[30,31]

$$D = 1.6 \exp(-138.5 \text{ kJ mol}^{-1}/RT) \text{ m}^2 \text{ s}^{-1} \qquad (2.3.5)$$

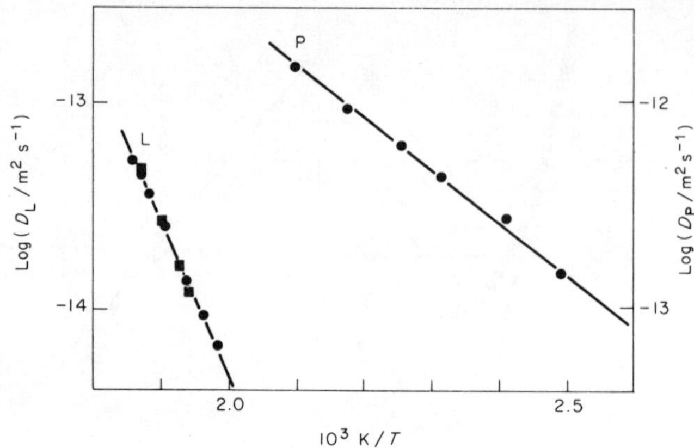

Figure 2.7 Arrhenius plots for lattice (L) and pipe (P) self-diffusion in adamantane[31]

$$D_p = 3.4 \times 10^{-5} \exp(-49 \text{ kJ mol}^{-1}/RT) \text{ m}^2 \text{ s}^{-1} \qquad (2.3.6)$$

We can now assess the influence of pipe diffusion on the total penetration process at higher temperatures i.e. on the initial linear portion of the penetration profile (Figure 2.6(a)).

At low penetrations, when the true lattice self-diffusion coefficient is considerably lower than that in the pipes and/or the pipe concentration is high, Hart[32] has shown that the measured diffusion coefficient (D_{obs}) can be compounded from the lattice and pipe coefficients such that

$$D_{obs} = (1 - g)D + gD_p \qquad (2.3.7)$$

where g is the fraction of the specimen comprising pipes. Thus the observation of a linear ln A against x^2 plot is not in itself evidence of pure lattice diffusion uninfluenced by the pipe process. Using the present data we can construct Figure 2.8 which shows the apparent enhancement of D by D_p for different values of g. For the samples used, $g \leqq 10^{-5}$. Thus we see that pipe self-diffusion will have little influence above 500 K. Above this temperature, the values of D determined from the ln A against x^2 curve will represent true lattice self-diffusion. This will be equally true (relatively) for most of the present materials.

In the very soft or highly plastic materials, e.g. cyclohexane, the situation is less well-defined. The penetration profiles here are highly curved (Figure 2.9). Since the values of lattice and pipe-self-diffusion are similar, relatively, to those of adamantane, the curvature must result from the presence of a very high concentration of pipe defects. This conclusion is rational if we associate plasticity with easy dislocation generation. Obviously, here, the temperature range of

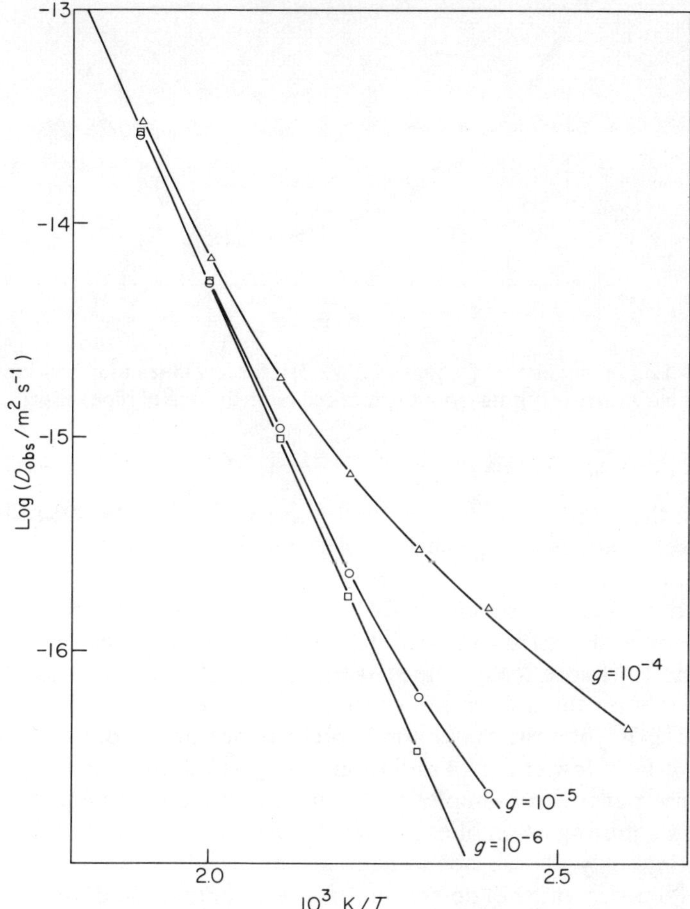

Figure 2.8 The influence of pipe self-diffusion on the observed diffusion coefficient of adamantane for different values of g (the pipe fraction)

accurate assessment of D will be smaller. It can be maximized however by careful preannealing to reduce the pipe defect concentration and by carrying out lengthy diffusion anneals on unstrained crystals so that lattice diffusion dominates the process at least near the surface ($x \leqq \sqrt{Dt}$). With such care in experiment design[33] reliable values of D can be obtained.†

†The reliability can be assessed from the values obtained for the isotope-mass-factor E_β^α as described in Section 2.3.4. For lattice self-diffusion in cyclohexane $E_\beta^\alpha = 0.78$. Simultaneous pipe-diffusion ($E_\beta^\alpha \sim 0$ to 0.3) will depress this value. Thus D's evaluated for experiments which yield $E_\beta^\alpha = 0.78$ should be reliable. This was the method used to test the data.[33]

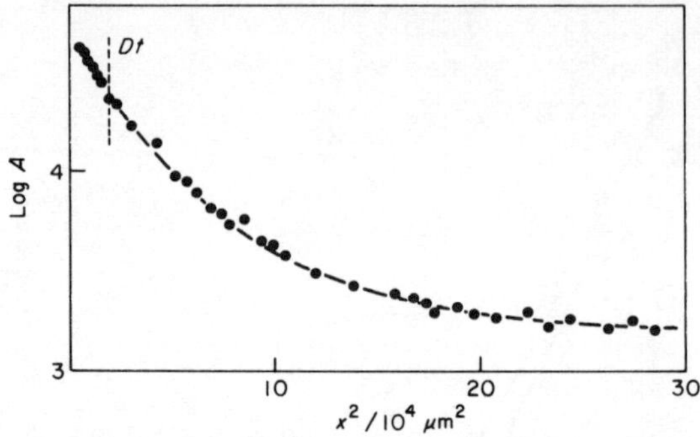

Figure 2.9 Log A against x^2 [equation (2.3.3)] for cyclohexane showing the pronounced curvature resulting from the marked contribution of pipe self-diffusion

2.3.1(b) Integrating Techniques

Amongst these are included those methods which follow the overall increase (or decrease) of activity at the surface of a solid during the course of a diffusion experiment.

Surface decrease experiments, in which the activity of a surface deposit is monitored over the diffusion period have been tried for naphthalene[34] and cyclohexane.[35] Theoretically, the sensitivity of such methods should be high since they rely on the absorption of the radiation emitted by the solid as the tracer diffuses beyond the maximum β range in the sample. Both carbon 14 and tritium emit very low energy β radiation. Also the relatively low experimental temperatures permit the samples to be mounted in sealed Geiger–Muller or scintillation counting assemblies so that the progress of diffusion can be monitored continuously. These advantages are outweighed, however, by experimental difficulties in the detection of the radiation and due to sample imperfection (see below). The result is that, at best, the method is useful as a check that decreases in surface activity are compatible with those measured by sectioning and hence that losses by vaporization are negligible.

The rate of exchange between an isotopically enriched solid and its molecules in solution or in the gas phase has been extensively used for ionic solids[36] and the rare gas solids.[37–39] Chadwick and his co-workers[39,40] have recently extended this technique to the more volatile of the organic solids including some plastic crystalline materials.

The solid, doped with a labelled species and grown as a single crystal, *in situ*, in a vacuum system is equilibrated with inactive vapour of the same material. Following initial rapid surface exchange the process becomes dominated by the rate of self-diffusion of radioactive molecules from the bulk of the specimen to the surface. Under these conditions the exchange kinetics are described by

$$d(n^2)/dt = 4S^2C_0^2D/\pi \tag{2.3.8}$$

where n is the number of tracer atoms in the gas phase at time t, S the surface area of the solid and C_0 the initial concentration of tracer in the solid. There are considerable difficulties in overcoming the problems of evaporation–condensation, but once equilibrium is established the experiment can be continued as a function of temperature on a single sample. This is advantageous because a major problem in the evaluation of D is the determination of S. In order to provide a reasonably accurate assessment it is essential to detach the specimen from the walls of the growth tube before the experiment is commenced. Then the total geometrical area is the best representation that one can achieve. This lack of definition of S does mean that D_{obs} may not be exactly equal to D. However, since for any one sample the difference between D and D_{obs} will be constant, subsequently derived parameters such as the activation energy should be reliable. Another important factor which is impossible to overcome is the influence of pipe self-diffusion. Whereas the sectioning technique does provide an indication (from the shape of the profile) of the degree of interference due to this process, the integrating techniques do not. For reasons given above (Section 2.3.1(a)) there is a high probability that $D_{obs} > D$ for this reason also. Since in this case the difference will increase with decreasing temperature, the activation energy will also be influenced. With careful sample preparation and experimentation these differences need not be excessive. At worst the results serve as a test of the serial sectioning data anticipating that the diffusion coefficients may not be exactly equivalent and that the activation energy evaluated from the exchange measurements may be lower due to the pipe diffusion contribution. Chadwick's results for cyclohexane (Figure 2.10) demonstrate the agreement achievable.

The technique comes into its own for materials such as methane where serial sectioning is not possible.

2.3.2 Results

Thus in carefully prepared pure samples reliable values of D can be measured and their dependence on temperature and hydrostatic pressure assessed. The range of measurement accessible is limited by two factors. Firstly, the pipe defect structure and its contribution to D_{obs}. This will increase at lower temperature and higher pressures leading to $D_{obs} > D$ and also to a lowering of the activation energy towards that for pipe diffusion. Secondly, the lower limit of the measuring technique, $D \cong 10^{-16}\ m^2\ s^{-1}$, for these materials. Within these limits the temperature dependence of the diffusion coefficients follows the expected Arrhenius type behaviour

$$D = D_0 \exp -E_d/RT \qquad (2.3.9)$$

Variation of hydrostatic pressure leads to the definition of an activation

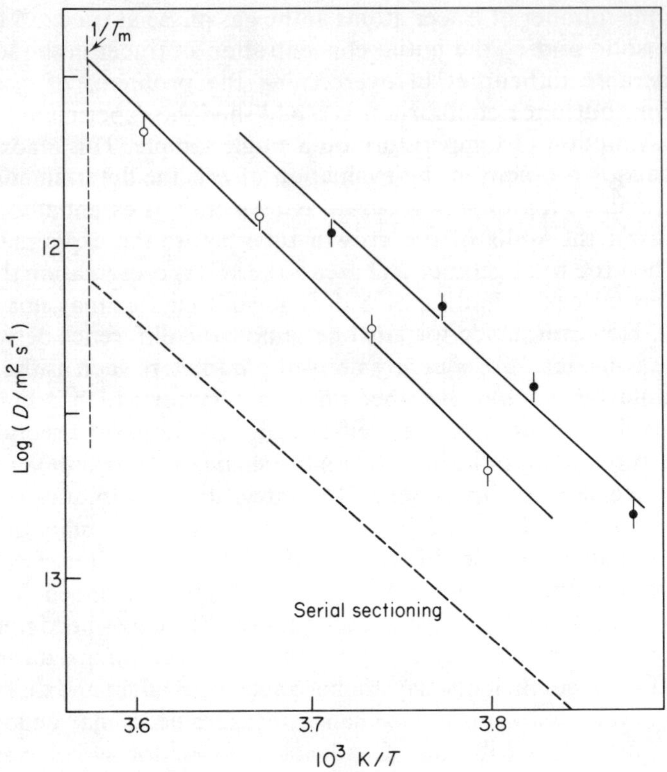

Figure 2.10 Self-diffusion coefficients for cyclohexane obtained for two different crystals by the exchange technique compared with the serial sectioning data. The difference reflects the difficulty in defining the surface area of the sample [equation (2.3.8)] (Reproduced by permission of Dr. A. V. Chadwick)

volume V^+. Writing the diffusion coefficient in the form[25]

$$D = \gamma a^2 v \exp(-G_d/RT) = \gamma a^2 v \exp S_d/R \cdot \exp -E_d/RT \quad (2.3.10)$$

where γ is a geometric factor, a the jump distance, v the lattice vibrational frequency and G_d the free energy for diffusion, it will be seen that

$$\left[\frac{\partial(\ln D)}{\partial P}\right]_T = \left[\frac{\partial(\ln \gamma a^2 v)}{\partial P}\right]_T - \frac{1}{RT}\left[\frac{\partial(G_d)}{\partial P}\right]_T \quad (2.3.11)$$

The second term is small ($<3\%$ of the first term)[41] and can be neglected within the experimental error. Thus

$$\left[\frac{\partial(\ln D)}{\partial P}\right]_T = -\frac{1}{RT}\left[\frac{\partial(G_d)}{\partial P}\right]_T = -\frac{V^+}{RT} \quad (2.3.12)$$

Values of E_d, D_0 and V^+ are given in Table 2.3. In all but one case these refer to self-diffusion. In this case—hexamethyldisilane[42]—the tracer was carbon-14 labelled hexamethylethane. Consideration of the result relative to other data[42] suggests that it mirrors self-diffusion adequately.

Departures from the Arrhenius equation are few. Deviations from linearity (log D against $1/T$) at low temperatures are usually ascribable to the influence of an excessive pipe defect structure. Figure 2.11(a) for cyclohexane shows the influence of the inherent defect structure resulting from its high plasticity. Increasing the annealing time (characterized by the increasing $\sqrt{Dt}$ values) for the experiment leads to a predominance of lattice diffusion at least in the surface region ($x < \sqrt{Dt}$) and the better definition of D. It will be noted that pipe-diffusion is still a major influence at low temperatures even at very long anneal times.

Figure 2.11(b) shows the enhancement of self-diffusion in polycrystalline and

Table 2.3 Values of the pre-exponential factor ($D_0/\mathrm{m^2\,s^{-1}}$), activation energy ($E_d/\mathrm{kJ\,mol^{-1}}$) and activation volume (V_d^+/Ω) for radiotracer self-diffusion in molecular solids

	$\Delta S_f/R$	D_0	E_d	E_d/L_s^a	V_d/Ω^b	Reference
Plastic Crystals						
Pivalic acid (F)	0.8	5×10^2	91.2	1.6	1.2	8,41
Cyclohexane (F)	1.0	1×10^5	92	2.0	1.0	33, 41
Cyclohexane^e		—	91	2.2		40
Phosphorus	1.0	4×10^5	114	1.9	1.4	6
t-Butyl thiol^e (F)	1.1	1×10^5	84	2.4	—	40
dl-Camphene (B)	1.2	2×10^4	96	1.9	1.0	10, 41
Norbornylene (H)	1.2	3×10^{-5}	49	1.5	0.8	43
Hexamethyldisilane (B)	1.3	2×10^{-3}	52	1.4		44
		9.1	71	1.9		42^d
Methane^e (F)	1.3	2.0	20	2.1	—	39
Succinonitrile (B)	1.5	4×10^{-5}	57	0.8	—	12, 41
Hexamethylethane (B)	2.4	2.2	86	2.2	1.7–1.8^c	15, 41
Adamantane (F)	2.5	1.6	139	2.1	1.2	30, 31
Normal Crystals						
Benzene (O)	4.2	1×10^5	97	2.1	—	45
Benzene^e		—	155	3.5	—	40
Naphthalene (M)	6.7	2×10^{11}	179	2.4	1.2	34, 36
Biphenyl (M)	6.4	4×10^{10}	169	2.3	—	17

$^a L_s$ = latent heat of sublimation.
^{b} = molar volume.
^{c} = varies with temperature.
^{d} = hexamethylene—^{14}C tracer.
^{e} = gas–solid exchange experiment, the remainder are serial sectioning experiments.

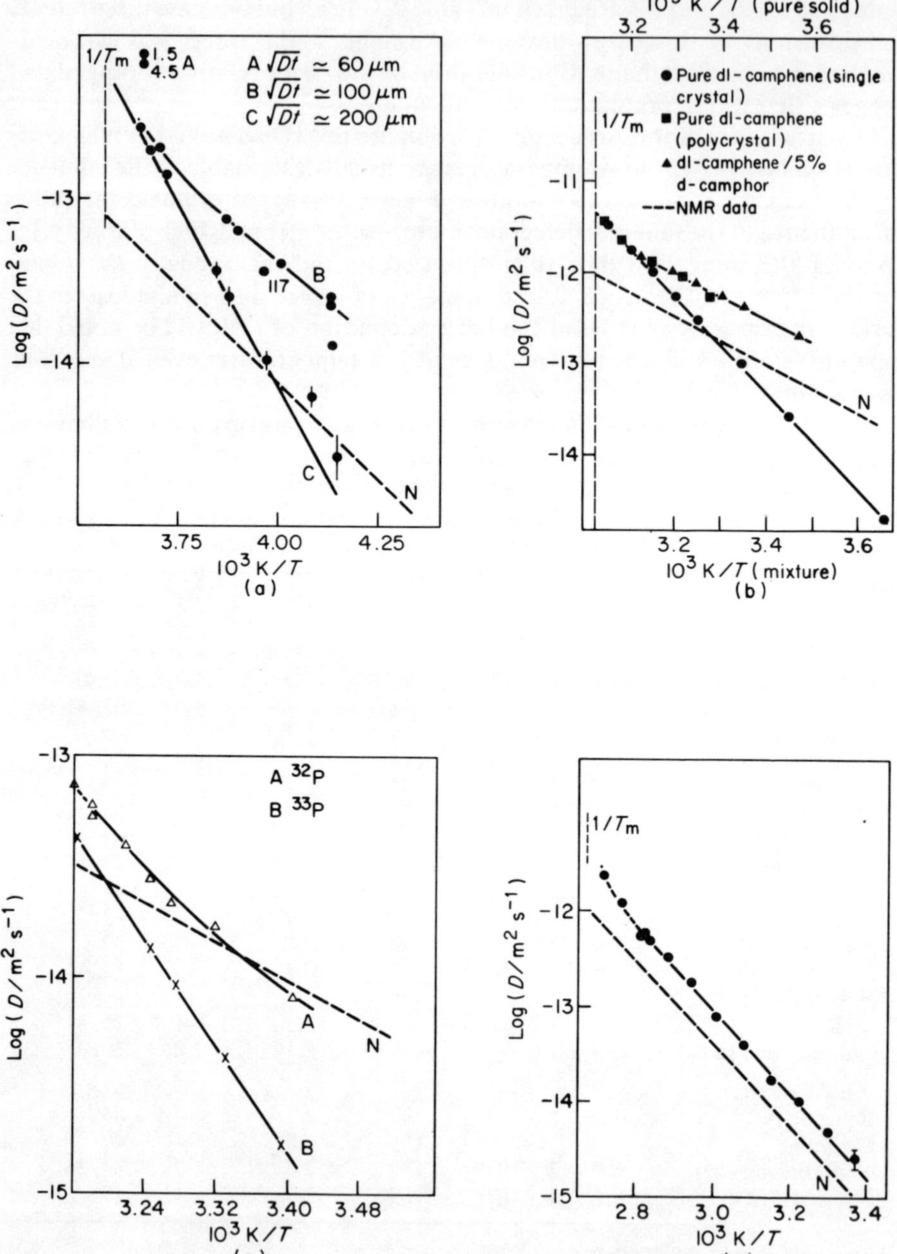

Figure 2.11 Arrhenius plots for several plastic crystals (a) Cyclohexane[9]; (b) *dl*-Camphene[10]; (c) α-Phosphorus[6]; (d) Hexamethylethane[15]. The dotted line indicates the relative position of the NMR data in each case

impure specimens of *dl*-camphene compared with the pure single crystalline specimen.

The variations noted in α-phosphorus (Figure 2.11(c)) result from radiation damage. Line A corresponds to measurements made using ^{32}P tracer and B, ^{33}P tracer. The former causes extensive lattice damage due to the high energy of the emitted β particle ($E_{\beta(\max)} = 1.71$ MeV). Repetition of the experiment using ^{33}P ($E_{\beta(\max)} = 0.24$ MeV) yielded reliable values. It is interesting to note, in this context, that the energies of the β particles emitted by carbon 14 and tritium are significantly lower than those of ^{33}P.

In all the above cases careful preparation, purification and annealing either eliminates or reduces the curvature.

Hexamethylethane (Figure 2.11(d)) exhibits a divergence at high temperatures. This is noteworthy since it implies the onset of a new mechanism of lattice self-diffusion. The change is reflected in the activation volume which is much higher than the values found for other solids and varies with temperature. This is consistent with two (or more) mechanisms of different V^{+}, one dominating at low temperatures and the other becoming more dominant with increasing temperature. This may be a consequence of the body-centred cubic lattice and although it is not observed for *dl*-camphene (b.c.c.), there is evidence from deformation experiments[44] for a similar behaviour in body-centred cubic hexamethyldisilane (Section 2.2.4). The lack of a more distinct curvature in the Arrhenius plot at lower temperature may simply indicate that, whereas the activation volumes for the two mechanisms differ significantly, the activation energies are similar.

For most curved plots, the low temperature, pipe-defect dominated portions yield activation energies similar to those expected for pipe diffusion (see equation (2.3.6)). The higher temperature portions and the linear plots found for less defective samples extrapolate to values of D_m (melting point) in the range 10^{-12}–10^{-13} m^2 s^{-1}. The body-centred cubic solids yield values at the upper end of the range, and the more close-packed solids, lower values. These values are characteristic of other types of solid of similar structure.[25] They are confirmed by NMR spin–echo studies, a technique which also assesses $R^2(t)$.[47,48] Unfortunately, the range of the spin–echo technique is limited to $D > 10^{-13}$ m^2 s^{-1}. Thus it cannot be used to assess the temperature dependence of D in these solids.

Examination of Table 2.3 shows that values of E_d obtained from the serial sectioning technique lie in the range 1.4–2.4 L_s and D_0 values cover an extremely wide range. Considering the plastic crystals alone, however, and neglecting pivalic acid, norbornylene, succinonitrile and one value for hexamethyldisilane, then for the remainder $E_d \sim 1.9$–$2.1\ L_s$ and $D_0 \sim 2$–10^5 m^2 s^{-1}. For the latter three materials $E_d = 0.8$–$1.4\ L_s$ and $D_0 \sim 10^{-4}$–10^{-3} m^2 s^{-1}. A convincing argument for accepting the reliability of these latter data is that for simple cubic metals, $D_0 \sim 10^{-5}$–10^{-4} m^2 s^{-1}, equivalent, within the error, to the above range.

Some other factors when taken into account throw some doubt on this proposition.

Succinonitrile[12] comprises two geometrical isomers [*gauche* and *trans* (18%)] of significantly different volume. Thus in this effectively impure system a different behaviour to the remaining solids is not surprising. The physical consequences of the solid solution which cause the difference are difficult to define. It may be that the *trans* isomer acts as an impurity point defect causing a lowering of the energy barriers to diffusion, with the observed result.

Hexamethyldisilane as purchased is extremely impure (15% total impurity) and is very difficult to purify.[42,44] Any residual impurity could be in solid solution (hence the difficulty in purification) or could segregate during crystal preparation to yield a high density of pipe defects. The former consequence could cause enhanced diffusion as suggested above for succinonitrile, the latter could produce a similar effect to that discussed in Section 2.3.1(a). Each could yield a $D_{obs} > D$ and consequently a lower E_d and D_0 than the true values. In this case repetition of the experiment with extensively purified material yielded a behaviour more compatible with the remaining data (Table 2.3).[42]

The similarity of the data for norbornylene[43] to these other materials suggests that similar problems may have arisen here, although considerable care was taken in the purification of the material. In this case the activation volume is also much lower than the values found for other solids and approaches that determined for succinonitrile by the deformation technique.

These solids apart, the rest give $E_d \cong 2L_s$ with the exception of pivalic acid. For this solid it is not possible to assess the reliability of the published value of L_s[49] and this may account for the discrepancy.

High values of D_0 compared with those for cubic metals are to be expected. D_0 contains an entropy term (S_d) (equation (2.3.10)) and it would not be surprising to find that this (and hence D_0) is much greater for the diffusion of polyatomic molecules than for simple metal atoms. The fact that D_0 is much greater for the lower entropy of fusion, and hence more disordered, solids than for those at the other extreme also seems reasonable.

The magnitude of D_0 can also be rationalized on the basis of a Zener analysis[50] which suggests that the entropy of diffusion is related to the activation energy by

$$S_d = \lambda\beta\frac{E_d}{T_m} \tag{2.3.13}$$

where λ is a constant (1 for b.c.c. solids and 0.55 for f.c.c. solids), T_m the melting temperature and

$$\beta = \frac{(\kappa/\kappa_0)}{(T/T_m)} \tag{2.3.14}$$

κ is the bulk modulus and κ_0 the value at 0 K. Zener reports that values of D_0 for metals, calculated using this equation, compare favourably with those determined experimentally. Detailed calculations for molecular solids are not possible due to a lack of data. Sufficient information exists however to calculate S_d for the two extreme solids: cyclohexane ($D_0 = 10^5\ \mathrm{m^2\,s^{-1}}$) and adamantane ($D_0 = 1.6\ \mathrm{m^2\,s^{-1}}$). For these we evaluate

Cyclohexane S_d(calculated) $= 150\ \mathrm{J\,mol^{-1}\,K^{-1}}$
S_d(experimental) $= 160\ \mathrm{J\,mol^{-1}\,K^{-1}}$
Adamantane S_d(calculated) $= 120\ \mathrm{J\,mol^{-1}\,K^{-1}}$
S_d(experimental) $= 140\ \mathrm{J\,mol^{-1}\,K^{-1}}$

The agreement is satisfactory.

2.3.3 Diffusion Mechanisms and Point Defects

Before returning to the basic theme of this section and comparing the measured diffusion parameters with the deformation data it is of interest to consider the nature of the predominant point defect in organic crystals and whether or not this, or some more mobile defect is responsible for self-diffusion.

In terms of self-diffusion by a point defect mechanism, equation (2.3.10) can be rewritten[25]

$$D = \tfrac{1}{6}a^2 v \exp(S_v + S_m)/R \exp -(E_v + E_m)/RT. \qquad (2.3.15)$$

S and E are entropy and energy terms relating to the formation (v) and migration (m) of the diffusing defect. The relationship between these equations and the Arrhenius equation is obvious.

One possible method of elucidating the mechanism is to evaluate or measure the formation and migration energies of the various defects and to compare the sums of these quantities with the experimental value.

2.3.3(a) Theoretical Calculations

Up to the present time there have been no theoretical calculations of either the formation or migration energies for point defects in organic solids. The best approximation to these values currently available are parallel calculations for the rare gas solids. For these materials a variety of calculations[51–63] using various intermolecular potentials to describe the systems yield values of the formation energy for a lattice vacancy which group closely round that for the lattice energy of the solid, L_s, (mainly 0.95–1.1 L_s) a value predicted by Jost[64] for all molecular solids. In comparison, the value calculated for the formation of interstitials is much higher ($>3L_s$).[61,63] Thus from theoretical view point lattice vacancies should predominate.

Fewer calculations have been made for vacancy migration but, combined with the vacancy formation energy these yield values of $E_d \sim 1.7\text{–}2L_s$[55,56,58,65] in excellent agreement with the tracer data for both rare gas and organic crystals. Coupled with the fact that E_d (interstitials) $> 3L_s$, this suggests that not only should vacancies dominate the point defect structure but that self-diffusion should occur by vacancy migration. Although this information does refer to simpler molecular systems than those presently under consideration, it does provide a strong indication of the mechanism.

Further support for this conclusion is derived from the few calculations[55] of activation volumes which yield V^+ (vacancy) $= 1.2\,\Omega$ and V^+ (divacancy) $=$ $1.8\,\Omega$. The former correlates well with values for the close-packed solids (Table 2.3). The divacancy value is strongly suggestive that this might be the type of mechanism which comes into play with increasing temperature in hexamethylethane.†

2.3.3(b) Experimental Determinations

(i) Differential Expansivity Measurements. The molar volume of a solid determined by measurement of the bulk dimensions (V) differs from that (V_x) calculated from the measured values of the x-ray lattice parameter. The difference can be used to measure directly the point defect concentration.

If the volume of formation of a vacancy is V_v this may be less than the volume of a vacancy V_m due to lattice relaxation around the vacancy.

i.e.
$$V_v = V_m(1 + x) \qquad (2.3.16)$$

x is a constant associated with the possible relaxation.

The real volume of the crystal can then be written

$$V = V^0 + n_v V_m(1 + x)$$

Where n_v is the number of vacancies. The value of the x-ray volume will not be equal to V^0 (The molar volume of the perfect crystal) due to the relaxation but will be

$$V_x = V^0 + n_v V_m x \qquad (2.3.17)$$

Therefore

$$n_v = (V - V_x)/V_m \qquad (2.3.18)$$

and provided that the total number of molecules $N_m > n_v$, the concentration

† The divacancy content will increase in parallel with the vacancy concentration. The relatively higher melting temperature and looser lattices of the b.c.c. system may allow this mechanism to become more apparent in this system than in the close-packed f.c.c. solids.

C_v is given by

$$C_v = (V - V_x)/V^0 \tag{2.3.19}$$

Where the crystal contains both vacancies and interstitials (number n_i) then

$$n_v - n_i = (V - V_x)/V_m \tag{2.3.20}$$

Thus if $V - V_x$ is positive it is a proof that vacancies are the dominant defect.
Equation (2.3.20) can be rewritten as

$$C_v = (L/L^0)^3 - (a/a^0)^3 \tag{2.3.21}$$

or

$$C_v = (\rho_x - \rho)/\rho_x \tag{2.3.22}$$

where ρ and ρ_x are the bulk and x-ray densities and L and a are the length and lattice parameter of the crystal respectively $[\Delta L = L - L^0, \Delta a = a - a^0]$. (The superscript indicates the value for the perfect crystal) Experiments of this type are extremely difficult and the bulk and x-ray measurements should be performed simultaneously. This practice serves to remove errors due to sample variation and thermometer calibration which could invalidate the comparison. The most reliable method is to determine C_v from simultaneous measurements of L and a on one crystal. This technique, originally developed by Simmons and Baluffi[66] has been used by Simmons and his later coworkers to examine the rare-gas solids.

No direct comparisons of expansivity data for organic solids have been attempted. Staveley and his coworkers[68] and Green and Wheeler[69] compared measured bulk densities with previously published x-ray determinations to yield maximum values of the vacancy concentrations for several plastic crystals and benzene. (Table 2.4).

A careful x-ray study of succinonitrile and cyclooctane[70] and comparison of the results with bulk densities provides the most convincing proof that the predominant point defects in plastic crystals are lattice vacancies. The results (Figure 2.12 and Table 2.4) show quite clearly that the bulk expansivity exceeds the x-ray measurements. Unfortunately the data were not sufficiently precise to allow the evaluation of an energy of vacancy formation but it was possible to define the vacancy concentration as $\sim 0.1\%$ near the melting point in accord with vacancy concentrations in other solids.

(ii) Excess Specific Heat Studies. In a manner parallel to that used to define a vacancy contribution to the bulk expansivity, a similar contribution can be inferred to the specific heat. Thus we can define[71,72,73] an excess specific heat, $\Delta C_p = C_p - C_p^0$ such that

$$\Delta C_p = n_v E_v^2/kT^2 \tag{2.3.23}$$

Table 2.4 Vacancy concentrations ($C_v/\%$) in organic crystals determined from differential thermal expansion measurements

	$T_{t.p.}/K^a$	$C_v/\%^b$	Method[c]	Reference
Plastic Crystals				
Carbon tetrachloride	250.17	2.0	p, p_x	68
t-Butyl chloride	247.8	2.0	p, p_x	68
t-Butyl bromide	257	2.0	p, p_x	68
Cyclohexane	279.71	2.0	p, p_x	68
	—	0.5	p, p_x	69
Cyclohexanol	298.31	2.0	p, p_x	68
		0.3	p, p_x	69
dl-Camphene	323	7.7	p, p_x	69
Succinonitrile	328	0.1	$\Delta L, \Delta a$	70
Cyclooctane	287.98	0.4 ± 0.1	$\Delta L, \Delta a$	70
Normal Crystals				
Benzene	278.69	0.2	p, p_x	68

[a]t.p. triple point.
[b]C_v at the triple point.
[c]See text equations (2.3.21), and (2.3.22)

Experimentally, the excess specific heat is seen as an 'anomalous' curvature in plots of C_p versus T at temperatures close to the melting point (Figure 2.13(a)) ΔC_p^0 is estimated by extrapolating the C_p from low temperatures and subtracting the extrapolated values from the measured C_p. Equation (2.3.23) then yields E_v (Figure 2.13(b)). A major point of uncertainty in this analysis is the extrapolation to obtain C_p^0.[74] C_p can be rising near the melting point for reasons other than vacancy formation (e.g. impurities and anharmonicity) and it has been suggested that one does not need to invoke vacancies to interpret the high temperature specific heat curve.[75]

Using existing specific heat data Baughman and Turnbull[70] have evaluated the vacancy formation energy for several organic solids. The results, shown in Table 2.5 yield $E_v \sim 0.5$–$1\ L_s$ and vacancy concentrations in the range 0.1–1%. The errors involved in the calculations will be large since the original data were not of the quality required for such an analysis. Estimates of the perfect lattice values would be difficult. Consequently, all that can be said is that within this error, the data are in satisfactory agreement with the theoretical calculation. It should perhaps be noted that calculations using carefully measured specific heat data[73,74,76,77] for the rare gas solids also yield values in the range 0.7–1 L_s. A final point of interest is the high value of S_v predicted for each compound. This would contribute to high D_0 values, as found experimentally (Section 2.3.2).

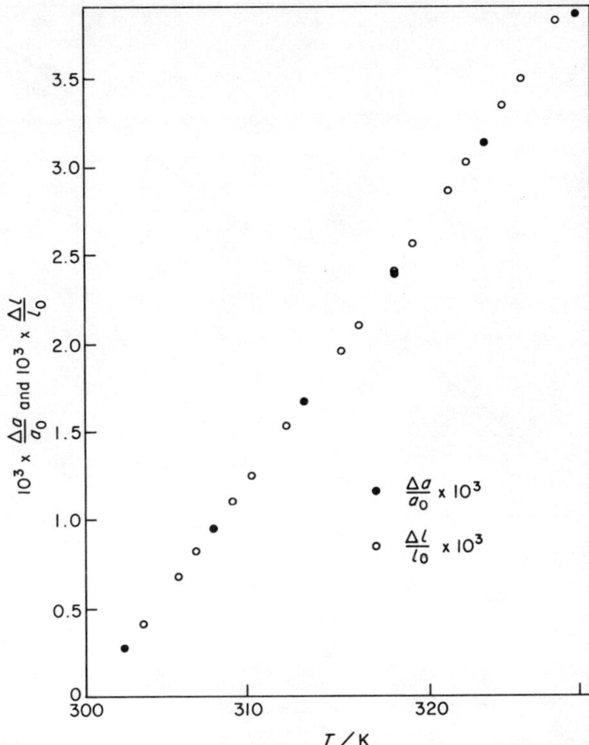

Figure 2.12 X-ray and bulk linear thermal expansion for succinonitrile[70] (Published by kind permission of the Editors of *J. Phys. Chem. Solids*)

2.3.3(c) Summary

In conclusion it can be said that lattice vacancies predominate in plastic crystals and that within a rather large error the calculated and evaluated formation energies are consistent with self-diffusion occurring by a vacancy mechanism. More careful and detailed experiments are necessary however to provide better assessments of C_v and E_v.

2.3.4 Isotope-mass-effect Experiments

Comparison of activation energies as described above provides an indication of the likely diffusion mechanism. Direct evidence is more satisfactory. Such evidence has been forthcoming for some other systems from the examination of the degree of correlation between successive atomic jumps in the diffusion process.

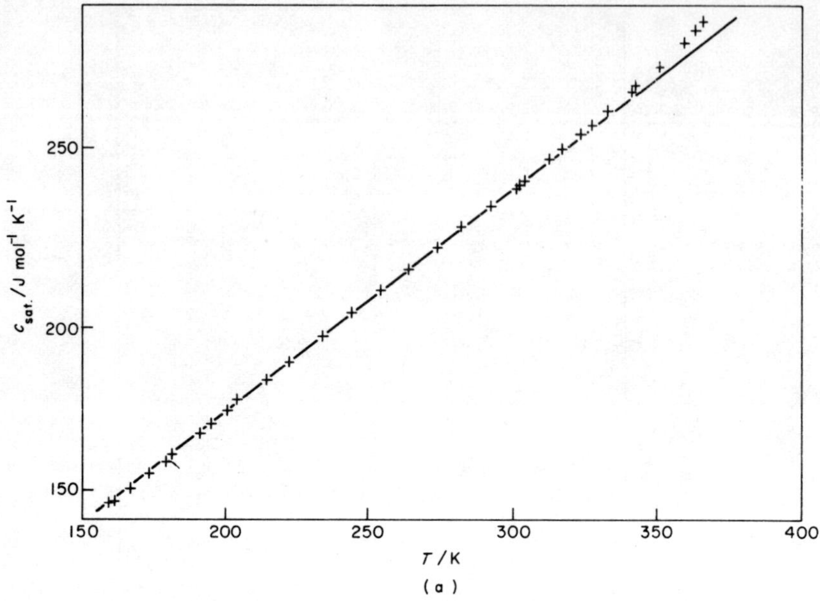

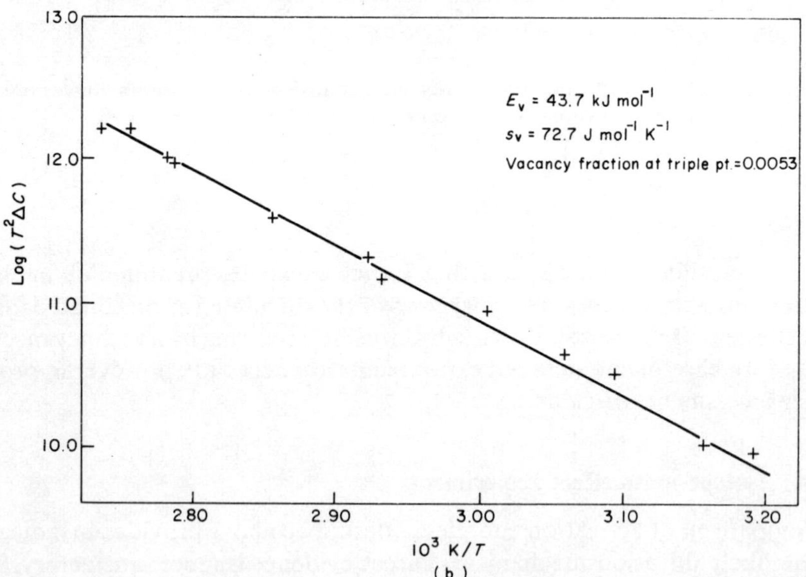

Figure 2.13 (a) Experimental heat capacity for hexamethylethane as a function of temperature.[10] (b) The function log $T^2 \Delta C_p$ against T^{-1} for hexamethylethane.[70] (Published by permission of the Editors of *J. Phys. Chem. Solids*)

Table 2.5 Vacancy formation parameters for organic crystals derived from specific heat data[70]

	$T_{t.p.}/K^{a}$	$C_v/\%^{b}$	S_v/J mol^{-1}	E_v/kJ mol^{-1}	L_s/kJ mol^{-1}
Plastic Crystals					
cis-1,2-Dimethylcyclopentane	219	0.46 ± 0.08	52.7 ± 14.2	21.4 ± 3.4	41.4
Cyclooctane	288.98	0.36 ± 0.11	54.8 ± 11.7	29.2 ± 4.2	46.2
1,1-Dimethylcyclohexane	239.81	0.22 ± 0.05	73.6 ± 12.1	29.8 ± 3.3	43.1
2,2,3-Trimethylbutane	248.57	0.76 ± 0.13	37.6 ± 5.0	19.4 ± 1.6	36.9
Perfluoropiperidine	274.12	0.33 ± 0.07	61.4 ± 14.6	29.8 ± 4.5	33.7
Tetra(fluoromethyl)methane	367.43	0.60 ± 0.25	74.4 ± 18.4	43.1 ± 8.4	39.3
Hexamethylethane	373.97	0.53 ± 0.20	72.7 ± 20.9	43.5 ± 9.2	39.0
Normal Crystals					
Benzene	278.691	0.09 ± 0.02	129.6 ± 17.1	52.3 ± 4.6	44.8

[a] T.p. triple point.
[b] C_v at the triple point.

As indicated previously, the diffusion coefficient can be written most generally in the form

$$D = \tfrac{1}{6}\,\overline{R^2(t)}\,t \tag{2.3.1}$$

where $\overline{R^2(t)}$ is the square of the vector displacement R of a diffusing molecule averaged over all paths. If during the course of the migration N jumps occur, then each R is the vector sum of N individual jump vectors a. If these are of equal length, which is the situation in a crystal, then provided N is large[78]

$$\overline{R^2(t)} = \overline{\sum^{N} a_i^2} = Na^2 + 2(\sum \overline{a_i a_{i+1}} + \sum \overline{a_i a_{i+2}} + \cdots) \tag{2.3.24}$$

$$= Na^2(1 + 2\,(\overline{\cos\theta_1} + \overline{\cos\theta_2} + \cdots). \tag{2.3.25}$$

Cos θ_j is the average value of the cosine of the angle between one jump and the jth following jump.

In the preceding argument it was assumed that all jumps occurred at random and were uninfluenced by the directions of earlier jumps. In this circumstance all the $\overline{\cos\theta_j}$ terms average out to zero and equation (2.3.25) becomes

$$D = \tfrac{1}{6}\Gamma a^2 \tag{2.3.2}$$

In some of the diffusion mechanisms however the jumps are not random but are correlated and the $\overline{\cos\theta_j}$ terms do not average out to zero. Writing

$$1 + 2\,(\overline{\cos\theta_1} + \overline{\cos\theta_2} + \cdots) = f \tag{2.3.26}$$

leads to

$$D = \tfrac{1}{6}\Gamma a^2 f \tag{2.3.27}$$

f is the correlation factor which defines the departure from random behaviour.

The assumption of random jumps is not valid for diffusion by a vacancy mechanism. Following the exchange of a molecule with a vacancy, the possible directions for the second jump are not of equal probability. The vacancy into which it could make an immediate second jump is in a position that would cause the molecule to reverse its direction. It cannot proceed, it can only reverse. Thus the probability of the reverse jump is greater than random and $\overline{\cos\theta_1} \neq 0$. Similarly for the second jump and so on. In contrast, for interstitial motion, the jump from one interstitial site to the next will always be at random and $f = 1$. The correlation factor is dependent only upon the lattice and the mechanism by which the diffusion process occurs. Calculations lead to the values quoted in Table 2.6. It will be seen that the values associable with specific mechanisms in particular types of solid are distinctive. If they can be measured, then comparison of the experimental value with the series of calculated values could define the mechanism.

Table 2.6 Correlation factors for self-diffusion.

Lattice	Mechanism	
F	Vacancy	0.7815
	Divacancy	0.475
	Interstitial	1
	Interstitialcy	0.8
		1
B	Vacancy	0.7272
	Interstitial	1
	Interstitialcy	0.7
		1
M	Vacancy	0.66
O	Vacancy	0.78

2.3.4(a) Isotope-effect Method

For a radiotracer study of self-diffusion by a vacancy mechanism the situation is as depicted in Figure 2.14.[78] The vacancy may exchange with the tracer (jump frequency ω_2), a nearest neighbour (ω_1) or next nearest neighbour (ω_3), (ω_4). Assuming that all other exchanges, vacancy/normal molecule, occur at the same frequency ω_0 as in the pure solvent and are unaffected by the presence of the tracer, allows the calculation of f. For an f.c.c. crystal

$$f = \frac{\omega_1 + F\omega_3}{\omega_2 + \omega_1 + F\omega_3} = \frac{u}{\omega_2 + u} \tag{2.3.28}$$

where F is a slowly varying function of (ω_0/ω_1).

The correlation factors for two tracers of different mass can now be written

$$f_\alpha = \frac{u}{\omega_2^\alpha + u} \quad \text{and} \quad f_\beta = \frac{u}{\omega_2^\beta + u} \tag{2.3.29}$$

u is the same for both since it contains only normal molecule jump frequencies.

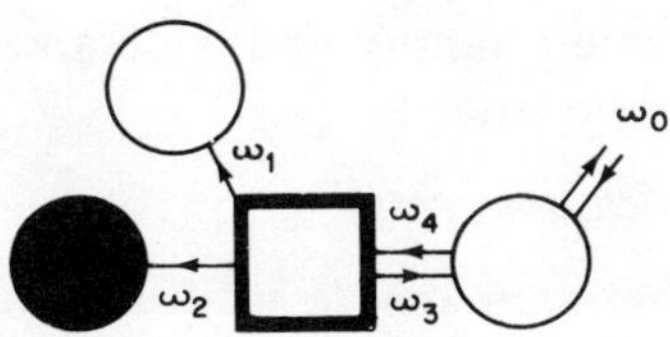

Figure 2.14 The exchange frequencies between a tracer molecule (●), a vacancy (□) and a normal lattice molecule (○)

From equation (2.3.27)

$$D = \tfrac{1}{6}a^2\Gamma f = \tfrac{1}{6}a^2(\omega_2 N_d)f \tag{2.3.30}$$

and

$$D_\alpha / D_\beta = \omega_2^\alpha f_\alpha / \omega_2^\beta f_\beta \tag{2.3.31}$$

where N_d is the defect fraction.
Eliminating **u** and one of the f's yields

$$(D_\alpha / D_\beta - 1) = f(\omega_2^\alpha / \omega_2^\beta - 1) \tag{2.3.32}$$

The distinction between f_α and f_β has been dropped since their difference is insignificant.

Now

$$\omega = \nu \exp G_{\mathrm{m}}/RT \tag{2.3.33}$$

and

$$\nu \propto m^{-1/2} \tag{2.3.34}$$

where m is the mass of the diffusing species. Thus

$$(D_\alpha / D_\beta - 1) = f((m_\beta / m_\alpha)^{1/2} - 1) \tag{2.3.35}$$

This equation is approximate since equation (2.3.33) considers only the diffusing molecule and ignores any motional disturbance in the rest of the lattice. If such disturbance occurs then the overall effect will be to produce some coupling between the moving molecule and the rest of the lattice. This will effectively reduce the mass effect by some fraction ΔK with the result that

$$E_\beta^\alpha = \frac{(D_\alpha / D_\beta - 1)}{(m_\beta / m_\alpha)^{1/2} - 1} = f\Delta K \quad (0 < \Delta K < 1) \tag{2.3.36}$$

If only the migrating atom is moving, $\Delta K = 1$ but if there is some concomitant motion of the surrounding molecules during the jump step, then $\Delta K < 1$. Such motions may occur for instance as a consequence of lattice relaxation around the vacancy. The relaxed molecules then have to expand to allow passage of the diffusing molecule and contract after its passage. E_β^α, the mass factor is the experimentally determinable quantity and is obtained by comparing the diffusion coefficients of the two tracers.

2.3.4(b) Experimental Measurements[45]

Since the difference between the two diffusion coefficients is small, the measurements must be made by the simultaneous diffusion of the two species. Only then are the basic errors due to sample variation, temperature stability and sectioning ruled out. Otherwise these errors are sufficient to invalidate the experiment. The experiment thus requires two tracers of different mass, differently radio-labelled so that the rates of penetration of each can be distinguished. The experiment is then performed by the serial sectioning method as described in Section 2.3.1(a).

The method used to produce distinctive tracers is to label protonated organic compounds with carbon-14 and deuterated compounds with tritium (or *vice versa*). For example, with benzene, the two tracers are $^{14}C_1{}^{12}C_5{}^1H_6$ and $^{12}C_6{}^2H_5{}^3H_1$. The masses are 78 and 85 respectively but the relative molar volumes are $\Omega(C_6H_6)/\Omega(C_6D_6) = 0.997$ suggesting that in spite of the electronic differences the lattice mis-match should not be significant. Both isotopes are β emitters but can be resolved by liquid scintillation counting and pulse height analysis.

After diffusion and sectioning the concentrations of each 'isotope' in the sections is assayed. Since each will diffuse according to equation (2.3.3) we have

$$\ln A_C = \left[Q/(\pi Dt)^{1/2}\right] \exp(-x^2/4D_C t) \tag{2.3.37}$$

$$\ln A_T = \left[Q/(\pi D_T t)^{1/2}\right] \exp(-x^2/4D_T t) \tag{2.3.38}$$

where the subscripts refer to carbon-14 and tritium respectively and A the specific activity. Since t is equivalent for each 'isotope' then we can compound the equations to give

$$\ln(A_C/A_T) = (D_C/D_I - 1) \ln A_C + \text{constant} \tag{2.3.39}$$

and $(D_C/D_T - 1)$ can be evaluated from a plot of $\ln(A_C/A_T)$ versus $\ln A_C$.

Figure 2.15 shows the precision attainable for the benzene experiments. Measurement of the slope leads to the assessment of the mass effect. The best fit to the experimental points is given by the dotted line. These are plotted against full lines corresponding to the slopes expected for diffusion by interstitial (I), vacancy (V) and divacancy (DV) mechanisms. There is no doubt that the experiment confirms vacancy self-diffusion ($f = 0.78$) with no contribution from the concomitant motion of other molecules ($\Delta K = 1$). Complicated interstitial type processes e.g. interstitialcy which yield a similar value of f (Table 2.6) are most unlikely in these close-packed structures.

The measurements have been extended to several other organic solids mostly with similar results. These are reported in Table 2.7. The outstanding differences are hexamethylethane[15] and stearic acid[79] both of which yield values of E_β^α lower than that expected for vacancy self-diffusion. In the former case there is also evidence of a temperature dependence of E_b^α (Figure 2.16). This is of considerable interest since it reflects the variation noted in the activation volume measurements and the curvature of the Arrhenius plots (see Section 2.3.2) and again suggests a variation in diffusion mechanism with temperature. There are two possible explanations. The first is that a mixed vacancy–divacancy mechanism holds. This would result in values of f between 0.72 and 0.48 the values characteristic of the two processes, the latter becoming more dominant at high

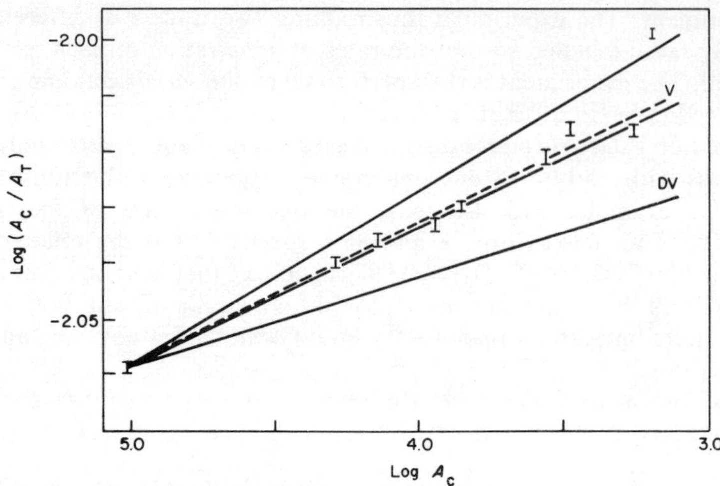

Figure 2.15 The results of a mass effect experiment in benzene single crystals. The data points are plotted against bold lines corresponding to the expected behaviour for interstitial (*I*), vacancy (*V*) and divacancy (DV) mechanisms. The dotted line is the regression line for the experimental points

temperatures. The experimental data do follow this pattern. Alternatively, lattice relaxation around the vacancy, which might be expected in this loose lattice, could result in a temperature dependent contribution via ΔK. Of these two alternatives, the first is most likely since this is in accord with the relatively high, temperature-dependent values of V^+ which would seem to rule out lattice relaxation. With no other data for stearic acid it is not possible to be specific about the result for this material. The lattice is however not close-packed and the mass factor could reflect a ΔK or divacancy contribution.

Table 2.7 Mass effect measurements for self-diffusion in organic crystals

Material	Tracers	E_β^α Measured	E_β^α Calculated	Reference
Adamantane	$^{12}C_9{}^{14}C_1{}^1H_{16}$, $^{12}C_{10}{}^2H_{15}{}^3H_1$	0.76 ±0.08	0.7812	31
Cyclohexane	$^{12}C_5{}^{14}C_1{}^1H_{12}$, $^{12}C_6{}^2H_{11}{}^3H_1$	0.75 ±0.05	0.7812	33
Hexamethyl-ethane	$^{12}C_7{}^{14}C_1{}^1H_{18}$, $^{12}C_8{}^1H_5{}^2H_{12}{}^3H_1$	0.30 –0.52[a]	0.7272	15
Benzene	$^{12}C_5{}^{14}C_1{}^1H_6$, $^{12}C_6{}^2H_5{}^3H_1$	0.78 ±0.01	0.78	45
Naphthalene	$^{12}C_9{}^{14}C_1{}^1H_8$, $^{12}C_{10}{}^2H_7{}^3H_1$	0.61 ±0.05	0.66	46
Stearic acid	$^{12}C_{17}{}^2H_{35}{}^{14}C_1OO{}^1H_1$, $^{12}C_{17}{}^1H_{34}{}^3H_1{}^{12}C_1OO{}^1H_1$	0.30 ±0.05	0.66	79

[a]Varies with temperature.

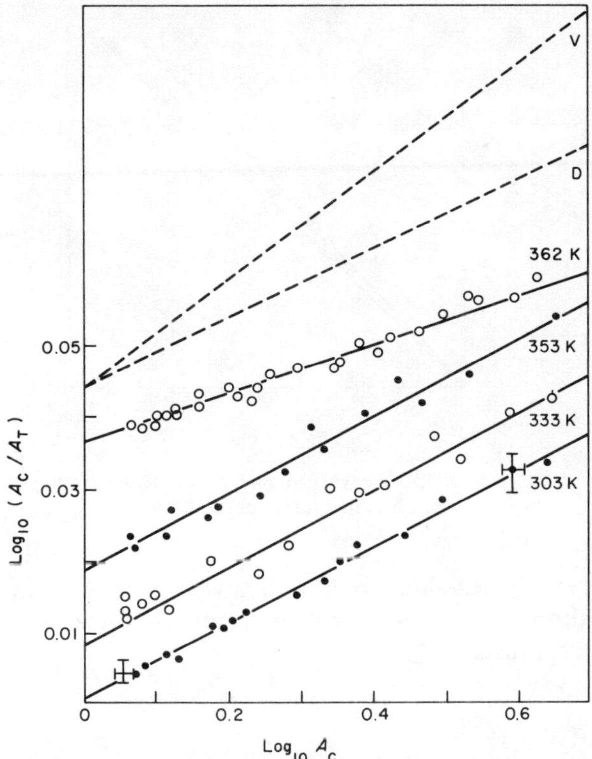

Figure 2.16 The isotope-mass-effect in hexamethylethane as a function of temperature. The dotted lines represent the predicted behaviour for vacancy (V) and divacancy (DV) mechanisms

In all cases examined, it is obvious that self-diffusion occurs by a vacancy process as predicted from considerations of the activation energies and activation volumes for self-diffusion.

2.4 THE CORRELATION BETWEEN SELF-DIFFUSION AND HIGH TEMPERATURE PLASTIC DEFORMATION

As indicated in Section 2.2 the stress dependence of the deformation rate follows the pattern expected for a self-diffusion-controlled mechanism (stress exponent $n \cong 5$). This is adequately confirmed by comparison of the measured activation energies with those for radiotracer self-diffusion. (Tables 2.2 and 2.3 and Figure 2.17.) A small departure from $E_c = E_d$ is predictable and reflects a temperature dependence of the elastic moduli which are contained in the constant A (equation (2.2.4)). This can make a contribution of about 5–10%.[80]

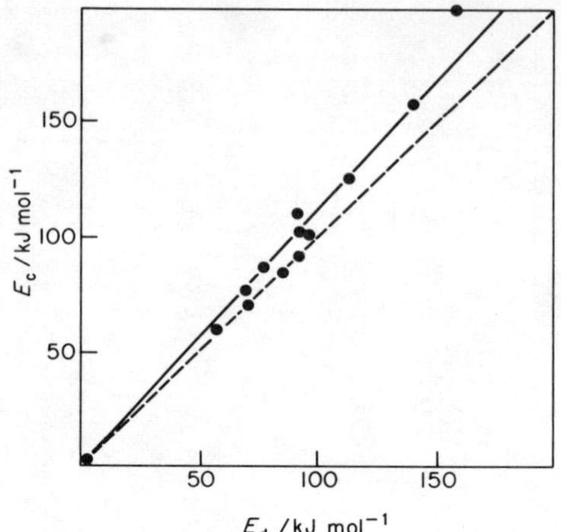

Figure 2.17 Comparison of the activation energies for creep (E_c) and self-diffusion (E_d) in plastic crystals

This similarity can be extended to the activation volumes for the two processes since, if self-diffusion is rate-controlling in the deformation, then equation (2.2.4) can be rephrased

$$\dot{\epsilon} = A'(\sigma)^{4.5}D \tag{2.4.1}$$

and

$$V^{+} = -RT[\partial(\ln D)/\partial P]_T = -RT[\partial(\ln \dot{\epsilon})/\partial P]_T \tag{2.4.2}$$

other terms in the expression being effectively pressure independent. Values are reported in Table 2.8. The similarity in magnitude and behaviour between the values for tracer self-diffusion and those derived from deformation studies is impressive.

The final definition of the rate-controlling nature of self-diffusion should come from a comparison of the rates of deformation with those predicted by the measured self-diffusion coefficient. The absolute calculation is not possible however because of the indefinite nature of the constant A′ in equation (2.4.1). This constant includes a number of structural factors such as the dislocation density, elastic moduli, dislocation energy, etc. If it is assumed that these factors are not too dissimilar or at least that their combination A′ is similar for all organic materials then an assessment can be made.

Table 2.9 tabulates the values of the diffusion coefficient at the melting point (D_m) for a number of plastic crystals and compares this with the stress required to yield a deformation rate of 10^{-6} s^{-1} at the same temperature. For carbon tetrabromide the comparison temperature is the transition point from the monoclinic phase to the face-centred cubic plastic phase. The relative stresses

Table 2.8 Comparison of activation volumes for high temperature plastic deformation (V_c^+/Ω) with those for radiotracer self-diffusion (V_d^+/Ω)

	$\Delta S/R$	V_c^+	V_d^+
Plastic Crystals			
Pivalic acid (F)	0.8	1.2	1.2
Cyclohexane (F)	1.0	1.0	1.0
Phosphorus	1.0	1.3–1.7	1.4
dl-Camphene (B)	1.2	1.1	—
Norbornylene (H)	1.2	—	0.8
Hexamethyldisilane (B)	1.3	1.4	—
Succinonitrile (B)	1.5	0.55	—
Tetra(fluoromethyl)methane (F)	1.7	1.2	—
Perfluorocyclohexane (F)	2.3	1.3	—
Hexamethylethane (B)	2.4	1.5–1.9	1.7–1.8
Adamantane (F)	2.5	1.2	1.2
Normal Crystals			
Naphthalene (M)	6.7	—	1.2

Table 2.9 Comparison of self-diffusion ($D_m/m^2\ s^{-1}$) with the stress ($\sigma_m/kN\ m^{-2}$ required to yield a deformation rate of $\dot{\epsilon} = 10^{-6}\ s^{-1}$, both at the melting point, for several organic solids

	D_m	σ_m	μ
Plastic Crystals			
Cyclohexane (F)	10^{-13}	10	0.26
Pivalic acid (F)	10^{-13}	70	0.64
Phosphorus	10^{-13}	80	1.8
dl-Camphene (B)	10^{-12}	12	0.29
Carbon tetrabromide (I) (F)	—	50	—
Adamantane (F)	10^{-13}	300	2.5
Normal Crystals			
Carbon Tetrabromide (II) (M)	—	600	—
Naphthalene (M)	10^{-15}	1000	3.0
Biphenyl (M)	10^{-15}	700	2.0

μ = Shear modulus/10^9 N m^{-2}

can then be compared directly with Michels values (Table 2.1). The plastic crystals of equivalent character to those examined by him are 2–10 times more plastic than the normal crystalline phases in excellent agreement with his predictions. Within any reasonable error the deformation rates parallel the self-diffusion rates. From equation (2.4.1), for constant $\dot{\epsilon}$

$$D \sim (\sigma)^{-4.5}$$

Taking for comparison adamantane with naphthalene and biphenyl the ratios of the diffusion coefficients are 100. The corresponding ratios for $(\sigma)^{-4.5}$ are, naphthalene 95 and biphenyl 120, leaving little doubt that the deformation is self-diffusion controlled. The agreement is similar for the other materials in Table 2.9 above within the limits of accuracy of the experiment and, more important, the values of the bulk modulus.

A noteworthy feature is that the plastic crystals show a gradation of plasticity not revealed in Michels' original measurements. The materials of low entropy of fusion are more plastic than those of higher entropy of fusion. This cannot reflect a variation in mechanism since the values of D are equivalent within experimental error. Also, where isotope-effect experiments have been carried out, these yield similar mass factors (e.g. for the extreme face-centred cubic crystals adamantane and cyclohexane $E_{\beta}^{\alpha} = 0.78 \pm 0.03$ corresponding to a simple vacancy mechanism). The variation can be rationalized approximately by normalizing with respect to μ. It must therefore reflect a variation of some fundamental structural property of the solid or perhaps more specifically, of the migrating dislocation. As will be seen below, many of the properties of dislocations are related to the elastic moduli of the solid. There is no doubt however from the preceding agreement that high temperature plastic deformation is self-diffusion controlled and that it occurs by a characteristically solid state process.

2.5 LOW TEMPERATURE/HIGH STRESS DEFORMATION—DISLOCATION SLIP

With such a clear definition of high temperature deformation and its theoretical association with dislocation motion it is of interest to consider what other evidence exists for dislocations in plastic crystals and whether or not this does indicate a higher mobility of these defects in plastic crystals relative to other organic crystals.

The principal alternative process to dislocation climb as a deformation mechanism is slip.[81] Slip occurs by the migration of dislocations under applied stress along specific directions in the crystal. Figure 2.18 depicts the situation for an edge dislocation, the slip plane being at right angles to the dislocation plane and the direction of the slip in line with the Burgers vector. A similar motion can also take place for screw dislocations but now, since the Burgers vector is parallel to the dislocation core, slip is not restricted to a unique plane. The dislocation can cross slip from the plane in which it lies into any equivalent plane parallel to its core. Much exploratory work has been carried out on dislocations in normal organic solids.[82–85] In contrast, information on the identification and characterization of dislocation slip processes in plastic crystals is very limited.

The first indication that the climb mechanism has a limited range of dominance can be deduced from stress–strain measurements of the type described

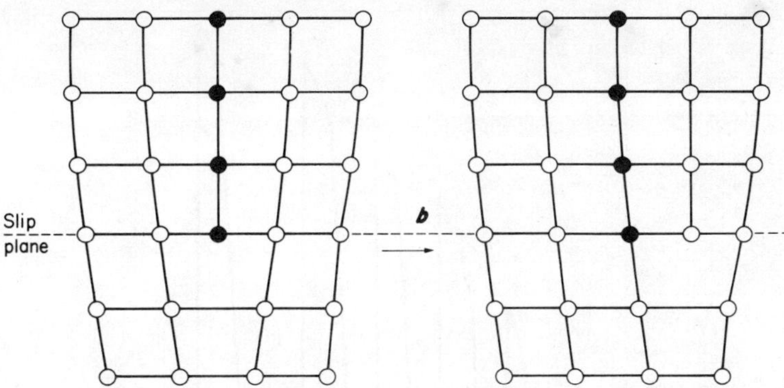

Figure 2.18 The movement of an edge dislocation by slip. (b = Burgers vector)

above.[33,86,87] The use of higher stresses and lower temperatures leads to a well-defined variation in the stress exponent and activation energy as shown in Table 2.10. The variations of n and E_c with stress and temperature are compatible with a change from a diffusion controlled mechanism to a slip controlled mechanism. Particularly noteworthy are the small stresses and minor increases in stress required to cause these changes in cyclohexane, again a factor of ~ 100 times lower than for adamantane.

The few tensile stress/strain experiments which have been carried out confirm this behaviour and the noted variation. They reveal (e.g. Figure 2.19) that all solids examined exhibit a small elastic deformation giving way to plastic flow at yield stresses of: adamantane,[85] 200–250 kN m^{-2}, *dl* camphor,[86] 40–50 kN m^{-2} and succinonitrile,[86] 20–30 kN m^{-2}. That such easy deformation is associable with dislocation motion has been demonstrated by microscopic examination of deformed and etched samples. Light deformation of the {111} faces of both *d*-camphor[87] and adamantane[88] yields the characteristic distribution of slip-traces (Figure 2.20(a)). Subsequent etching of adamantane yielded equi-

Table 2.10 The influence of decreasing temperature/increasing stress on the plastic deformation of highly plastic crystals

	T/K	$\delta/\mathrm{kN\,m^{-2}}$	n	$E/\mathrm{kJ\,mol^{-1}}$
Cyclohexane[33]	270–279	1–3	5	112
	270–279	4–10	3	52
Adamantane[85]	473–543	~ 300	5	151
	373–453	~ 700	3	75
	293–373	~ 1000	2	42

Melting points: cyclohexane, 299 K, adamantane 543 K.

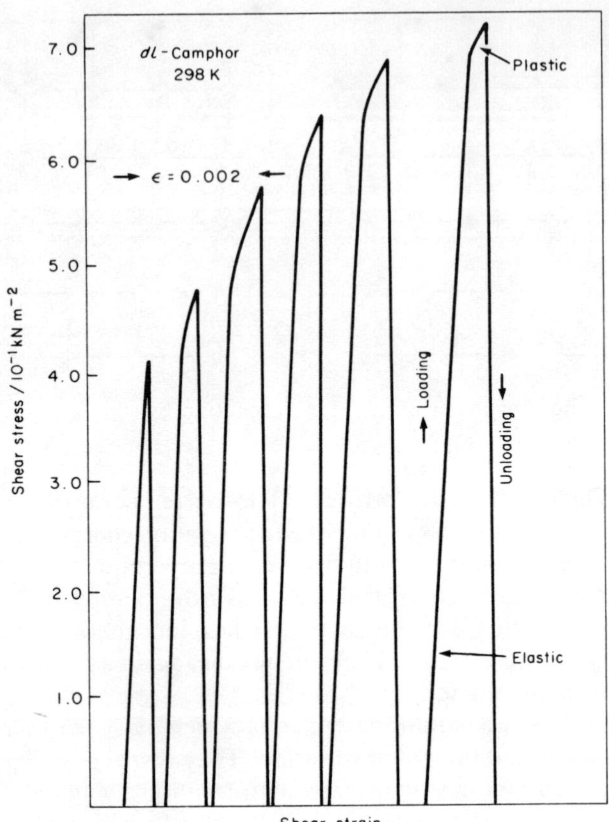

Figure 2.19 Shear stress against shear strain curve for *dl*-camphor deformed in tension at 298 K (strain rate 0.002 min^{-1}) showing the effect of loading and unloading.[87] (Reproduced by permission of the Editors of *Mol. Cryst. Liq. Cryst.*)

valent distributions of etch-pits (Figure 2.20(b)). From this it can be concluded that the deformation of these two solids occurs by the slip of $\{111\}\langle 10\bar{1}\rangle$ dislocations, generally characteristic of face-centred cubic solids. Similar slip patterns have been reported on the more plastic crystals cyclohexane[33] and *dl*-camphene.[10]

A further point of evidence for both the characteristic solid-like behaviour and high-plasticity of these materials is furnished by the observation by Ogilvie and Robinson[87] of easy twinning in *dl*-camphor. Light indentation yielded the patterns of $\langle 10\bar{1}\rangle$ twins shown in Figure 2.21, a result produced only by shock straining in harder face-centred cubic materials.

The evidence, although sparse, is sufficient to indicate that dislocation slip is likely to be the dominant deformation mechanism at high stresses and low

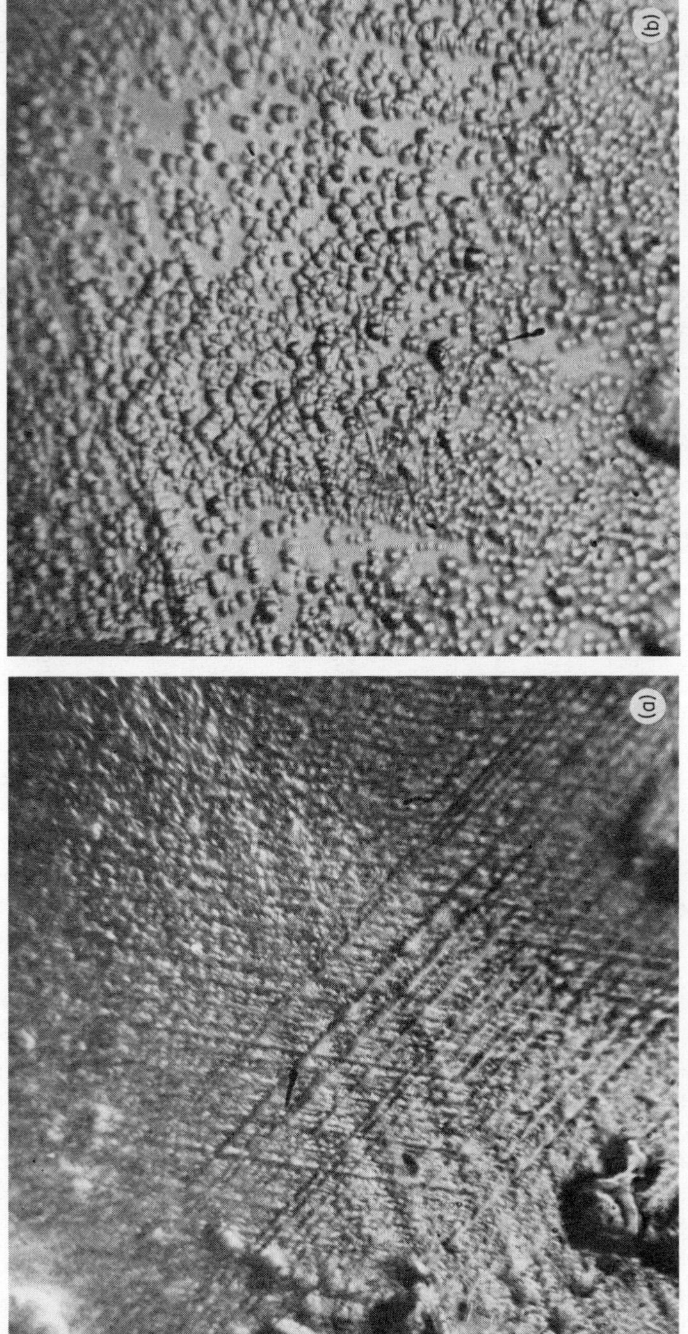

Figure 2.20 (a) Photomicrograph of a {111} surface of an adamantane crystal deformed in compression showing characteristic ⟨10$\bar{1}$⟩ slip lines. (b) The same area as shown in (a) etched to reveal dislocation etch pits

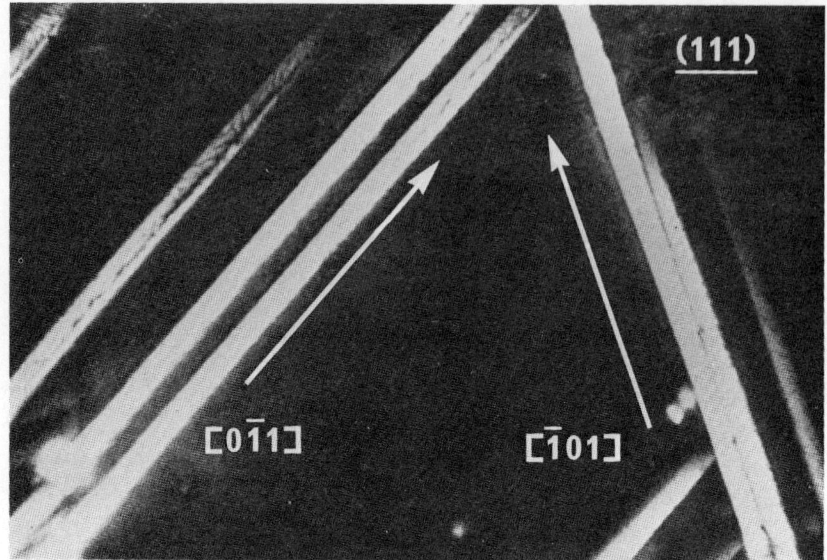

Figure 2.21 Deformation twins produced by rapid indentation, on the (111) surface of a vapour grown single crystal of *dl*-camphor Polarized transmitted light, × 100.[87] (Reproduced by permission of the Editors of *Mol. Cryst. Liq. Cryst.*)

temperatures. The high plasticity of these materials can then be accounted for by the fact that they have a high number of relatively low energy slip systems in comparison with the normal organic materials where slip is highly anisotropic and the energies are high.[83,84]

This can be quantified. The total energy of a dislocation is the sum of contributions from the core energy U_c and the elastic energy U_{el}.[81] In general $U_c < 0.1\ U_{el}$ and can be ignored. $U_{el} \cong \mu b^3$ per molecule plane threaded by the dislocation: b is the Burgers vector. Values of μb^3 are tabulated in Table 2.11 for a series of molecular crystals and two metals of cubic structure for comparison. The values quoted for the non-cubic molecular crystals refer to the most commonly observed (and hence the easiest) dislocation slip system as evidenced by x-ray topography and etching techniques.[83,84,89] The distinction between the normal and plastic crystals is well marked. It is interesting to note that the dislocation line energies for the two harder cubic organic crystals are equivalent to those calculated for the two metals. The values for the line energies of the two solids of low entropy of fusion are considerably lower.

2.6 THE DEFECT STATE OF PLASTIC CRYSTALS

On the basis of the above data it is possible to make some reasonable speculations on the defect structure of these materials relative to other solids.

Table 2.11 Calculated values of the line energy for common dislocation slip systems

	Dislocation system	$\mu/10^9\,\text{Nm}^{-2}$	$\mathbf{b}/10^{-10}\,\text{m}$	U_{el}/eV
Plastic Molecular Crystals				
Cyclohexane (F)	$\{111\}\langle 10\bar{1}\rangle$	0.25	6.2	0.38
Adamantane (F)	$\{111\}\langle 10\bar{1}\rangle$	2.5	6.7	4.7
dl-Camphene (B)	$\{110\}\langle 111\rangle$	0.29	6.1	0.6
Normal Molecular Crystals				
Hexamethylenetetramine (B)	$\{110\}\langle 111\rangle$	5.0	6.9	7.4
Benzophenone (O)	$(001)[001]$	1.8	7.9	6.4
Sulphur (O)	$(1\bar{1}1)[110]$	7.2	8.3	25.7
Cubic Metals				
Silver (F)	$\{111\}\langle 10\bar{1}\rangle$	30	2.9	4.6
Iron (B)	$\{110\}\langle 111\rangle$	83.	2.5	8.1

A well prepared and annealed crystal will contain the equilibrium number of lattice vacancies. At the melting point these will be present in concentrations of $C_v \cong 10^{-3}$ and will decrease in concentration with temperature ($E_v \cong 1L_s$).

Dislocations on the other hand present considerable problems. At the best, these solids have the mechanical characteristics of the simple cubic metals which themselves are difficult to prepare in a dislocation free state. It is possible to do this however, provided that the metal crystal is grown under equilibrium conditions and is not subjected to strain.[90] Any thermal or mechanical shock causes irreperable damage to the specimen. This behaviour is also characteristic of the hardest of the present materials, adamantane. Careful growth from solution at temperatures close to room temperature can yield a sample of very low dislocation content and low mosaic spread (Table 2.12). Alternative methods of growth yield less perfect samples due to adhesion of the growing

Table 2.12 Influence of growth methods, growth temperature and stress on the perfection of adamantane crystals[26]

	T/K	Dislocation count /cm^{-2}	Mosaic[a] spread
Melt growth	543	10^8	1000
Vapour growth	383	10^7	200
	323	10^7	60
Solution growth	300	0	50
+ 1 g load		$> 10^5$[b]	200
+ 10 g load		$> 10^5$[b]	1000

[a]In seconds of arc.
[b]Limit of resolution of x-ray topography.

crystal to the walls of the growth vessel (vapour and melt growth) and thermal constraints (melt growth). Application of very minor mechanical constraints to the most perfect crystals results in considerable changes in dislocation content and general perfection. This emphasizes that dislocations readily multiply in the crystal and then polygonize to yield a high degree of disordering. Judged from the self-diffusion experiments described above, careful handling and annealing, although it cannot eliminate the dislocations, does reduce them sufficiently for intrinsic behaviour to be observed.

To achieve and maintain such perfection will become progressively more difficult as one moves to the other extreme of the range of these solids. A casually handled highly plastic crystal such as cyclohexane will become extremely defective. It is most unlikely that annealing will remove much of this disorder and even then the minor mechanical strains involved in making measurements could regenerate vast numbers of dislocations, etc. The presence of impurity precipitates could exacerbate this situation. Since it is well established that in a normal crystal containing 10^{10}–10^{12} dislocations per cm^2, all molecules will be elastically distorted from the equilibrium position,[82] it is possible to see that a highly imperfect plastic crystal of moderate purity could be in a semi-crystalline state. Again the magnitude of the problem can be assessed from the self-diffusion measurements where even after careful preparation and prolonged annealing, intrinsic self-diffusion in ultra-pure cyclohexane could only be observed over a limited temperature range, the influence of disorder being significant.

Such variations as these could have a profound influence on experimental measurements particularly when carried out using casually prepared samples. In this context it is interesting to recall that (a) the differences between the radiotracer and NMR studies of self-diffusion arise for the very soft materials,[91] and (b) for these materials the temperature dependence of the NMR diffusion coefficient parallels that for radiotracer self-diffusion in the defective state (see for example Figure 2.11). In a later chapter Boden discusses in detail the NMR measurements and concludes that these represent bulk behaviour and not pipe diffusion effects as discussed in Section 2.3.1 (a) of this Chapter. On the basis of the present information however one could foresee that the bulk of a casually handled sample could differ significantly from that of a well annealed carefully prepared crystal.

2.7 CONCLUSIONS

On the basis of the experimental evidence there seems little doubt that the plasticity of these materials can be described in terms of the basic processes which govern this property in most other crystalline materials. Consequently these materials must be considered as an extreme of the crystalline state.

REFERENCES

1. J. Timmermans, *J. Phys. Chem. Solids*, **18**, 1 (1961).
2. A. Michels, *Bull. Soc. Chim. Belg.*, **57**, 575 (1948).
3. J. E. Dorn, *J. Mech. Phys. Solids*, **3**, 85 (1955).
4. F. R. N. Nabarro, *Report of Conference on Strength of Materials*, Physical Society, London, p. 75 (1947), C. Herring, *J. Appl. Phys.*, **21**, 437 (1950).
5. J. H. Weertman, *J. Appl. Phys.*, **28**, 362 (1957).
6. E. M. Hampton, P. McKay, and J. N. Sherwood, *Phil. Mag.*, **30**, 853 (1974).
7. N. C. Lockhart and J. N. Sherwood, *Unpublished Work*.
8. H. M. Hawthorne and J. N. Sherwood, *Trans. Faraday Soc.*, **66**, 1783 (1970).
9. E. M. Hampton and J. N. Sherwood, *J. C. S. Faraday I*, **72**, 2398 (1976).
10. N. T. Corke, N. C. Lockhart, R. S. Narang, and J. N. Sherwood, *Mol. Cryst. Liq. Cryst.*, **44**, 45 (1978).
11. P. W. Salthouse and J. N. Sherwood, *J. C. S. Faraday I*, **73**, 1843 (1977).
12. H. M. Hawthorne and J. N. Sherwood, *Trans. Faraday Soc.*, **66**, 1792 (1970).
13. P. Bladon, N. C. Lockhart, and J. N. Sherwood, *Mol. Cryst. Liq. Cryst.*, **19**, 315 (1973).
14. P. Bladon, N. C. Lockhart, and J. N. Sherwood, *Mol. Phys.*, **20**, 577 (1971).
15. N. C. Lockhart and J. N. Sherwood, *Faraday Symposia C. S.*, **6**, 57 (1972).
16. H. A. Resing, N. T. Corke, and J. N. Sherwood, *Phys. Rev. Letters*, **20**, 1227 (1965).
17. N. T. Corke and J. N. Sherwood, *J. Mats. Sci.*, **6**, 68 (1971).
18. D. N. Bolshutkin, *Soviet Phys. Solid State*, **7**, 2111, (1965).
19. D. N. Bolshutkin, L. I. Borisova, and A. V. Leonteva, *Soviet Phys. Solid State*, **10**, 1248 (1968).
20. A. I. Prokvatilov, *Soviet Phys. Solid State*, **9**, 2160 (1967).
21. D. N. Bolshutkin, A. I. Prokvatilov, T. V. Silvestrova, and J. I. Startsev, *Soviet Phys. Solid State*, **7**, 2255 (1966).
22. A. D. Prokvatilov, T. N. Silvestrova, and D. N. Bolshutkin, *Soviet Phys. Solid State*, **9**, 546 (1967).
23. G. A. Geach and A. A. Wolff, *Powder Metallurgy*, Editor W. Leszynski, Interscience, New York, p. 210, (1961).
24. S. N. G. Chu and J. C. M. Li, *J. Mats. Sci.*, **12**, 2214 (1977).
25. Y. Adda and J. Philbert, *La Diffusion dans les Solides*, Dunod, Paris, (1966).
26. R. M. Hooper and J. N. Sherwood, *J. Crystal Growth*, to be published.
27. R. T . P. Whipple, *Phil. Mag.*, **45**, 1225 (1954).
28. T. Suzuoka, *Trans. Jap. Inst. Metals*, **2**, 25 (1961).
29. A. D. Le Claire, *Brit. J. Appl. Phys.*, **14**, 351 (1963).
30. E. M. Hampton, N. C. Lockhart, and J. N. Sherwood, *Chem. Phys. Letters*, **23**, 191 (1971).
31. J. Bleay, P. W. Salthouse, and J. N. Sherwood, *Phil. Mag.*, **36**, 885 (1977).
32. E. W. Hart, *Acta. Metall.*, **5**, 597 (1957).
33. E. M. Hampton and J. N. Sherwood, *J. C. S. Faraday I*, **72**, 2398 (1976).
34. J. N. Sherwood and D. J. White, *Phil. Mag.*, **15**, 745 (1967).
35. G. M. Hood and J. N. Sherwood, *Mol. Cryst.*, **1** , 97 (1966).
36. L. W. Barr and J. A. Morrison, *J. Appl. Phys.*, **31**, 617 (1960).
37. A. Berne, G. Boato, and M. M. de Paz, *Nuovo Cimento*, **46B**, 182 (1966).
38. A. V. Chadwick and J. A. Morrison, *Phys. Rev.*, **B1**, 2748 (1970).
39. A. H. Brooks and A. V. Chadwick, To be published.
40. M. Y. Al Shakar and A. V. Chadwick, To be published.
41. P. McKay and J. N. Sherwood, *J.C.S. Faraday I*, **71**, 2331 (1975).

42. P. W. Salthouse and J. N. Sherwood, *J.C.S. Faraday I*, 1845 (1977).
43. A. V. Chadwick and J. W. Forrest, *J.C.S. Faraday I*, In press.
44. A. V. Chadwick, J. M. Chezeau, R. Folland, J. W. Forrest, and J. H. Strange, *J.C.S. Faraday I*, **71**, 1610 (1975).
45. R. Fox and J. N. Sherwood, *Trans. Faraday Soc.*, **67**, 3364 (1971).
46. E. M. Hampton and J. N. Sherwood, *J. C. S. Faraday I*, **71**, 1392 (1975).
47. E. L. Hahn, *Phys. Rev.*, **80**, 580 (1950).
48. J. E. Tanner, *J. Chem. Phys.*, **56**, 3850 (1972).
49. H. Suga, M. Sugusaki, and J. Seki, *Mol. Cryst.*, **1** , 337 (1966).
50. C. Zener, *Imperfections in Nearly Perfect Crystals*, Edited by W. Shockley Wiley New York, p. 289 (1952).
51. H. Kanzaki, *J. Phys. Chem. Solids*, **2**, 24–36 (1957).
52. G. F. Nardelli and A. Repanai Chiarotti, *Nuovo Cimento*, **18**, 1053–1071 (1960).
53. G. F. Nardelli and N. Terzi, *J. Phys. Chem. Solids*, **25**, 815–826 (1964).
54. G. L. Hall, *J. Phys. Chem. Solids*, **3**, 210–222 (1957).
55. J. J. Burton and G. Jura, *J. Phys. Chem. Solids*, **28**, 705–710 (1967).
56. H. R. Glyde and J. A. Venables, *J. Phys. Chem. Solids*, **29**, 1093–1098 (1968).
57. K. Mukherjee, *Phys. Lett.*, **25A**, 439–440 (1967).
58. J. J. Burton, *Phys. Rev.*, **182**, 885–891 (1969).
59. D. R. Squire and W. G. Hoover, *J. Chem. Phys.*, **50**, 701–706 (1969).
60. S. R. Druger, *Phys. Rev. B*, **3**, 1391–1396 (1971).
61. A. R. Allnatt and L. A. Rowley, *J. Phys. Chem. Solids*, **30**, 2187–2199 (1969).
62. R. M. J. Cotterill and M. Doyama, *Phys. Lett.*, **25A**, 35–36 (1967).
63. M. Doyama and R. M. J. Cotterill, *Phys. Rev.*, **B1**, 832–833 (1970).
64. W. Jost, *Diffusion* (New York, Academic Press) p. 116 (1960).
65. R. Fieschi, G. F. Nardelli, and A. Repanai Chiarotti, *Phys. Rev.*, **123**, 141–147 (1961).
66. R. O. Simmons and R. W. Balluffi, *Phys. Rev.*, **117**, 52–61 (1960).
67. D. L. Losee and R. O. Simmons, *Phys. Rev.*, **172**, 934 (1968).
68. P. F. Higgins, R. A. B. Ivor, L. A. K. Staveley, and J. J. des C. Virden, *J. Chem. Soc.*, 5762–5768 (1965).
69. J. R. Green and D. R. Wheeler, *Mol. Cryst. Liq. Cryst.*, **6**, 13–21 (1969).
70. R. H. Baughman and D. Turnbull, *J. Phys. Chem. Solids*, **32**, 1375–1394 (1971).
71. R. A. Swalin, '*Thermodynamics of Solids*', Wiley, New York (1961) .
72. R. E. Howard and A. B. Lidiard, *Rep. Prog. Phys.*, **27**, 161–240 (1964).
73. R. H. Beaumont, H. Chihara, and J. A. Morrison, *Proc. Phys. Soc.* (*London*), **78**, 1462–1481 (1961).
74. A. J. E. Foreman and A. B. Lidiard, *Phil. Mag.*, **8**, 97–103 (1963).
75. M. L. McGlashan, *Disc. Faraday Soc.*, **40**, 59–68 (1965).
76. D. L. Martin, in '*Melting, Diffusion and Related Topics*', Report of 2nd Symposium N.R.C., p. 31, Ottawa (1957).
77. J. Kubler and M. P. Tosi, *Phys. Rev.*, **137**, 1617–1620 (1965).
78. A. D. le Claire, *Physical Chemistry—An Advanced Treatise*, Edited by W. Henderson, H. Eyring, and W. Jost, Academic Press, New York, Vol. 10, Chapter 5 (1970).
79. R. S. Narang and J. N. Sherwood, To be published.
80. O. D. Sherby and P. M. Burke, *Progress Mats. Sci.*
81. J. Friedel, *Dislocations*, Pergamon, London (1964).
82. J. M. Thomas, *Phil. Trans. Roy. Soc.*, **277**, 251 (1974). *Adv. Catalysis*, **19**, 202 (1969).
83. J. M. Thomas and J. O. Williams, *Progr. Solid State Chemistry*, **6**, 271 (1971).
84. E. M. Hampton, R. M. Hooper, B. S. Shah, J. N. Sherwood, B. Escaig, and J. Di Persio, *Phil. Mag.*, **29**, 743 (1974).

85. P. M. Robinson and H. G. Scott, *Acta Metals*, **15**, 1581 (1967), *Mol. Cryst. Liq. Cryst.*, **11**, 13 (1970).
86. B. S. Shah and J. N. Sherwood, Unpublished work.
87. G. J. Ogilvie and P. M. Robinson, *Mol. Cryst. Liq. Cryst.*, **12**, 379 (1971).
88. B. S. Shah and J. N. Sherwood, *Trans. Faraday Soc.*, **67**, 1200, (1971).
89. R. S. Narang and J. N. Sherwood, To be published.
90. B. K. Tanner, *Z. Naturforsch*, **28a**, 676 (1973).
91. P. Bladon, N. C. Lockhart, and J. N. Sherwood, *Mol. Phys.*, **20**, 577 (1971); *Mol. Cryst. Liq. Cryst.*, **19**, 315 (1973).

3

Diffuse X-Ray Scattering by Orientationally-Disordered Crystals

R. Fouret

3.1 INTRODUCTION

As a consequence of the very high Debye–Waller factors, orientationally disordered molccular crystals yield only a few diffracted x-ray beams at low Bragg angles. Consequently, structural determinations are imprecise. The problem is exacerbated by the presence of an intense diffuse scattering as first noted by Finbak[1] in 1938. This diffuse scattering is highly structured and it is possible that its detailed study could yield further information both to supplement the structural data and to provide details of the dynamics and orientation of molecules in the disordered crystal lattice.

Such analyses have been successfully carried out for organic molecular solids which comprise non-reorientating molecules. In these cases the diffuse x-ray scattering arises only from thermal vibrational motions of the molecules. Although an apparently simpler situation the problem is complicated but with reasonable approximations a satisfactory description of the source of the scattering can be obtained. The methods used[2-5] for these previous analyses have been based on a harmonic approximation to the crystal dynamics. Dynamical studies[6,7] indicate that the present materials are highly anharmonic and that concepts developed for harmonic crystals are in general unsuitable. Consequently it is doubtful if the previous descriptions could be applied directly and some better approach should be considered.

A novel and ideal approach would be via a computerized molecular dynamics calculation which could then be arranged to include all molecular motions. Such calculations have been used successfully to describe[7] coherent neutron scattering from orientationally disordered crystals. They represent a rather too sophisticated approach to the present problem however and within the experimental limitations a more modest approach using simpler approximate models such as those of Pauling[4] and Frenkel[8] is satisfactory.

The Pauling model[4] in which rotations are almost free but are restricted by weak anisotropic potentials with no pronounced minima is of rather limited application. Phases in which molecules approximate to free rotors are few and only occur at low temperatures e.g. H_2 and CH_4 (II). The Frenkel model[8] is more generally applicable and will form the basis of our theoretical approach. In this model[8] the crystal is considered to be disordered; in different sites the molecules occupy one of D possible orientations and perform librations in the potential well accompanied by jumps from one equilibrium position to another. The diffuse scattering arises from both the orientational disorder of the molecules and the thermal motions. In many cases the contribution of the latter will be significant and we do not feel, as do some authors, that it can be neglected. Also by taking account of this contribution we can more precisely determine the influence of disorder alone on the diffuse scattering. This enables us to confirm that the equilibrium positions of molecules as indicated by the structural analyses are correct. Furthermore, even if we do not have long range order in the crystal there will be a local order due to correlations between the orientations of neighbouring molecules. These correlations will have a considerable influence on the observed diffuse diagram. An assessment can be made of these correlations and their range when the thermal contributions have been subtracted.

Only in one circumstance does the influence of the thermal motions become negligible; when the crystal is slightly above its critical temperature. Here, large fluctuations in molecular orientations occur and the scattering due to orientational disorder has a critical behaviour. This has been observed experimentally for methane (CH_4, CD_4),[9] sodium nitrite,[10] and sodium nitrate.[11] Although not applied widely to molecular crystals as yet, the technique provides a possible means of probing the forces which decide the phase transition.

The principal aim of this article is to present the underlying theory for normal and critical scattering and to describe the interpretation of the experimental results for the few materials studied. We hope that this will serve to stimulate further theoretical and experimental interest in this currently little examined area of the structural study of orientationally disordered solids.

Before proceeding to the detailed analysis of the diffuse scattering figures let us first consider briefly how they are recorded experimentally.

3.2 EXPERIMENTAL METHODS

The prime experimental essentials are a strictly monochromatic source of radiation and a monocrystalline specimen.

The simplest procedure is to use a Laue configuration and photographic recording as depicted in Figure 3.1. The resulting diffuse diagram on the film corresponds to the location of the reciprocal lattice points on the Ewald sphere. The method, which is admirable for the rapid, qualitative study of diffuse scattering has been widely used by the Orsay group.[12]

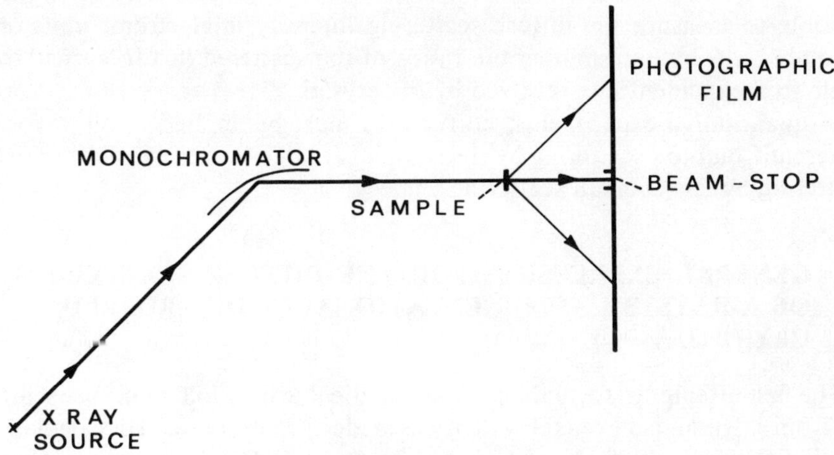

Figure 3.1 Monochromatic Laue setting

A more elaborate system uses a moving film camera of the Weissenberg or precession type together with a graphite monochromator. With the precession camera in particular subsequent photometric measurements are easier because the film records only one undistorted reciprocal lattice plane. A good example of this method is the pivalic acid study of Longueville, Fontaine, Descamps, and Coulon.[13]

As an alternative to photographic recording, intensity measurements can be made using a counter or diffractometer (Figure 3.2). The diffuse scattering intensity is measured point by point for different values and orientations of the scattering vector $\mathbf{X} = (\mathbf{s} - \mathbf{s}_0)/\lambda$ where $\mathbf{s}_0$ and $\mathbf{s}$ are unit vectors in the incident and scattering directions respectively and λ the wavelength of the monochromatic radiation. This technique is particularly useful for the examination of critical diffuse scattering which occurs on particular points in reciprocal space.[10,11] It has also been used for the study of succinonitrile.[14] Using this method it is

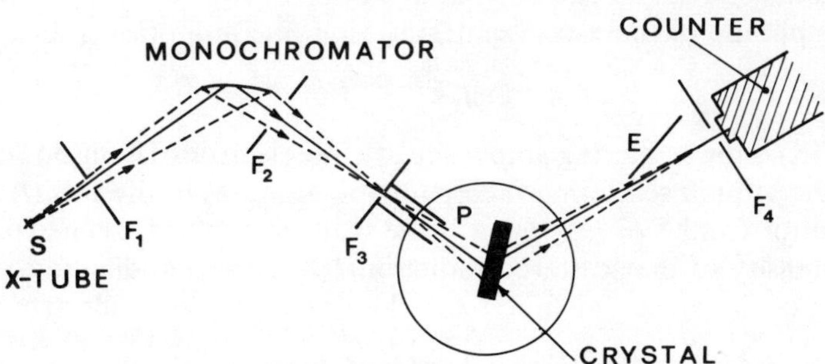

Figure 3.2 Diffractometer arrangement for measuring the diffuse intensity

possible to measure the diffuse scattering intensity in electronic units or in absolute units by determining the ratios of the scattered flux in a small solid angle to the incident flux received by the crystal.

In quantitative experiments, corrections must be applied to allow for incoherent Compton scattering by the sample (at large scattering angles) and for scattering by air (at small scattering angles).

3.3 GENERAL EXPRESSION FOR THE DIFFUSE SCATTERING OF X-RAYS BY AN ORIENTATIONALLY DISORDERED CRYSTAL

The first attempt to formulate a mathematical expression for diffuse scattering from a dynamical Frenkel system was made by Matsubara.[15] His calculation, applied theoretically to N_2, KCN, NaCN, and $C(NO_2)_4$, took no account of thermal scattering. The result was not compared with experiment. We shall take the thermal scattering into account and base our analysis on the method developed by Descamps[14] following Fournet[16] and Guinier.[17]

3.3.1 Approximation of the Independent Vibration of Molecules

Let us suppose that the crystal contains one molecule per unit cell. This is a frequent situation and greatly simplifies the notation. The orientation of this molecule on a site will then be defined by a parameter ω_l. This parameter can be, for instance, the set of Euler angles which define the orientation of 3 axes attached to the molecule relative to the axis of the laboratory. It will take D values corresponding to the D allowed orientations of the molecules i.e. $\omega_l = 1, 2, \ldots D$.

At a given moment, an atom (lk) of this molecule (l) will be at the end of the vector $\mathbf{R}_l + \mathbf{r}_{lk} + \mathbf{u}_{lk}$. $\mathbf{r}_{lk}$ marks the equilibrium position of the atom (k) in the cell, the origin of which corresponds to the end of the vector $\mathbf{R}_l$. $\mathbf{r}_{lk}$ will depend on the orientation ω_l of the molecule in the site l. $\mathbf{u}_{lk}$ represents the displacement of this atom (k) from its equilibrium position.

The amplitude of the x-ray scattering from the atom (lk) is then given by:

$$A_e(\mathbf{X})\, f_k(\mathbf{X}) e^{i2\pi \mathbf{X}(R_l + \mathbf{r}_{lk} + \mathbf{u}_{lk})} \tag{3.3.1}$$

where $A_e(\mathbf{X})$ is the scattering amplitude of a free electron (Thomson Scattering), $f_k(\mathbf{X})$ is the atomic scattering factor for the atom (k) in the cell (l) and is independent of l and $\mathbf{X} = (\mathbf{s} - \mathbf{s}_0)/\lambda$ is the scattering vector, defined previously.

The intensity of the scattered radiation $I(\mathbf{X})$ corresponding to a given configuration of the crystal at time t can then be expressed in electronic units as

$$I(\mathbf{X}) = \sum_l \sum_{l'} F_l(t)\, F_{l'}^*(t) e^{i2\pi \mathbf{X}\cdot\mathbf{R}_{ll'}} \qquad 3.3.2$$

where

$$F_l(t) = \sum_k f_k{}^{(\mathbf{X})} e^{i2\pi \mathbf{X}(\mathbf{r}_{lk} + \mathbf{u}_{lk})} \tag{3.3.3}$$

is the instantaneous structure factor for the cell (l) and

$$\mathbf{R}_{ll'} = \mathbf{R}_l - \mathbf{R}_{l'}$$

The observed intensity will then be the average over a very long time of this function $I(\mathbf{X})$. To calculate this intensity we suppose that the lifetime of a configuration is long compared with the period of the vibrations of the molecules in the crystal. Let us first take the average of the observed intensity for the vibrations of the molecules in a given configuration. $\{\omega_1, \omega_2, \ldots \omega_N\}$, and then average over all configurations. The former can be expressed as

$$I_{\text{obs}} = \sum_l \sum_{l'} \overline{\langle F_l(t)\, F_{l'}^*(t)\rangle}\, e^{i2\pi \mathbf{X}\cdot\mathbf{R}_{ll'}}. \tag{3.3.4}$$

$\overline{\langle A\rangle}$ corresponds to the thermal average of the function A ($\langle A\rangle$) averaged over all allowed configurations.

The thermal average of $F_l F_{l'}^*$ can then be written:

$$\langle F_l(t)\, F_{l'}^*(t)\rangle = \sum_k \sum_{k'} f_k f_{k'} e^{i2\pi \mathbf{X}\cdot\mathbf{r}_{lk,l'k'}} \langle e^{i2\pi \mathbf{X}(\mathbf{u}_{lk} - \mathbf{u}_{l'k'})}\rangle \tag{3.3.5}$$

$$\mathbf{r}_{lk,l'k'} = \mathbf{r}_{lk} - \mathbf{r}_{l'k'}$$

For a harmonic crystal the calculation of the thermal average (equation (3.3.5) is performed classically. For a strongly anharmonic crystal we can write the following approximate expression:

$$\langle e^{i2\pi \mathbf{X}(\mathbf{u}_{lk} - \mathbf{u}_{l'k'})}\rangle \cong e^{-2\pi^2\langle |\mathbf{X}\cdot(\mathbf{u}_{lk} - \mathbf{u}_{l'k'})|^2\rangle} \tag{3.3.6}$$

or in a different form:

$$\langle e^{i2\pi \mathbf{X}(\mathbf{u}_{lk} - \mathbf{u}_{l'k'})}\rangle \cong e^{-[B_k(\omega_l) + B_{k'}(\omega_{l'})]} \cdot e^{D(lk;\, l'k')} \tag{3.3.7}$$

The Debye–Waller factor for the atom (k), $B_k = 2\pi^2\langle (\mathbf{X}\cdot\mathbf{u}_{lk})^2\rangle$ depends on the orientation ω_l of the molecule in the given configuration. $D(lk; l'k')$ introduces the correlation between the vibration of the atom (lk) and that of the atom ($l'k'$)

$$D(lk;\, l'k') = 4\pi^2\langle (\mathbf{X}\cdot\mathbf{u}_{lk})(\mathbf{X}\cdot\mathbf{u}_{l'k'})\rangle \tag{3.3.8}$$

In their calculation for ordered molecular crystals, Amoros and Amoros [5] assume that the vibrations of the molecules are independent and do not take this correlation into account.

Let us first assume that $D(lk;\, l'k') = 0$ for $l \neq l'$. The thermal average of the intensity then becomes

$$\langle I(X)\rangle = \left\{ \sum_{\substack{l \\ l\neq l'}} \sum_{l'} F'_l F'^{*}_{l'} \mathrm{e}^{\mathrm{i}2\pi \mathbf{X}\cdot\mathbf{R}_{ll'}} + \sum_{l} \sum_{kk'} f'_k(\omega_l) f'_{k'}(\omega_l)\, \mathrm{e}^{D(lk;\, lk')} \right\} \mathrm{e}^{\mathrm{i}2\pi \mathbf{X}\cdot\mathbf{r}_{lk,l'k'}} \tag{3.3.9}$$

$f'_k(\omega_l)$ is the atomic scattering factor of the atom (k) weakened by the Debye–Waller factor. F'_l is the static structure factor of the cell (l) with account taken of the Debye–Waller factors of each atom.

To calculate the average over the different configurations, we shall introduce two probability terms.

$P_1(\omega_l)$, the probability that the molecule (l) is in site (l) and orientation ω_l. This corresponds to the average of the second term in equation (3.3.9).

$P_2(\omega_l \omega_{l'})$ the double probability that the molecule in site (l) is in the orientation ω_l and the molecule in site (l') is in the orientation $\omega_{l'}$. This corresponds to the average of the first term in equation (3.3.9).

If there is no correlation between the orientations of the molecules then

$$P_2(\omega_l, \omega_{l'}) = P_1(\omega_l) \cdot P_1(\omega_{l'}) \tag{3.3.10}$$

If, on the contrary these correlations exist, we should write

$$P_2(\omega_l | \omega_{l'}) = P_1(\omega_l) \cdot P_2(\omega_{l'} | \omega_l) \tag{3.3.11}$$

$p_2(\omega_{l'} | \omega_l)$ is the conditional probability that the molecule on site l' has the orientation $\omega_{l'}$ and that on site l the orientation ω_l. When $\mathbf{R}_{llL} \to \infty$ the correlations between the orientations of the molecules vanish and

$$p_2(\omega_{l'} | \omega_l) \text{ tends to } P_1(\omega_{l'}).$$

In an orientationally disordered crystal, the difference $g_2(\omega_{l'} | \omega_l) = p_2(\omega_{l'} | \omega_l) - P_1(\omega_{l'})_i$ will characterize the short range orientational order: for $(\omega_l = i, \omega_{l'} = j)$ the molecule (l') will tend or not tend to take the orientations j when the molecule (l) has the orientation i according to the sign of this difference.

Finally, the average intensity or observed intensity for x-rays scattered by an orientationally disordered crystal can be written

$$I_{\text{obs}} = \left\{ N^2 |\bar{F}'|^2 \Delta(X) + N(I'_1 - |\bar{F}'|^2) + N \sum_{l'-l\neq 0} C(R_{ll'}; X) \mathrm{e}^{\mathrm{i}2\pi \mathbf{X}\cdot\mathbf{R}_{ll'}} \right\} \tag{3.3.12}$$

The first term is the intensity diffracted by a mean crystal having N cells and a structure factor

$$\bar{F}' = \sum_{\omega_l} P_1(\omega_l)\, F'_l(\omega_l). \tag{3.3.13}$$

$\Delta(\mathbf{X})$ differs from 0 and equals 1 only if $\mathbf{X}$ is a reciprocal lattice vector. The second term $N(I_1^r - |F'|^2)$ gives in electronic units the intensity scattered when there are no correlations. The term I_1^r is calculated by assuming that the molecules are rigid and have a centre of mass translation $\mathbf{u}_l$ and a libration θ_l around the equilibrium position:

$$I_1^r = \sum_{kk'} \sum_{\omega_l} f_k f_{k'} P_1(\omega_l)\, e^{-2\pi^2 \langle (\mathbf{X}, \theta_l, \mathbf{r}_{lk,lk'})^2 \rangle}\, e^{i2\pi \mathbf{X} \cdot \mathbf{r}_{lk,lk'}} \tag{3.3.14}$$

If the librations were neglected, this term would simply equal the average of the square of the structure factor modules $|\overline{F}|^2$ without the weakening of the Debye–Waller factor.

Making this new approximation we can write for this second term:

$$I_{\text{no corr.}} = N(I_1^r - |\overline{F'}|^2) = N(|\overline{F}|^2 - |\overline{F}|^2) + N(|\overline{F}|^2 - |\overline{F'}|^2) \tag{3.3.15}$$

This shows that diffuse scattering without correlations is the sum of the random disorder scattering plus the thermal disturbance. It represents the generalization of the Amoros calculation to allow for disorder.[5] (The so-called Difference Fourier Transform).

Provided that the crystal structure is known, then according to equation (3.3.15), it should be possible to calculate the distribution of diffuse scattering. If the positions of the calculated maxima do not agree with the observed ones, the crystal presents correlations. The last term of equation (3.3.12)

$$I_{\text{corr}} = N \sum_{l'-l} C(\mathbf{R}_{ll'}, \mathbf{X})\, e^{i2\pi \mathbf{X} \cdot \mathbf{R}_{ll'}} \tag{3.3.16}$$

where

$$C(\mathbf{R}_{ll'}, \mathbf{X}) = \sum_{\omega_l} \sum_{\omega_{l'}} P_1(\omega_l) g_2(\omega_{l'} | \omega_l) F_l'(\omega_l) F_{l'}'(\omega_{l'}) \tag{3.3.17}$$

takes the correlations into account.

The value of this term can only be obtained if the conditional probability that orientations $\omega_{l'}$ and ω_l exist simultaneously can be calculated. Theoretically, the calculation could be made from a knowledge of the potential energy of the molecules for a given configuration of the crystal. Practically it can only be made in particular cases and with simplifying assumptions.

In spite of these simplifications and the fact that the assumption of independent molecular vibrations is only a very rough approximation to real behaviour, the theory does allow the prediction of the positions of the scattering maxima for specific systems. This has been qualitatively demonstrated by Cochran and Pawley[3] for a crystal comprising rigid molecules. In order to achieve quantitative agreement however the correlations between the atomic vibrations must be included.

3.3.2 Inclusion of Correlations Between Atomic Vibrations

The most important source of scattering of thermal origin is due to acoustic waves in the crystal. Neutron scattering experiments indicate that the acoustic vibrations in orientationally disordered crystals will exist in many cases anywhere within the Brillouin zone.

Let us call $\mathbf{u}_l$ the displacement of the centre of mass of the molecule (l) which we shall assume to be a rigid body. Then, if we neglect the vibration–libration and libration–libration correlations, we can write:

$$D(lk;\, l'k') = 4\pi^2 \langle (\mathbf{X}\cdot\mathbf{u}_l)(\mathbf{X}\cdot\mathbf{u}_{l'})\rangle = D(l, l') \qquad (3.3.18)$$

Inserting this expression into equation (3.9) and using a similar procedure as before to calculate the observed scattered intensity I_D, we find:

$$I_D = N(I_1^r - I_2^r) + \sum_{l,l'} \overline{F_l^r F_{l'}^{r*} D_u(ll')}\ \mathrm{e}^{i2\pi\mathbf{X}\cdot\mathbf{R}_{ll'}} + I_{corr} \qquad (3.3.19)$$

The first term now corresponds to the random disorder diffuse intensity where I_2^r equals:

$$I_2^r = \sum_{\omega_l} \sum_{\omega_{l'}} P_1(\omega_l)\, P_1(\omega_{l'}) f_k f_{k'}\, \mathrm{e}^{-2\pi^2 \langle (\mathbf{X},\theta_l,\mathbf{r}_{lk,l'k'})^2 \rangle} \cdot \mathrm{e}^{i2\pi\mathbf{X}\cdot\mathbf{r}_{lk,l'k'}} \qquad (3.3.20)$$

If the libration θ_l of the molecules is not taken into account this term reduces to $N(|\overline{F}|^2 - |\overline{F}|^2)$.

The second term corresponds to the average of the thermal diffuse scattering caused by the vibration waves in the disordered crystal. In a molecular crystal with only one molecule per cell, the vibration waves correspond to the acoustic waves. If we calculate this term approximately using cyclic boundary conditions and on the basis that the acoustic waves can be expanded as a function of the phonon creation and annihilation operator for the mean crystal, we find the same result as for an ordered crystal:

$$I_{Th.Av.} = N \sum_{\mathbf{q}j} |\overline{F}'|^2 |\mathbf{X}\mathbf{U}_j|^2 \frac{E_j(\mathbf{q})}{v_j^2(\mathbf{q})} \Delta(\mathbf{q} + \mathbf{X}) \qquad (3.3.21)$$

Here, E is the mean energy of the acoustic wave j of wave vector $\mathbf{q}$ frequency $v_j(\mathbf{q})$ and unit polarization vector $\mathbf{U}_j$. In the orientationally disordered phase it will be equal to $k_B T$ where k_B is the Boltzmann constant and T the absolute temperature of the crystal. Neglecting the vibration–libration and libration–libration correlations, we can calculate the frequencies which are present in $I_{Th.Av.}$ on the basis that the crystal can be considered as comprising atoms of mass m, with force constants determined by the elastic constants.

The last term in equation (3.3.19), I_{corr}, corresponds to the space correlation term which can be either positive or negative. This term is important in defining the contrast of the diffuse diagram. It can be expressed as

$$I_{corr.} = \sum_{\substack{l,l' \\ l \neq l'}} \sum_{\omega_l} \sum_{\omega_{l'}} P_1(\omega_l)\, g_2(\omega_{l'}/\omega_l)\, F'_l F'^*_{l'} \left[1 + D(ll')\right] e^{i2\pi \mathbf{X}\cdot\mathbf{R}_{ll'}}$$

An approximation of this term can be obtained by neglecting $D(ll')$.

The next stage in the development of the theory would be to include libration–libration and vibration–libration correlations. However the number and the degree of the approximations necessary to obtain a solution are such that no significant improvement would result. This, then represents the limit of the present theory and we must await the development of more detailed molecular dynamics calculations.

Consideration of the preceding analysis indicates that, in order to calculate the diffuse scattering, we need to know the conditional probability ($p_2(\omega_{l'}/\omega_l)$) and the orientational probability ($p_1\omega_l$). The latter is obtained from a knowledge of the crystal structure factor. Determination of the former is the main problem in the study of short range order in orientationally disordered crystals. It is essential to develop a model for the disordered solid based on the crystal and molecular structure (and often the experimentally observed scattering). From this the conditional probability is calculated and hence the intensity and intensity distribution of the diffuse scattering. This can then be compared with experiment. The mode of calculation of this parameter forms the theme of the following sections.

3.4 DIFFUSE SCATTERING WITH STRONGLY ANISOTROPIC CORRELATIONS

If correlations between the orientations of molecules are established in a particular direction, we can visualize the order within the crystal as comprising ordered clusters of variable length fluctuating around a mean value determined by the correlation length. This means that for the preceding calculation, the double probability $p_2(\omega_{l'}\omega'_l)$ is unity if l and l' are inside the same domain with ω_l and $\omega_{l'}$ the characteristic orientations of the molecules in the domain. Such correlations have been studied by Levelut and Lambert[12] in the hexasubstituted derivatives of benzene.

The particular materials examined by them were:

Hexachlorobenzene
1,2-dichloro-3,4,5,6-tetramethylbenzene (DCTMB)
1,2,3-trichloro-4,5,6-trimethylbenzene (TCTMB)

In all three solids, the substituted benzene rings form linear stacks parallel to the **b** axis of the monoclinic network.

3.4.1 Hexachlorobenzene

As shown in Figure 3.3(a), experiments on the high temperature phase yield a diffuse scattering localized in the (010) scattering planes and consistent with a molecular disorder.

The (010) scattering planes correspond in the crystal to periodic rows of benzene rings stacked parallel to the **b** axis. In spite of the similarity of the six substituents the order between these rows must be limited enough to lead to an incoherent scattering whilst the rows themselves must be predominantly ordered. This is difficult to rationalize if the molecule is planar and Levelut and Lambert [12] propose that it is not. They base their proposition on the results of neutron scattering study of hexamethylbenzene by Hamilton *et al.*[30] in which it is shown that the CH_3 groups lie alternately above and below the plane of the benzene ring. The similarity of size between $-CH_3$ and $-Cl$ suggests a parallel effect in hexachlorobenzene. This would result in a change from apparent six-fold symmetry to a three-fold one. The molecule can then adopt two distinct orientations related to each other by a $\pi/3$ rotation around the symmetry axis. Levelut and Lambert[12] propose that the packing of molecules imposes an order within the rows but that adjacent rows are randomly, dynamically, disordered between the two orientations ($\pm r$ about the mean position) as shown in Figure 3.4(a). On this basis and allowing a limit in correlation of N_2 elements along the rows but no correlation between the orientations of two adjacent rows, they calculate the scattered intensity to be

$$I(X) = (F - F^*)^2 N_1 N_2 \sin^2 2\pi \mathbf{r}\, \mathbf{X} \frac{\sin^2 N_2 \pi \mathbf{X}\,\mathbf{b}}{\sin^2 \pi \mathbf{X}\,\mathbf{b}}$$

where F is the structure factor corresponding to the three different positions of the 1,3,5-chlorine atoms (Figure 3.4b) and N_1, N_2, N_3 are the number of cells in the a, b, and c direction. This equation predicts that the diffuse scattering is localized in the $\mathbf{X} \cdot \mathbf{b} = \mathbf{k}$ planes i.e. (010) of the reciprocal lattice, as is observed. The factor $\sin^2 N_2 \pi \mathbf{X}\,\mathbf{b}/\sin^2 \pi \mathbf{X}\,\mathbf{b}$ accounts for the observed broadening of the lines. Photometric measurement of this broadening indicates that the extent of correlations is $N_2 \cong 5$ molecules at room temperature. The equation also accounts for the absence of scattering in the plane passing through the origin (Figure 3.4(a)).

Although an apparently satisfactory explanation there is no experimental evidence to confirm that the disorder is dynamic. Starting from new x-ray diffraction data Brown and Strydom[20] have shown that the molecule of hexachlorobenzene is planar with rather weak Debye–Waller factors. Thus, if

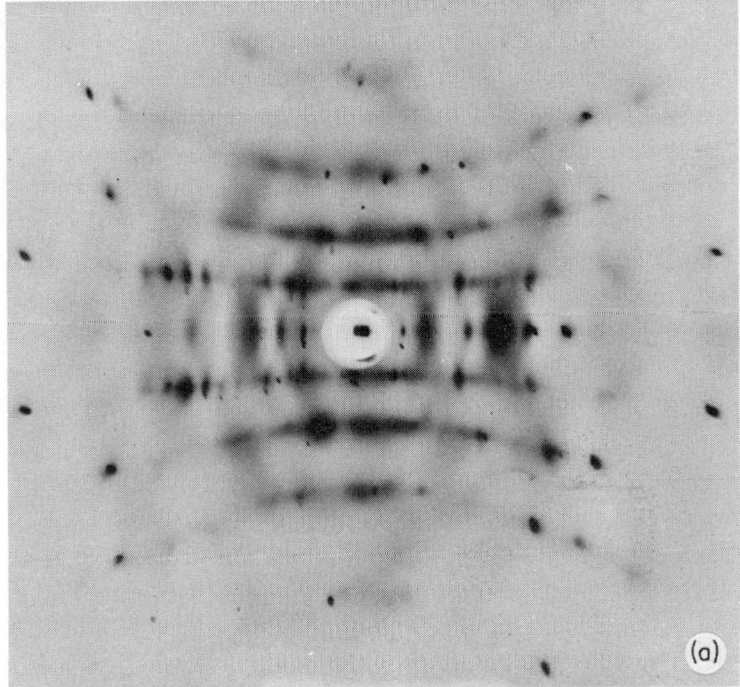

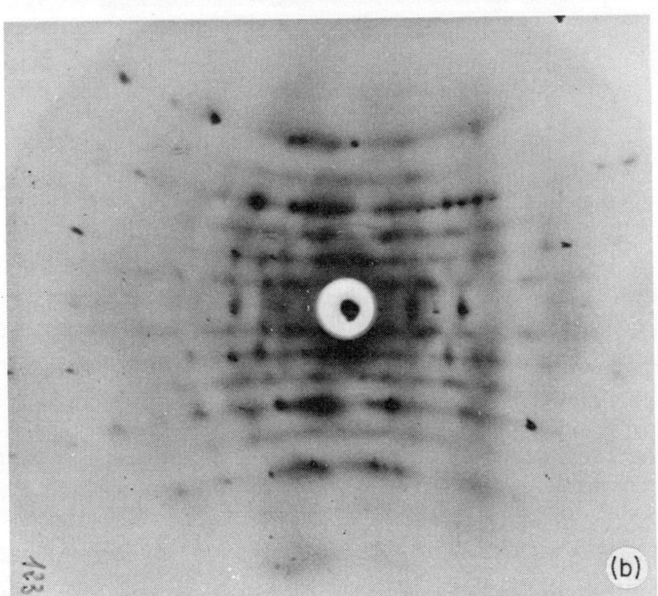

Figure 3.3 Laue photographs taken at 20 °C with MoKα radiation: the **b** axis is vertical. (a) hexachlorobenzene; (b) T.C.T.M.B.

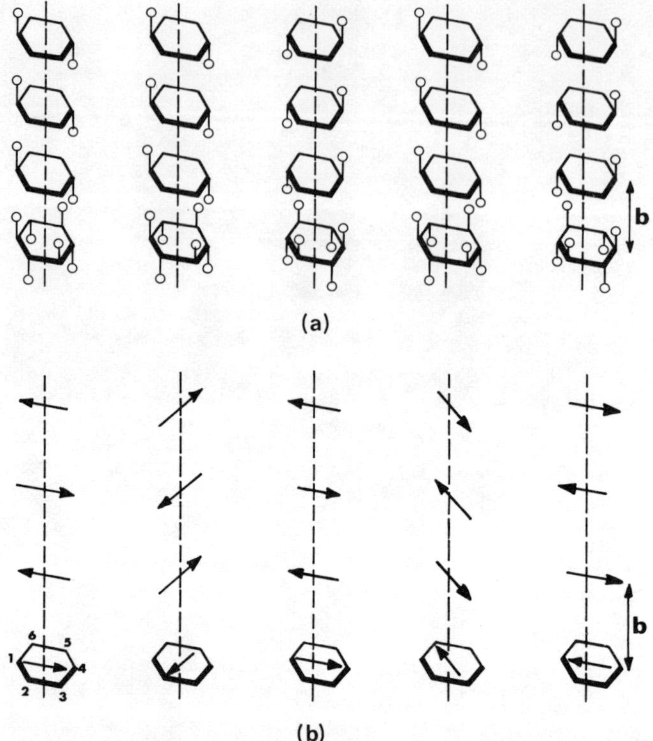

Figure 3.4 (a) Hexachlorobenzene chains—from one chain to another the displacement of the chlorine atoms is disordered—(all the chlorine atoms are not drawn). (b) Disorder in trichlorotrimethyl benzene

dynamic disorder does exist, the lifetime is certainly long enough to allow weakly anharmonic motions and small temperature factors.

On the basis of Brown and Strydom's data Reynolds[19] has contested the above explanation and suggested that the orientational disorder is static. In his model the molecules are planar and a fraction of them are disordered. The scattering defect corresponds to a stack of n misoriented molecules along the [010] direction starting and terminating at a lattice vacancy. The type of disorder visualized for a two dimensional crystal is shown in Figure 3.5. A and B correspond to ordered crystal structures of slightly different free energy ($A < B$). He shows qualitatively that this model also explains the observed diffuse scattering.

Unfortunately x-rays are not able to distinguish whether or not the phenomenon is static or dynamic. There is presently a lack of other experimental data to clarify the situation. We note however that with a similar model Levelut and Lambert have been able to interpret the diffuse scattering of the

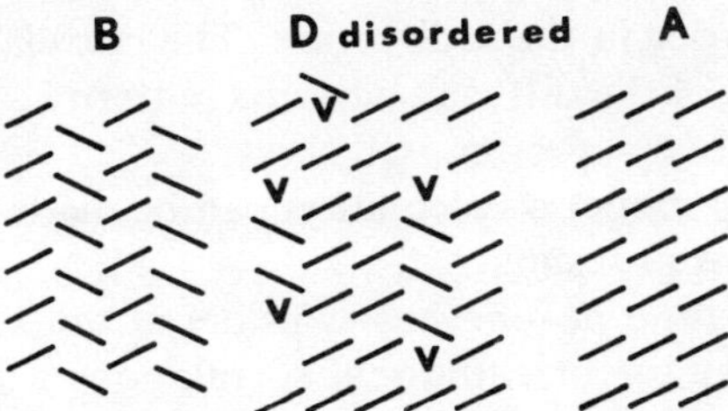

Figure 3.5 Stacking vacancy disorder in a two dimensional crystal after Reynolds. *A* and *B* are two possible structures close in Gibbs free energy

other hexasubstituted derivatives of benzene where the dynamical character of the disorder is well defined.

3.4.2 Dichlorotetramethylbenzene (DCTMB) and Trichlorotrimethylbenzene (TCTMB)

It is well established that DCTMB and TCTMB show orientational phase transitions at 170 K and 260 K respectively. There is no doubt that the orientational disorder in the high temperature phase is dynamic.[21–22]

All the phases are monoclinic. At low temperatures the benzene rings are stacked face to face along the **b** axis in a antiferroelectric arrangement. Above the transition point the molecules become free to rotate around the hexad axis and take one of six equivalent positions. The antiferroelectric order disappears and the lattice parameter is halved in the **b** and **c** directions. We shall confine our attention to TCTMB. The conclusions are the same for DCTMB.

Laue diagrams for the two solids are equivalent (Figure 3.3(b)) and show a diffuse scattering localized in the (010) reciprocal plane. In detail we note that

(a) in the (010) plane passing through the origin, the intensity is zero.
(b) diffuse scattering also exists in planes corresponding to points $(h, k + \frac{1}{2}, l)$ of the low temperature phase.

The localization of scattering in the (010) planes is reminiscent of hexachlorobenzene and can be accounted for in the same manner with anisotropic correlations in the **b** direction of range, at room temperature, 5 molecules. The absence of the plane passing through the origin is characteristic of a displacement type of disorder.

Scattering in the intermediate planes is related to a strongly anisotropic local order in periodic rows parallel to **b** but with a repeat period double that of the crystal. This scattering indicates that the rings change their orientations locally so as to preserve, in part, the antiferroelectric order. The structure indicated in Figure 3.4(b) is adequately descriptive of this. In the plane of the rings there is an equal probability for each of six orientations coupled with the preservation of antiferroelectric order in the **b** direction. The correlation length, as determined from the width of the diffuse scattering pattern is again about five times the distance between two superimposed rings.

On the basis of this model a calculation can be made[12] which qualitatively explains the experimental results.

Up to the present time no-one has considered the alternative (and quite likely) possibility, that the orientational correlations will develop in parallel planes. Since the arrangement in successive planes is completely random this would yield lines of diffuse scattering corresponding to the Fourier transform of these planes. In the perpendicular plane these lines would have a range roughly the inverse of the correlation length.

These solids represent the few examples examined where local order is anisotropic. More generally the correlations are in three dimensions and studies are more complicated.

3.5 DIFFUSE SCATTERING FROM SYSTEMS WITH THREE-DIMENSIONAL CORRELATIONS ACCORDING TO MATSUBARA

To solve the problem of the order propagation in three dimensions, Matsubara[15] uses the Zernike approximation.[30]

In Section 3 the conditional probability was written:

$$p(\omega_{l'} \mid \omega_l) = P(\omega_{l'}) + g(\omega_{l'} \mid \omega_l) \tag{3.5.1}$$

In a homogeneous system, this probability depends mainly on $\mathbf{r}_{ll'}$. If we make the assumption that the orientation is ω for the point chosen as the origin, for the point at the end of $\mathbf{r}_l$ with orientation ω' the conditional probability will be $P_l(\omega' \mid \omega)$. Matsubara expresses this as

$$p_l(\omega' \mid \omega) = \sum_{\omega_1 \ldots \omega_Z} p_1(\omega_1 \mid \omega) \cdot p_2(\omega_2 \mid \omega) \ldots p_Z(\omega_Z \mid \omega) \cdot W(\omega_1 \ldots \omega_Z : \omega') \tag{3.5.2}$$

where the Z nearest neighbour cells of the cell (l) are in the states $\omega_1, \omega_2 \ldots \omega_z$. The cell ($l$) being in the state ω'. Substituting equation (3.5.2) into equation (3.5.1) yields

$$g_l(\omega' \mid \omega) = \sum_{\mu=1}^{Z} \sum_{\omega_\mu} A_\mu(\omega', \omega_\mu) \cdot g_\mu(\omega_\mu \mid \omega) + h_l(\omega', \omega). \tag{3.5.3}$$

Here $A_\mu(\omega', \omega_\mu)$ is the probability that cell (l) and its nearest neighbour μ are respectively in the states ω' and ω_μ. $h_l(\omega', \omega)$ is a non-linear term comprising the sums of terms involving two or more values of $g(\omega', \omega)$. For the calculation of the total scattered intensity we require to know

$$S(\omega', \omega) = \sum_{l} P_0(\omega) \cdot g_l(\omega' \mid \omega) \cdot e^{-i2\pi \mathbf{X} \cdot \mathbf{R}_l}. \tag{3.5.4}$$

$P_0(\omega)$ is the orientational probability of a molecule in the origin cell. Writing

$$B(\omega', \omega) = \sum_{\mu=1}^{Z} A_\mu(\omega', \omega) \cdot e^{-i2\pi \mathbf{X} \cdot (\mathbf{R}_l - \mathbf{R}_\mu)} \tag{3.5.5}$$

and using a matrix notation (with matrices of size D where D is the number of allowed orientations) equation (3.5.3) can be re-expressed in the form

$$S = H(E - B)^{-1} \tag{3.5.6}$$

Where

$$H(\omega', \omega) = \sum_{l} P_0(\omega) \cdot h_l(\omega', \omega) \cdot e^{-i2\pi \mathbf{X} \cdot \mathbf{R}_l} \tag{3.5.7}$$

If we use a column matrix $F(\omega)$, $\omega = 1, 2, \ldots D$, the scattered intensity then becomes:

$$I = N(F^+ S F) \tag{3.5.8}$$

F^+ is the adjoint matrix of F and the calculation of the scattered intensity becomes a matrix calculation. An example how to use this method follows.

3.5.1 Tetranitromethane[18]

This compound crystallizes in a body-centred cubic lattice (space group $I\bar{4}3\,m$). The central carbon atoms of the molecules in the unit cell lie at the positions (0, 0, 0) and $\frac{1}{2}, \frac{1}{2}, \frac{1}{2}$ and the eight nitrogen atoms on the four body diagonals (Figure 3.6). The orientation of the pairs of oxygen atoms are disordered around their respective C–N axes.

The simplest model is one in which all the pairs of oxygen atoms rotate freely

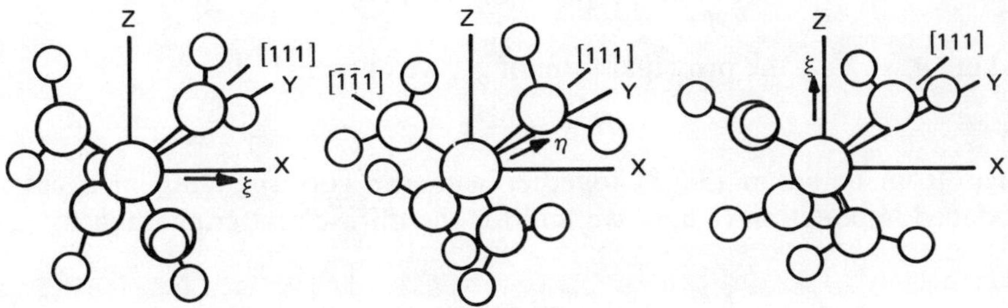

Figure 3.6 The three orientations of tetranotromethane molecules in the crystal

around the C–N axis. A calculation carried out on this basis accounts for both the Bragg intensities and the circular diffuse halo, the maximum of which occurs at 32° for CuKα. It does not explain the spot-like diffuse scattering which has been observed. Matsubåra supposes that the molecule has the symmetry $\overline{4}2m$ and that, according to a model of Stosick[23] a pair of oxygen atoms jump by a rotation around the C–N axis from one equilibrium position to another. The other pairs of oxygen atoms must simultaneously rotate since the repulsive energy between oxygen atoms of different pairs must be a minimum. The single $\overline{4}$ axis of the molecule goes from one fourfold axis to another of the cubic crystal during this motion. There are, therefore, three possible orientations of a molecule on its site and $P_1(\omega) = \frac{1}{3}$ for any value of ω. We suppose that $A(\omega_l, \omega_\mu)$ is proportional to the Boltzman factor, $\exp\{-V(\omega_l, \omega_\mu)/RT\}$ where $V(\omega_l, \omega_\mu)$ is the interaction energy between the neighbouring molecules l and μ. If the orientations are specified by 1, 2, 3, the symmetry gives us:

$$V(1, 1) = V(2, 2) = V(3, 3)$$
$$V(1, 2) = V(2, 3) = V(3, 1)$$

We may assume that:

$$A(1, 1) = A(2, 2) = A(3, 3) = \frac{1 + 2\alpha}{3}$$

$$A(1, 2) = A(2, 3) = A(3, 1) = \frac{1 - \alpha}{3}$$

α is a function depending upon potential energy and temperature. From equation (3.5.5), $B(\omega', \omega)$ becomes:

$$B(\omega', \omega) = 8(\cos \pi h.\ \cos \pi k.\ \cos \pi l)\ \mathrm{A}(\omega', \omega).$$

h, k, 1 are the numerical coordinates of X in reciprocal space.

The evaluation of H (equation (3.5.7)) requires additional approximations. In equation (3.5.7), let us retain only the term $l = 0$. Then, as $P_0(\omega) = \frac{1}{3}$, we have

$$H = \tfrac{1}{3}h_0 = g_0 - \sum_{\mu=1}^{z} A g_\mu$$

Taking $g_\mu = A g_0$, the principal term of g_μ, we obtain:

$$H = g_0 - z\,A^2 g_0$$

Then from equation (3.5.8) together with the corresponding matrices introduced in equation (3.5.6), we find for the diffuse scattering intensity:

$$I_{\text{diff}} = N(1 - 8\alpha^2)\,\frac{\overline{|F|^2} - |\overline{F}|^2}{1 - 8\alpha \cos \pi h \cos \pi k \cos \pi l}. \qquad (3.5.9)$$

F is the molecular structure factor.

$\alpha \neq 0$, would correspond to diffuse scattering without correlation and this correctly predicts the occurrence of the observed diffuse halo. When there are correlations $\alpha \neq 0$, but we may assume that $8|\alpha|1$. The sign of α will depend on the interaction energy between two neighbouring molecules. For example, if $V(1, 1) > V(1, 2)$ we have $\alpha < 0$ and if $V(1, 1) < V(1, 2)$ then $\alpha > 0$.

The numerator of equation (3.5.9) is maximum in the diffuse halo while the denominator will be a minimum when h, k, l are integers with $h + k + l$ odd $(2n + 1)$ when $\alpha < 0$ and even $(2n)$ when $\alpha > 0$. Since the additional diffuse spots surround reciprocal lattice points such as $h + k + l = 2n$: (110), (200), (112), (220) then $\alpha > 0$. This then accounts for the general features. A more precise study would be needed however to assess if the diffuse spots arise from molecular correlations or from thermal acoustical waves. It is known that thermal waves do scatter the x-rays significantly near reciprocal lattice points.

3.6 DIFFUSE SCATTERING WITH STRONG CORRELATIONS

In some plastic crystals, the correlations are determined by an all or nothing law. This is the case when steric hindrance restricts the random orientations of the molecules or when there is dimer formation. The resulting diffuse scattering is strongly influenced by these correlations. We shall consider two cases in detail.

(a) Succinonitrile—studied by Descamps.[14,24]
(b) Pivalic acid—studied by Coulon, Descamps, Fontaine and Longueville.[13]

We shall attempt to show how the calculation of the conditional probability can be generalized to obtain better agreement with the experimental data.

3.6.1 Succinonitrile

Succinonitrile (1,2-dicyanoethane, $(CH_2–CN)_2$) has a body-centred cubic rotator phase between its transition temperature $T_t = 233\ K$ and its melting temperature $T_f = 331\ K$. The structure, studied first by Finbak[1] and again by Fontaine[25] shows that we have a thermodynamic equilibrium between the three following molecular configurations (Figure 3.7).

A trans-isomer of symmetry C_{2h} (20%). The cyanide groups CN are in the symmetry plane of the molecule.

Two gauche isomers G_1 and G_2 of C_2 symmetry (40% each). G_1 and G_2 are derived from the trans isomer by a $\pm 2\pi/3$ rotation around the central C–C bond.

On the lattice site, the equilibrium positions of the molecule are as follows. The middle C–C bond of the molecule lies along one of the four, three fold axes of the cubic cell. Around this axis, the molecule may occupy three equilibrium positions related to each other by a $2\pi/3$ rotation around this axis (Figure 3.8). Each isomer can orientate itself according to twelve positions which define the

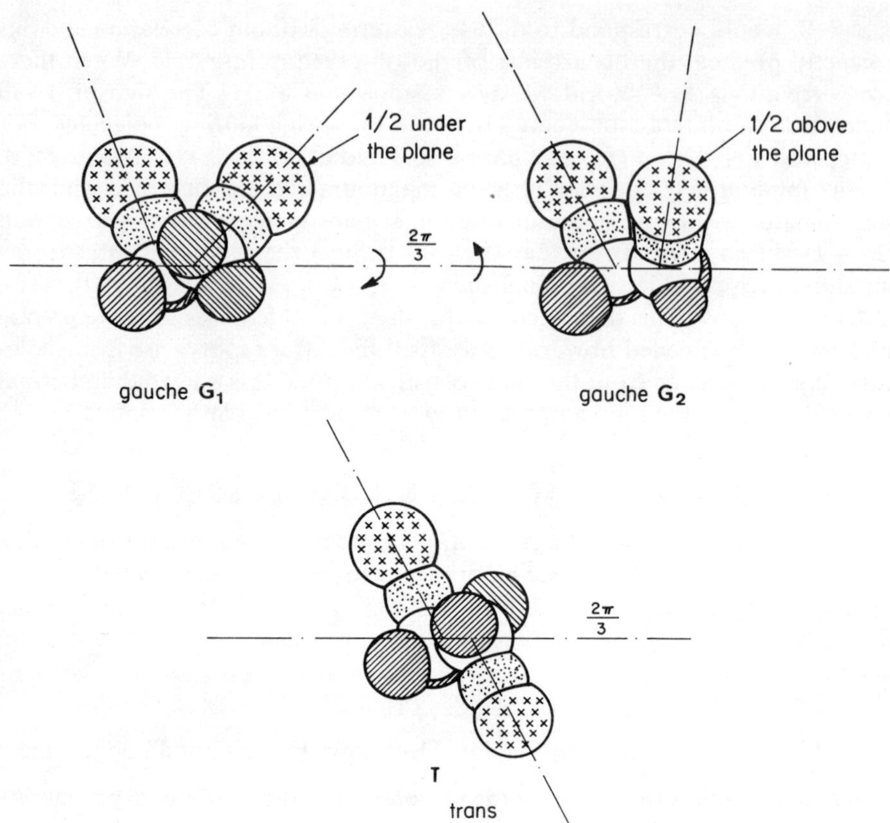

Figure 3.7 The three isomers of succinonitrile

same mean molecular configuration. From the intensity of the Bragg reflections, we obtain an isotropic Debye–Waller factor for the translation vibrations corresponding to a mean square amplitude $\overline{u^2} = 0.089$ Å^2 and an anisotropic Debye–Waller factor connected to the librations around the three fold axis with a mean square amplitude $\overline{\theta^2} = 0.37$ rad^2.

Laue photographs (Figure 3.9) taken using MoKα monochromatic radiation on single crystalline slabs show intense diffuse zones in the shape of structured halos. These halos are located at scattering angles $2\theta = 12°; 20°; 33°$. They are not coincident with the diffraction spots. This diagram coupled with intensity measurements performed with a counter has been interpreted by Descamps[24] by taking into account correlations produced by steric hindrance in the molecule. Figure 3.10 compares the diffuse scattering intensity along one crystallographic direction with the intensity calculated assuming diffuse scattering without correlations (equation (3.3.15)). The calculated diffuse x-ray scattering presents two maxima at $|\mathbf{X}| = 0.1$ Å^{-1} and 0.3 Å^{-1}. In contrast the

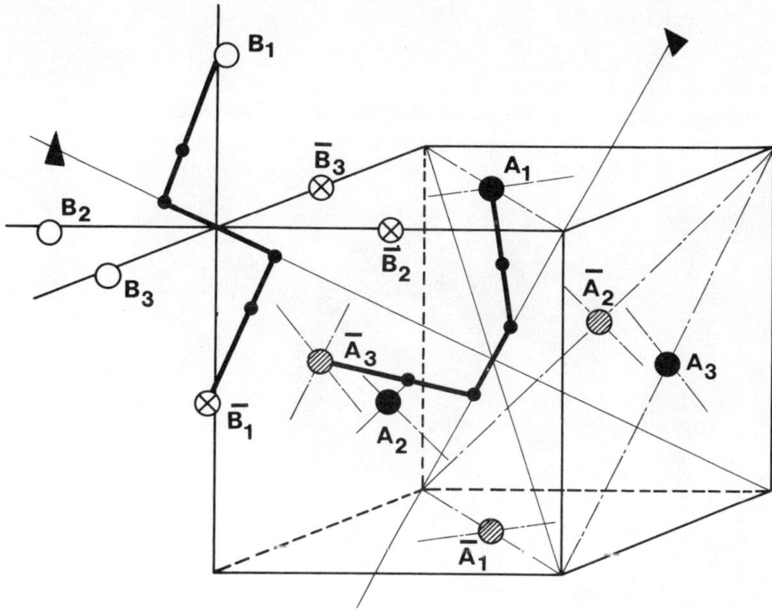

Figure 3.8 Equilibrium positions of gauche and a trans-molecule in the cubic unit cell. A_1, A_2, A_3, ... B_1, B_2, B_3 ... are the positions of the nitrogen atoms

experiment shows that there is only one maximum at 0.3 Å^{-1} corresponding to the Bragg diffraction. The spatial correlations associated with steric hindrance account for the position of the diffuse halo. When we consider a molecule on its site in any equilibrium position, the nitrogen atoms of the –CN groups lie in the middle of one edge of the cubic cell the origin of which coincides with the middle of the central C–C bond of the molecule. Steric hindrance takes place only between next nearest neighbouring molecules. As will be seen from Figure 3.11 we cannot have two consecutive trans molecules along the [001] direction. To show more clearly how the correlations occur let us first consider a simplified case before approaching the more complicated succinonitrile situation.

Consider a cubic crystal, unit cell parameter *a* and one linear and centrosymmetric molecule of length *a* per unit cell. If this molecule has its centre of mass on a lattice point it can take three orientations along the directions Ox_1, Ox_2, Ox_3 of the cube edges. The structure factors for these orientations are F_1, F_2, F_3. Descamps makes an approximate calculation of the conditional probabilities in the following manner. Firstly he defines them separately as the probabilities of Markov chains along the three directions 1, 2, 3 and then relates them by the condition that the molecules located at the ends of the cube diagonal of the crystal have independent behaviour. The development of $p_2\,(1/1; na)$ is given in Figure 3.12 for different values of *n* and yields a correlation length $\xi =$

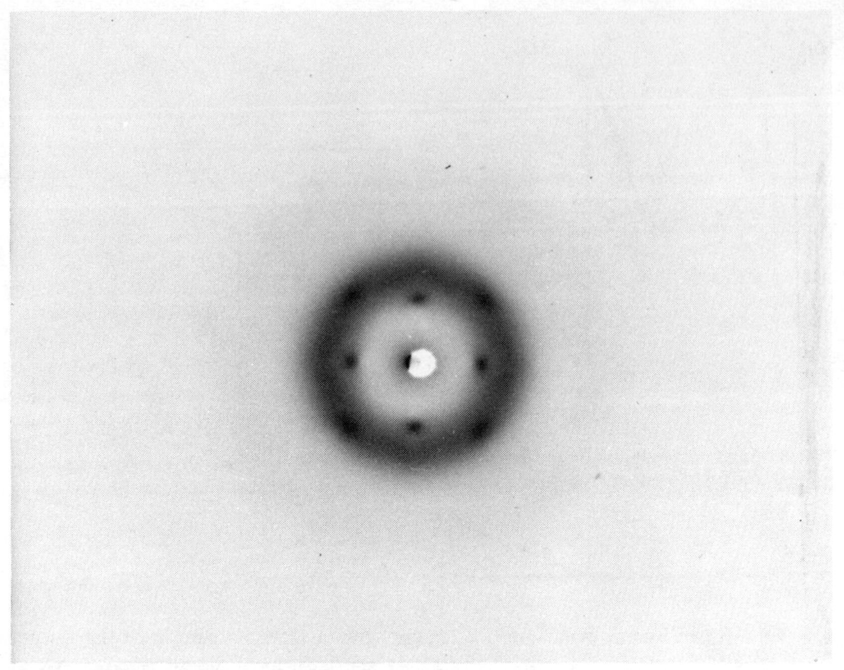

Figure 3.9 Monochromatic Laue photograph of succinonitrile taken with MoKα radiation. $d = 60$ mm. The [100] axis is along the incident direction

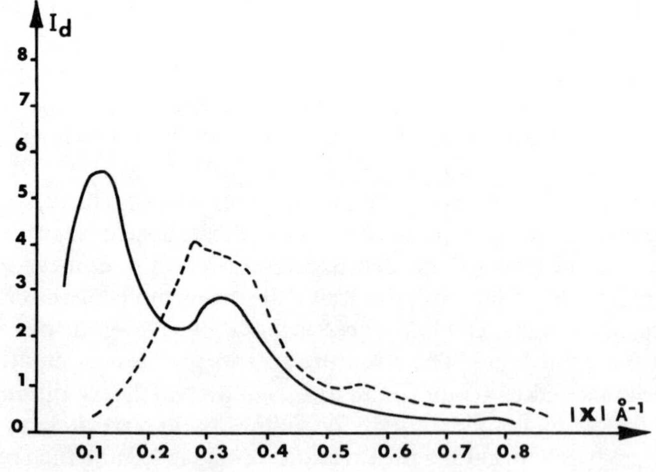

Figure 3.10 The diffuse scattering intensity along the (111) direction for succinonitrile. ———— Calculated intensity; – – – – – – – – – – Experimental diffuse power

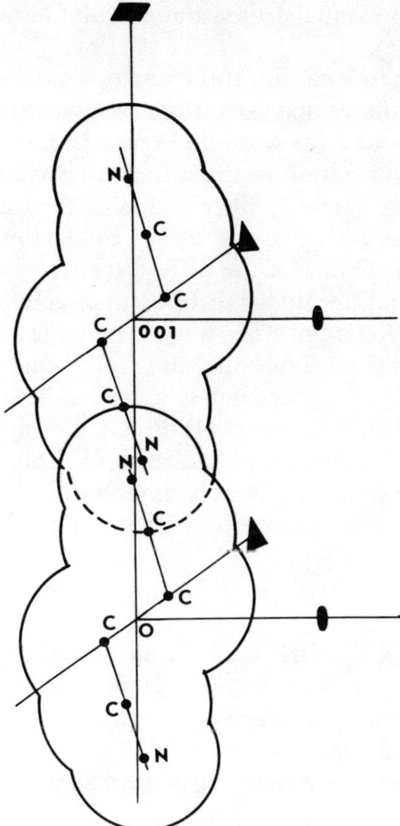

Figure 3.11 An incompatible situation for two trans-molecules along the [001] direction

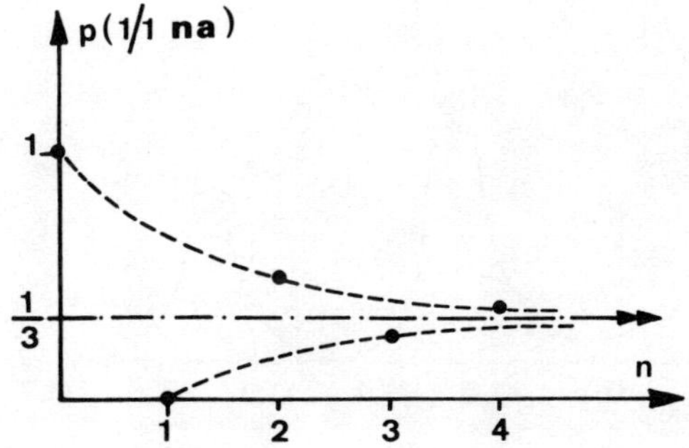

Figure 3.12 Conditional probability along the [100] direction of succinonitrile for the three orientations model

9.3 Å assuming an exponential decrease and a cubic lattice parameter $a = 6.4$ Å (that for succinonitrile).

For a molecule with four carbon atoms and two nitrogen atoms such as trans-succinonitrile he calculates that the diffuse x-ray scattering intensity with and without correlations varies as shown in Figure 3.13.

The correlation contribution to the diffuse scattering intensity has a significant influence on the intensity distribution of the total scattering, yielding minima at points where the non-correlated calculation alone gives maxima.

To perform the calculation of the diffuse scattering for succinonitrile itself the conditional probabilities along the different directions are established using the total probability theorem. This was carried out as indicated previously using the independant vibrations approximation (equation 3.3.13). Figure 3.14 shows that a satisfactory agreement is achieved between the experimental results and the calculation for the (111) direction. The agreement is less satisfactory for the other crystallographic directions. This result proves however that the spatial correlations profoundly modify the diffuse scattering diagram of a plastic crystal. Their inclusion does help to explain the observed scattering.

3.6.2 Pivalic Acid

Pivalic acid, $(CH_3)_3C.COOH$, exists in an orientationally disordered phase between the transition temperature, $T_t = 280$ K, and the melting temperature, $T_f = 309$ K. It crystallizes in a face-centred cubic lattice, unit cell parameter $a_0 = 8.82$ Å at room temperature.

The dielectric properties[31] and the structural analysis[32] show that in the solid

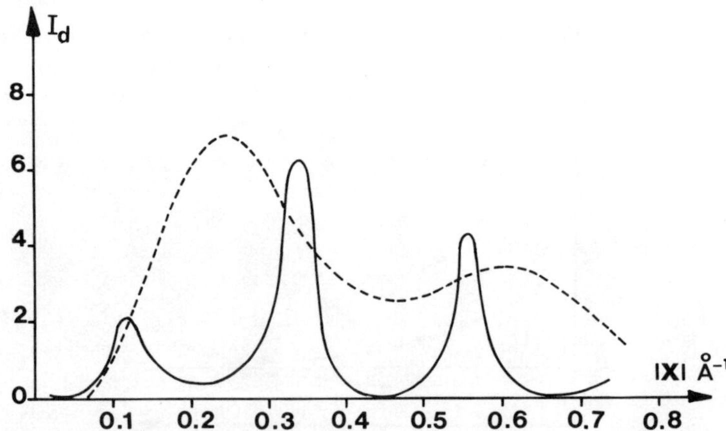

Figure 3.13 The distribution of diffuse x-ray scattering from succinitrile resulting the simplified model ([111] direction): -------- without correlations; ———— with correlations

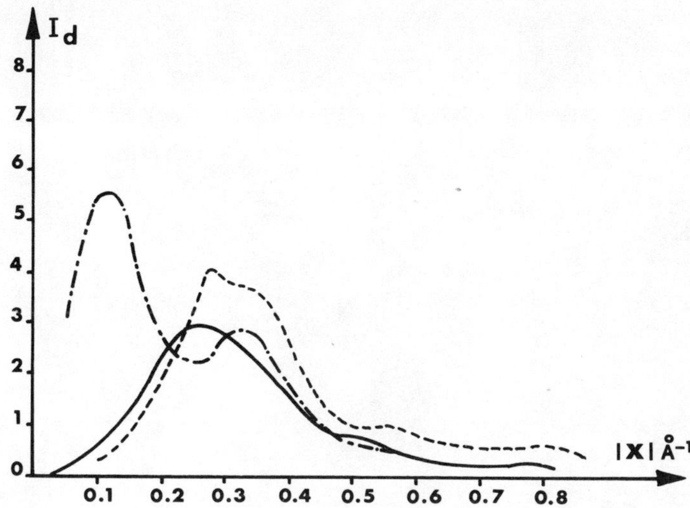

Figure 3.14 The diffuse scattering intensity along the (111) direction for succinonitrile --------- experimental intensity; –o–o–o–o– calculated intensity without correlation; ______________ calculated intensity with correlations

state the molecules are arranged as dimers orientated along one of the twelve [110] directions of the unit cell. The formation of cyclical dimers coupled by hydrogen bonds implies that the centres of mass of the two molecules forming the dimer are on two nearest neighbour sites and that their C–C axes are orientated along [110] directions. At room temperature the structural analysis gives a mean square amplitude of translation $\overline{u^2} = 0.11$ Å^2 and a mean square amplitude of libration $\overline{\theta^2} = 0.032$ rad^2.

Coulon *et al.*[13] undertook a diffuse x-ray scattering study in order to confirm both the dimer formation and the orientation of the dimer axis. Laue and precession photographs were taken with monochromatic MoKα radiation using small monocrystals (mms in diameter) kept in a spherical glass container to prevent sublimation. Figure 3.15 shows the precession photograph of the (hkO) reciprocal plane. Diffraction spots corresponding to the 100 and 110 reflexions are obvious. The 220 reflexion is absent because the structure factor is almost zero. Three structured halos are noted; the first at **X** = 0.12 Å^{-1}, the second near 0.27 Å^{-1} and the third in the neighbourhood of 0.37 Å^{-1}. The halo at 0.27 Å^{-1} is due to the glass envelope. Microphotometric examination of the intensity along the [010] distribution axis is shown on Figure 3.16: the contribution of scattering from the glass vessel has been subtracted). A similar study along the [100] direction gave only weak scattering.

Using the structural data, the diffuse scattering neglecting the correlations can be calculated following equation (3.3.13). The result, Figure 3. 16(a), curve I, does not agree with the experiment since it predicts a single broad halo. When

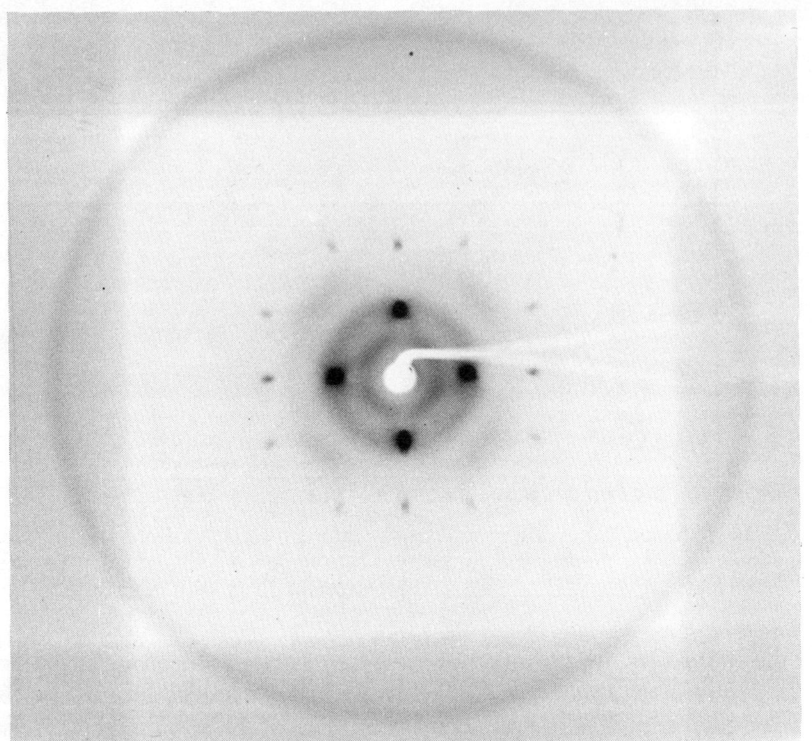

Figure 3.15 Precession photograph of the (khO) reciprocal plane for pivalic acid taken using MoKα monochromatic radiation

we take account of correlations however curve II results. This calculation is performed on the basis that two molecules set on the same diagonal, one on a vertex, the other in the centre of a face, are each surrounded by twelve molecules. Each then has a probability of orientation 1/12. When they make a dimer, the remaining eleven molecules surrounding them have only a 1/11 probability of making a dimer with another molecule. We see that the introduction of this correlation results in the prediction of the two halos at 0.12 Å^{-1} and 0.37 Å^{-1}.

Curve III (Figure 3.16(b)) gives the scattered intensity calculated according to equation (3.20). The diffuse scattering by the thermal motions has been calculated by equation (3.21) and substituted for the vibration term of equation 3.15). The halos are narrower and the agreement between experiment and calculation is satisfactory. There is some disagreement with regard to the halo intensities. Some improvement might result from a new evaluation of the thermal diffuse scattering and the Debye–Waller factor. Better agreement can only be expected however following a more elaborate calculation of the conditional probability, $p(\omega_{l'} | \omega_l)$ or, what is more directly related to the theory, the double probability $P(\omega_{l'}|\omega_l)$. A possible method for doing this will be outlined below. We note however that even at this stage, the study confirms the hypotheses made following the original structure determination: that molecules form dimers, the axes of which are orientated along the [110] directions.

3.6.3 Evaluation by Graphs of the Conditional Probability for a System of Dimers or with Steric Hindrance

Better agreement between theory and experiment, can only be expected following the calculation of the conditional probability $p(\omega_{l'}|\omega_l)$ or, what is in more direct relation with the theory, the double probability $P(\omega_{l'}|\omega_l)$.

Let us assume that the potential energy V between molecules is merely the sum of potentials due to molecular pairs. That is

$$V = \sum_{1 \ll i<j \ll N} v_{ij}(\omega_i, \omega_j)$$

We will restrict this sum to the interactions between the nearest neighbouring molecules. At the equilibrium position, the probability that we have a configuration characterized by the molecules with orientations $\omega_1, \omega_2 \ldots \omega_D$ is:

$$\begin{aligned} p(\omega_1, \omega_2 \ldots \omega_D) &= \frac{1}{Z} e^{-\beta V} \\ &= \frac{1}{Z} \prod_{1 \leqq i<j \leqq N} e^{\beta v_{ij}(\omega_i, \omega_j)} \end{aligned} \tag{3.6.1}$$

where

$$Z = \sum_{\omega_1} \ldots \sum_{\omega_D} \prod_{1 \leqq i<j \leqq N} e^{-\beta v_{ij}(\omega_i, \omega_j)} \tag{3.6.2}$$

is the partition function for this problem. The probability that molecules l and l' have orientations ω_l and $\omega_{l'}$ is then:

$$P(\omega_l, \omega_{l'}) = \frac{1}{Z} \sum_{\substack{\omega_1 \\ \omega i \neq \omega l \\ \neq \omega l'}} \ldots \sum_{\omega_D} \prod_{1 \leqq i<j \leqq N} e^{-\beta v_{ij}(\omega_i, \omega_j)} \tag{3.6.3}$$

In this expression, we fix the orientation of the molecules l and l' and sum over the orientations of all other molecules.

The calculation of $P(\omega_l|\omega_{l'})$ can be performed by the cluster expansion method of Ursell and Mayer[34] which consists of expanding the product according the power of $e^{-\beta v_{ij}} - 1 = f_{ij}$. To enumerate all the terms $f_{ij} f_{kl} \ldots$ in the expansion we associate each term with a graph drawn on the lattice. By using the cluster sums, an evaluation of the double probability is also possible but lengthy.

A particular example of this calculation is the graphical series approximation method developed by Nagle[34] which is applicable when we have hydrogen bond formation or a dimer problem. This method was used by Descamps to calculate the entropy and Kirkwood Dielectric factor of succinonitrile.[35] The steric hindrance is equivalent to a hard core interaction energy. When two molecules located on nearest neighbours sites have sterically hindered orientations ω_i and ω_j we may write

$$v_{ij}(\omega_i, \omega_j) = \infty$$

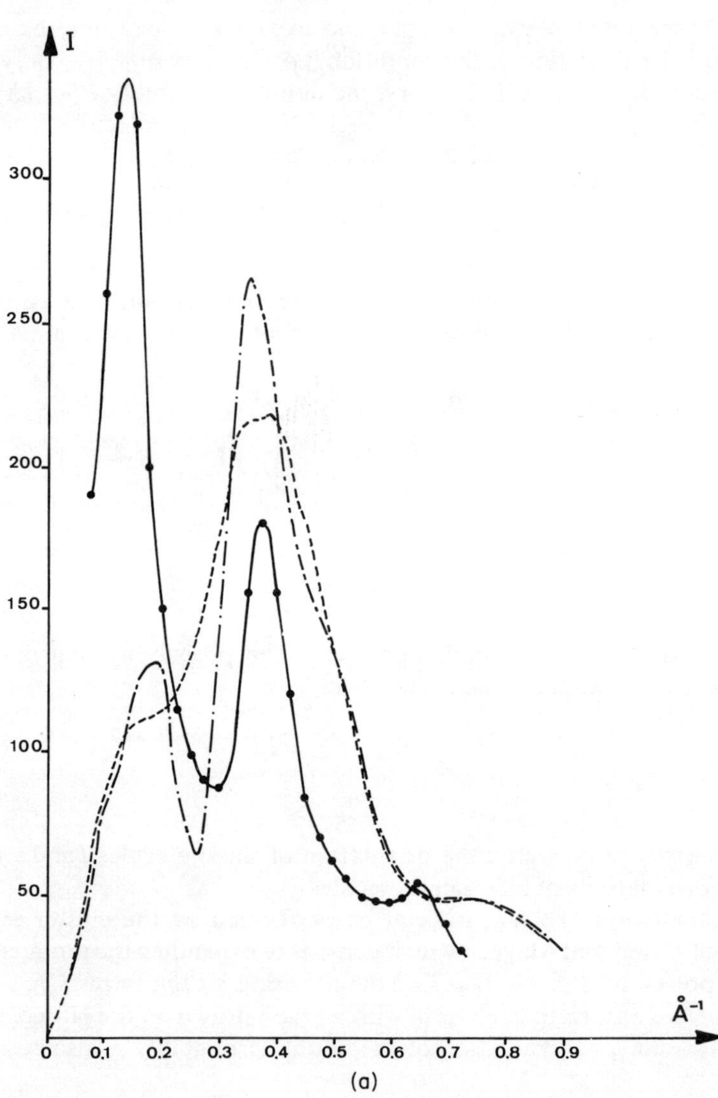

(a)

and for other situations:

$$v_{ij}(\omega_i, \omega_j) = 0$$

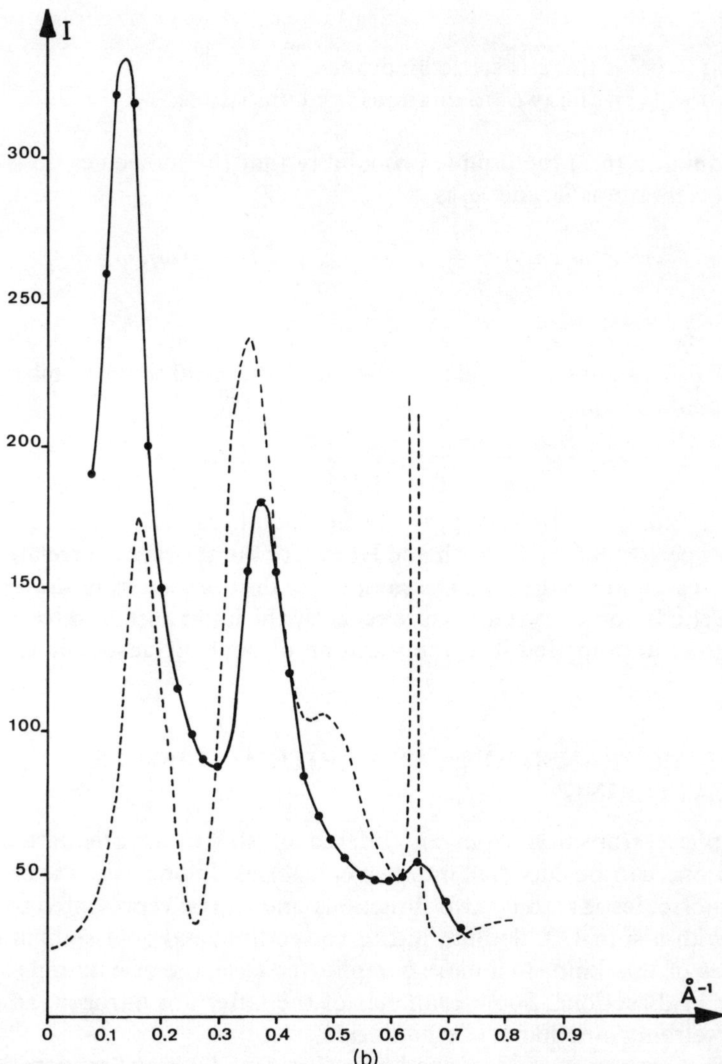

Figure 3.16 Diffused intensity along (110) for pivalic acid. (a) ______________ experimental measurements; --------- calculated intensity without correlation (curve I); -o-o-o-o- calculated intensity with correlation (curve II). Curves calculated from equation (33.15) and (33.16). (b) ______________ experimental measurements; --------- calculated intensity (curve III); Curve calculated from equation (33.19)

Then following Nagle we define a compatibility function for the *ij* lattice edge

$$A(\omega_i, \omega_j) = e^{-\beta\omega_{ij}}$$

so that

$A(\omega_i, \omega_j) = 0$ if there is steric hindrance
$A(\omega_i, \omega_j) = 1$ if the two orientations are compatible.

From equation (6.3) the double probability that the molecules on sites *l* and *l'* have orientations ω_l and $\omega_{l'}$ is

$$P(\omega_l, \omega_{l'}) = \frac{1}{Z} \sum_{\substack{\omega_1 \\ \omega_i \neq \omega_l \\ \neq \omega_{l'}}} \cdots \sum_{\omega_D} \prod_{1 \leqq i < j \leqq N} A(\omega_i, \omega_j) \tag{3.6.6}$$

where Z is the configurational partition function equal to the number of compatible configurations.

$$Z = \sum_{\omega 1} \cdots \sum_{\omega_D} \prod_{1 \leqq i < j \leqq N} A(\omega_i, \omega_j) \tag{3.6.7}$$

Equation (3.6.6) can be solved by the method of graphs. A calculation of this form was performed by Gobush and Hoeve for the dielectric correlation factor of cubic ice [36] and Coulon and Descamps [37] for diffuse scattering and local order in a smectic B liquid crystal. [38] More recently the latter authors have developed this method and applied it to the situation of neutron elastic scattering from ice. [38]

3.7 PHASE TRANSITIONS AND CRITICAL DIFFUSE SCATTERING

The phase transition from an ordered crystal to an orientationally disordered one can be classified into two categories. In one case, the orientation of the molecules take only two directions and can be represented by an Ising model with a spin 1/2. Sodium nitrite and sodium and potassium nitrates are examples of this kind. In a more complicated case, the orientation can take a number of directions. Some examples of the latter are nitrogen, ammonium nitrate, tetranitromethane, caesium azide.

The first case is easily described as an Ising model in the random phase approximation [27] and we will consider the results of the diffuse scattering of sodium nitrate. [11] In the second case, a coherent statistical theory has been worked out recently by Naya. [28] We will explain the main outline of this theory. No attempt at experimental verification has been tried up to the present.

3.7.1 Critical Diffuse Scattering for a Two-Orientation System—Sodium Nitrate

The low temperature phase of sodium nitrate belongs to the space group $R\overline{3}c$ and involves two molecules per unit cell. In the high temperature phase, the parameter $\mathbf{c}$ is reduced by half and the space group becomes $R\overline{3}m$ with one molecule per unit cell. The nitrate ions, in antiparallel positions in the low temperature phase, become randomly distributed among the two stable orientations. It is of interest to investigate the interaction energy which causes this phase transition.

The order–disorder transition can be adequately represented by an Ising model with short range interactions. For each site of the high temperature phase we introduce an Ising variable $\sigma(\mathbf{R}_l)$ in order to specify the two orientations (Figure 3.17).

$$\begin{aligned} \sigma(\mathbf{R}_l) &= 1 \qquad \text{for } A \text{ orientation} \\ &= -1 \qquad \text{for } B \text{ orientation} \end{aligned}$$

At a given temperature, there will be statistical equilibrium between orientations, the equilibrium properties of which will be determined by the following hamiltonian:

$$H = -\frac{1}{2} \sum_{l,l'} J(\mathbf{R}_{ll'})\, \sigma(\mathbf{R}_l)\, \sigma(\mathbf{R}_{l'}) \tag{3.7.1}$$

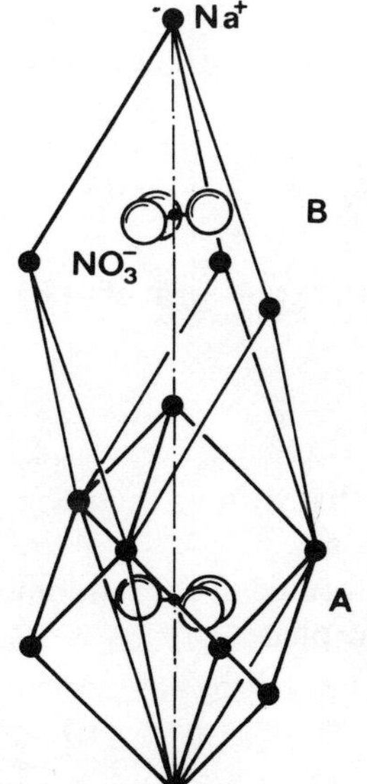

Figure 3.17 Crystal structure of $NaNO_3$. A rhombohedral bimolecular unit cell in the low temperature phase is drawn by thin solid lines. Thick solid lines indicate the unimolecular unit cell in the high temperature phase

With the hamiltonian in this form, thermal vibrations are neglected. Shinnaka[26] has shown experimentally that the corresponding diffuse scattering represented by the last term is centred around the reciprocal lattice points of the high temperature phase.

The diffuse x-ray scattering which yields information on $J(\mathbf{R}_{ll'})$ is determined by the structure factor F_l of the unit cell l of the high temperature phase. With the origin on the centre of one nitrogen atom and using as a reference a thermal average distribution of electrons where each orientation has a weight 0.5, the electronic density of the unit cell l can be written

$$\rho_l(\mathbf{R}) = \langle \rho(\mathbf{r}) \rangle + \Delta\rho(\mathbf{r})\,\sigma(\mathbf{R}_l) \tag{3.7.2}$$

$\langle \rho(\mathbf{r}) \rangle$, the average electronic density, is the symmetric part of the electronic distribution $\Delta\rho(\mathbf{r})$ is the antisymmetric part of the electronic density because we have $\Delta\rho(-\mathbf{r}) = -\Delta\rho(\mathbf{r})$.

The structure factor Fourier transform of the electronic density is, therefore, given by:

$$F_l = F_s + \mathrm{i}\sigma(\mathbf{R}_l)\,F_\alpha \tag{3.7.3}$$

F_s and F_α correspond to symmetric and antisymmetric parts of the electronic distribution. From the square modulus of the scattering amplitude we obtain a total intensity which comprises the diffracted intensity and the diffuse intensity I_d.

$$I_\mathrm{d} = NI_\mathrm{e}(\mathbf{K})\,|F_\alpha|^2\,\chi(\mathbf{K}) \tag{3.7.4}$$

where $\chi(\mathbf{K})$ is the Fourier transform of the spin correlation function (equivalent to the conditional probability):

$$\chi(\mathbf{K}) = \sum_{l-l'} \langle \sigma(\mathbf{R}_l)\,\sigma(\mathbf{R}_{l'}) \rangle\, \mathrm{e}^{\mathrm{i}\mathbf{K}\cdot\mathbf{R}_{ll'}} = \langle\, |\sigma(\mathbf{K})|^2 \rangle \tag{3.7.5}$$

The periodicity in the reciprocal lattice of $\sigma(\mathbf{K})$ the Fourier transform of $\sigma(\mathbf{R})$ gives

$$\chi(\mathbf{K}) = \chi(\mathbf{q}) \text{ with } \mathbf{K} = 2\pi\mathbf{M} + \mathbf{q}$$

M being a reciprocal lattice vector. We can choose **q** so that with its origin on the lattice origin, its extremity is lying in the Brillouin zone.

The calculation of the thermal average $\langle |\sigma(\mathbf{q})|^2 \rangle$ performed in the random phase approximation[27] gives for the high temperature phase

$$\chi(\mathbf{q}) = \frac{1}{I - \beta J(\mathbf{q})}. \tag{3.7.6}$$

$J(\mathbf{q})$ is the Fourier transform of the coefficients $J(\mathbf{R}_{ll'})$. We have for the hamiltonian of the system

$$H = -\frac{1}{2} \sum_{q} J(\mathbf{q}) \,|\, \sigma(\mathbf{q}) \,|^2 \tag{3.7.7}$$

The energy (thermal average of H) will be minimum when $J(\mathbf{q})$ will be maximum. We obtain:

$$J(\mathbf{q}) = 2J_1\{\cos \mathbf{qa} + \cos \mathbf{qb} + \cos \mathbf{qc}\} \\ + 2J_2\{\cos \mathbf{q}(\mathbf{a} - \mathbf{b}) + \cos \mathbf{q}(\mathbf{b} - \mathbf{c}) + \cos \mathbf{q}(\mathbf{c} - \mathbf{a})\} \tag{3.7.8}$$

a, **b**, **c** are the fundamental translation vectors of the high temperature cell, J_1 is the interaction energy of the ion with one of its first neighbours. In the configuration where these ions are parallel, there will be an overlap of the charges and J_1 must be considered as negative.

J_2 is the interaction energy of the ion with one of its second neighbours. It arises from the octopole–octopole interaction between two NO_3^- ions and must be considered as positive.

With these signs for J_1 and J_2, $J(\mathbf{q})$ becomes maximum for $\mathbf{q}_0 = \pi(\mathbf{a}^* + \mathbf{b}^* + \mathbf{c}^*)$ which corresponds to the spin arrangement in the low temperature phase.

When the temperature is very high $\chi(\mathbf{q})$ approaches unity. This indicates that the crystal is completely disordered. When the temperature decreases, $\chi(\mathbf{q})$ becomes infinite when $J(\mathbf{q}_0)$, the maximum value of $J(\mathbf{q})$, equals kT_c, i.e. at the critical temperature.

We see, therefore, that when we substitute equation (3.7.6) in equation (3.7.4) the diffuse x-ray scattering becomes infinite at $T = T_c$ on the reciprocal space point $\mathbf{q} = \mathbf{q}_0$. Near $\mathbf{q}_0$ as $J(\mathbf{q})$ is a regular function of q, the intensity distribution will be more or less spherical around this point of the super-lattice. From a detailed study of the temperature dependence of the diffuse scattering intensity we can determine J_1 and J_2.

Observations are made both on and near a superlattice point. The results for the $(\frac{3}{2}, \frac{1}{2}, -\frac{1}{2})$ reflexion are shown on Figure 3.18. When we approach T_c, the intensity rises rapidly as expected for a superlattice reflexion. This is accompanied by an additional abnormal increase of the intensity characteristic of critical scattering in second order transitions.

A more precise study of the diffuse intensity distribution near $(\frac{3}{2}, \frac{1}{2}, -\frac{1}{2})$ in the plane containing $\mathbf{a}^* - \mathbf{c}^*$ and $\mathbf{a}^* + \mathbf{b}^* + \mathbf{c}^*$, at a temperature slightly higher than $T_c(T = T_c + 5\text{ K})$ allows us to determine J_1 and J_2. The iso-scattering curves indicated on Figure 3.19 are ellipses the axes of which are directed along $a^* - c^*$ and $\mathbf{a}^* + \mathbf{b}^* + \mathbf{c}^*$. The measured intensities along these directions versus $\mathbf{q}$ vary inversely with q^2 (Figure 3.20) as predicted by equations (3.7.6) and (3.7.8).

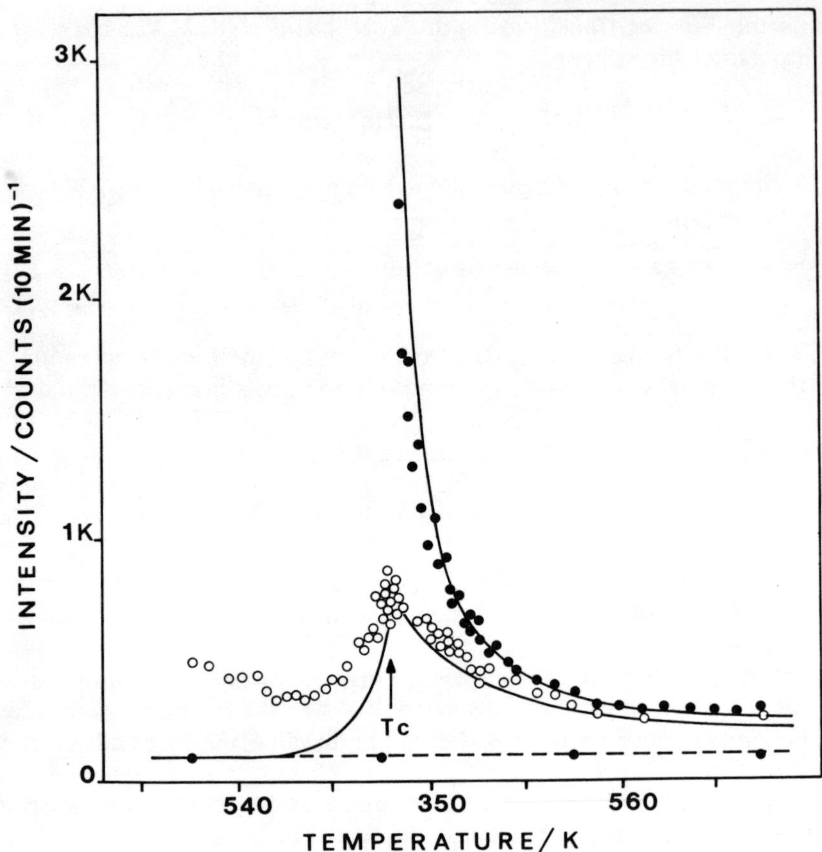

Figure 3.18 Temperature dependences of the diffuse scattering intensities observed at: 3/2, 1/2, -1/2 0 0 0 0 0, (3/2 + 1/12, 1/2 + 1/12, -1/2 + 1/12) 0 0 0 0 0 0 The broken line indicates the background level estimated from the intensity observed at 7/4, 1/2, −3/4. The solid curves show the results of calculations which should be compared with observed points

Calculation yields:

$$J_1 = -53.6k\ J_2 = 37.6k \text{ at } T = T_c + 5\text{ K}$$

Which values are in adequate agreement with values derived from theoretical force models using repulsive and octopolar attraction.

3.7.2 Critical Diffuse Scattering for a System of Multiple Orientations

In many cases [$C(NO_2)_4$, NH_4NO_3, $CsHF_2$, CsN_3] the molecule takes more than two orientations. A calculation of the x-ray diffuse scattering has

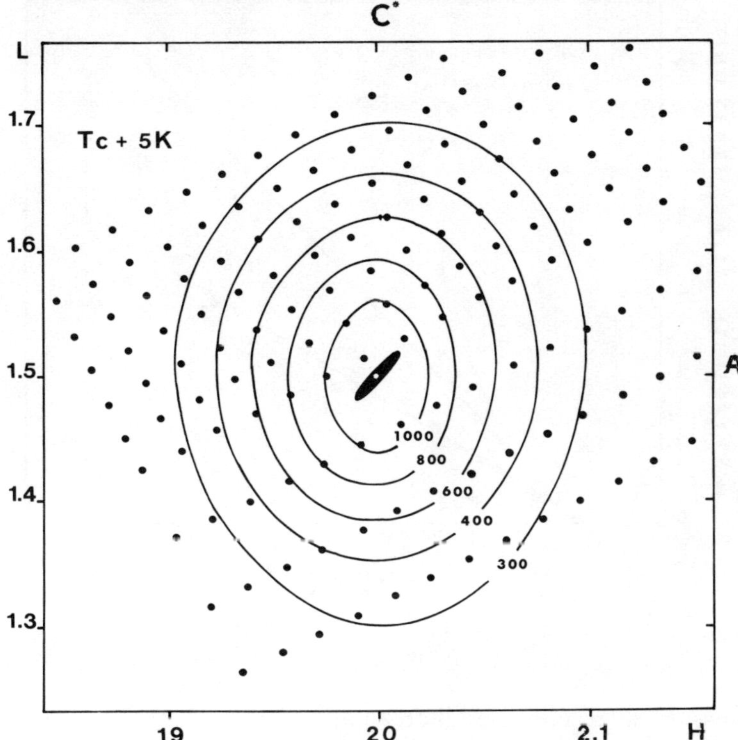

Figure 3.19 Observed distribution of diffuse scattering intensity in reciprocal space around (3/2, 1/2, −1/2)

been made by Naya[28] using the random phase approximation but there has been no experimental verification of the result. In his theory no account is taken of thermal agitation. This can be introduced easily as we have indicated in Section 3.3.1 by taking first, a thermal average over the vibrations of the molecules in the potential wells followed by averaging over all molecular orientations.

To mark the orientation of the molecule in its site, we use the set of values $\omega_l^1 \omega_l^2 \cdots \omega_l^D$ These parameters take the value 1 if the molecule is in the corresponding orientation and the value 0 if not. The orientation state of the molecule will be characterized by the vector ω_l with D components $(\omega_l^1, \omega_l^2 \ldots \omega_l^D)$.

The statistical distribution of the orientations of molecules at a given temperature will be determined by the interaction hamiltonian between the molecules in equilibrium positions. If we call $J_{ll'}^{ss'}$ the pair interaction energy between the l-th molecule with orientation s and l' th molecule with orientation s', we

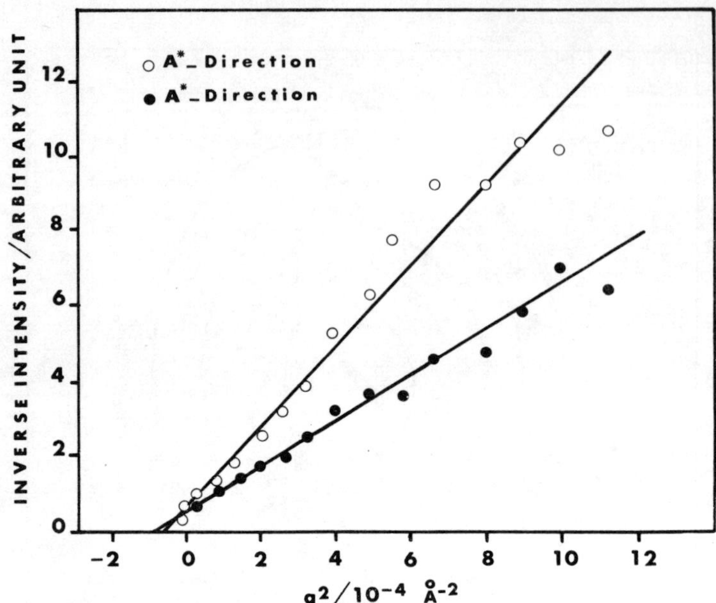

Figure 3.20 Plots of the inverse intensities against q^2 along two principal axes at $T_c + 5$ K

can express the interaction hamiltonian as:

$$\mathscr{H} = \frac{1}{2} \sum_{l} \sum_{l'} \sum_{s=1} \sum_{s'=1} \omega_l^s \cdot J_{ll'}^{ss'} \cdot \omega_{l'}^{s'}$$

$$= \frac{1}{2} \sum_{l} \sum_{l'} \tilde{\omega}_l \cdot \vec{\vec{J}}_{ll'} \cdot \omega_{l'}$$

(with $\vec{\vec{J}}_{ll} \equiv 0$)

If we introduce the Fourier transforms of the orientational variables and of the interaction energy matrix:

$$\omega(\mathbf{q}) = \frac{1}{\sqrt{N}} \sum_{l} \omega_l \cdot e^{i\mathbf{q}\mathbf{R}_l} \tag{3.7.10}$$

$$\vec{\vec{J}}(\mathbf{q}) = \sum_{l-l'} J_{ll'} \, e^{i\mathbf{q}\cdot\mathbf{R}_{ll'}} \tag{3.7.11}$$

where **q** is a vector of the first Brillouin zone we can write for the hamiltonian:

$$\mathscr{H} = \frac{1}{2} \sum_{q} \mathscr{H}(\mathbf{q}) \text{ where } \mathscr{H}(\mathbf{q}) = \tilde{\omega}(\mathbf{q}) \cdot \vec{\vec{J}}(\mathbf{q}) \cdot \omega(\mathbf{q}) \tag{3.7.12}$$

Neglecting the librations, we then obtain for the average scattered intensity over all vibrations in electronic units

$$I(\mathbf{K}) = \sum_{l} \sum_{l'} \mathrm{F}'_l \mathrm{F}'^*_{l'} \cdot \mathrm{e}^{\mathrm{D}_{ll'}} \cdot \mathrm{e}^{\mathrm{i}\mathbf{K}\cdot\mathbf{R}_{ll'}} \qquad (3.7.13)$$

In evaluating $I(\mathbf{K})$ we assume that we can separately average $F'_l\, F'^*_{l'}$ and $\mathrm{e}^{D_{ll'}}$ over the orientations i.e.

$$\overline{F'_l F'^*_{l'} \cdot \mathrm{e}^{D_{ll'}}} = \overline{F'_l F'^*_{l'}} \cdot \overline{\mathrm{e}^{D_{ll'}}}$$

and restrict the second term on the right hand side to $(1 + \overline{D}_{ll'})$. $\overline{D}_{ll'}$ takes account of first order diffuse power. For simplicity we shall subsequently neglect this but note that this contribution will be significant. For comparisons with experimental data at least an approximate value should be determined.

Let F' be a vector the components of which are the scattering factors for the molecule in its different orientations $(F'^1, \ldots F'^s \ldots F'^D)$ weakened by the Debye–Waller factor. Then we have $F' = \omega_l F'$ and

$$F'_l F'^*_{l'} = \mathrm{F}' + (\omega_{l'} \tilde{\omega}_l)\mathbf{F}'$$

$\vec{\mathrm{F}}'^+$ is the row matrix adjoint of the column matrix $\mathbf{F}'$

Using equation (3.7.13), we can write the radiation intensity form the disordered crystal as follows:

$$I(\mathbf{K}) = \sum_{l,l'} \mathbf{F}'^* \cdot \overline{(\omega_l \cdot \tilde{\omega}_{l'})} \cdot \mathbf{F}' \cdot \mathrm{e}^{\mathrm{i}\mathbf{K}\cdot\mathbf{R}_{ll'}}$$

or

$$I(\mathbf{K}) = N \cdot \mathrm{Tr}\, S \cdot \overline{\omega(\mathbf{q})} \cdot \overline{\tilde{\omega}(\mathbf{q})} \text{ with } \mathbf{K} = 2\pi\, \mathbf{M} + \mathbf{q}$$

where $\mathbf{q}$ is a vector of the first Brillouin zone and $\mathbf{M}$ a reciprocal lattice vector.

Here S is the diffusion matrix defined by $S = \mathbf{F}' \cdot \mathbf{F}'^*$ (where the product is a dyadic or tensorial one) and $\omega(\mathbf{q}) \cdot \tilde{\omega}(\mathbf{q})$ is the average over the orientations of the orientational fluctuation matrix.

This quantity is calculated by Naya in the random phase approximation and gives:

$$\omega(\mathbf{q}) \cdot \tilde{\omega}(\mathbf{q}) = M\{E + \beta M \cdot J(\mathbf{q})\}^{-1}$$

where M is the matrix of the second moment of $\omega(\mathbf{q})$ considered as a random variable:

$$M^{ss'} = \frac{1}{D}\delta_{ss'} + \frac{1}{D^2}\{N\Delta(\mathbf{q}) - 1\} \qquad (3.7.14)$$

Naya then obtains both the intensity scattered by the mean crystal and a diffuse scattered intensity

$$I_d(\mathbf{K}) = N \cdot \mathrm{Tr}\,\{S\,M[E + \beta M \cdot J(\mathbf{q})]\}^{-1} \tag{3.7.15}$$

The calculation of this intensity can be performed by changing the matrix $MJ\,(\mathbf{q})$ into a diagonal one by the matrix U:

$$I_d(\mathbf{K}) = N \cdot \sum_{i} \frac{(\mathrm{USMU}^{-1})_{ii}}{1 + \beta\lambda_i} \tag{3.7.16}$$

where λ_i is an eigenvalue of $MJ(\mathbf{q})$.

Equation (3.7.16) is very similar to that obtained by Matsubara. In such a model, the critical temperature T_c will be determined by the limit temperature when the orientational fluctuations become infinite.

It will be determined by the highest temperature corresponding to:

$$E + \beta MJ(\mathbf{q}_0) = 0$$

where $\mathbf{q}_0$ is given by this condition.

For $\mathbf{q}_0$ and for $T \to T_c$ the intensity tends to infinity: this corresponds to critical scattering.

In most examples which are suitable for study e.g. CBr_4, NH_4NO_3, $C(NO_2)_4$ the phase transition is known to be of first order and the transition temperature $T_t > T_c$. Hence we shall not see the critical fluctuations. However a detailed study over all the reciprocal lattice could reveal significant diffuse scattering near the superlattice reflections and an elaborate calculation from a model could enable one to calculate the pair interaction responsible for the phase transition. The theoretical work of Naya[28] which he has applied to situation of a simple cubic crystal in which molecules orientate along one of the three directions of the cube axes would be an approximate but useful starting point in the interpretation of experimental data.

3.8 CONCLUSIONS

We have shown using data from the few experiments that have been carried out how diffuse scattering measurements can be used to confirm structural analyses and also the ideas we have developed on local order in the orientationally disordered crystals and its influence on diffuse x-ray scattering. The analysis of the experimental data becomes complex when the disorder scattering is superimposed on the thermal fluctuation scattering but it is essential to take account of the latter.

There is at present no theory which allows us to make a direct treatment of the experimental data and the formulation of a model is essential. However, the technique is potentially useful and the extension both experimentally to other systems and theoretically could yield much information on the orientationally disordered state. In this context we note that the extension of experiments to include neutron elastic diffuse scattering would be extremely useful since with this type of experiment, measurements can be made with a zero energy transfer and the scattering due to thermal vibrations will be insignificant within the experimental resolution.

3.9 ACKNOWLEDGEMENTS

The author would like to thank the members of the 'Equipe de Dynamique des Cristaux Moleculaires' who make this work possible, and more particularly Mlle. Coulon, MM. Descamps, Fontaine and Longueville for helpful discussions and M. Damien for his help when trying to write English.

REFERENCES

1. C. Finbak, *Ark. Math. Naturvidenskab*, **B. 42**, 71 (1938).
2. C. von Hoppe, *Z. Krist.*, **107**, 406 (1956).
3. W. Cochran and G.S. Pawley, *Proc. Roy. Soc.*, **A 280**, 22 (1964).
4. L. Pauling, *Phys. Rev.*, **36**, 430 (1930).
5. J. Amoros and M. Amoros, *Molecular crystals, their transform and diffuse scattering,* Wiley, New York (1968).
6. S. Aubry, Thèse, Université de Paris (1975).
7. J.J. Weiss and M. L. Klein, *J. Chem. Phys.*, **63,** 2869 (1975).
8. J. Frenkel, *Acta Phys. Chemica U.S.S.R.*, **3**, 23 (1935).
9. A. Hüller and W. Press, *in Anharmonic Lattices, Structural Transitions and Melting*, Riste-Nordhoff-Leiden ed., 185 (1974).
10. Y. Yamada and T. Yamada, *J. Phys. Soc. Japan*, **21**, 2167 (1967).
11. H. Terauchi and Y. Yamada, *J. Phys. Soc. Japan*, **33**, 446 (1972).
12. A.M. Levelut and M. Lambert, *Mol. Cryst. and Liq. Cryst.*, **23**, 111 (1973).
13. G. Coulon, M. Descamps, H. Fontaine, and W. Longueville, *Third European Crystallographic Meeting*, Zürich, p. 395 (1976).
14. M. Descamps, *Solid State Communications*, **14**, 71 (1974).
15. Matsubara, *x-Rays, Japan*, **5**, 102, (1949); 6, 15 (1950), Reprinted in the *Bulletin of the Institute for Chemical Research, Kyoto University*, **41**, 131 (1963).
16. G. Fournet. Thèse, Paris (1950).
17. A. Guinier, *Théorie et Technique de la Radiocristallographie*, Dunod, Paris (1964).
18. T. Oda and T. Matsubara, *J. Chem. Soc. Japan*, **27**, 273 (1954).
19. P. A. Reynolds, *Acta Cryst.*, **A 31**, 386 (1975).
20. G. M. Brown and O. A. W Strydom, *Acta Cryst.*, **B 30**, 801 (1974).
21. C. Brot and I. Darmon, *J. Chem. Phys.*, **53**, 2272 (1970).
22. C. Brot, I. Darmon, and N. Dat Xuong, *J. Chem. Phys.*, **64**, 1061 (1967).
23. A. J. Stosick, *J. Amer. Chem. Soc.*, **61**, 1127 (1939).
24. M. Descamps, *Thèse de 3ème cycle*, Lille (1973).
25. H. Fontaine, *Thèse de Doctorat*, Lille (1973).
26. Y. Shinnaka, *J. Phys. Soc. Japan.* **16**, 1281 (1964).
27. R. H. Brout, *Phase Transitions*, Benjamin, New York. (1965).
28. S. Naya, *J. Phys. Soc. Japan*, **37**, 2 (1974).
29. W. C. Hamilton, J. W. Edmonds, A. Tippe, and J. J. Rush, *Faraday Disc.*
30. F. Zernicke, *Physica*, **7**, 565 (1940).

31. S. Kondo and T. Oda, *Bull. Chem. Soc. Japan*, **27**, 567 (1954).
32. W. Longueville and H. Fontaine, *Mol. Cryst. Liq. Cryst.*, **32**, 73 (1976).
33. K. Huang, *Statistical Mechanics*, Wiley, New York (1963).
34. J. Nagle, *J. Math. Phys.*, **7**, 1484 (1966).
35. M. Descamps, *Chemical Physics*, **10**, 199(1975).
36. W. Gobush and C. Hoeve, *J. Chem. Phys.*, **57**, 3416(1972).
37. G. Coulon and M. Descamps, *Solid-State Communications*, **20**, 379(1976).
38. M. Descamps and G. Coulon, *Chemical Physics, 25,* 170(1977).

4

Dielectric and Acoustic Studies

R. A. Pethrick

4.1 INTRODUCTION

The theoretical studies of Pauling[1] and Frenkel[2] in the thirties indicated the possible existence of a mesophase of molecular crystals in which the molecules could be rotationally disordered. In 1938 Timmermans[3] noted and classified a group of molecular solids with the structural and thermodynamic characteristics expected of such systems. The pioneering dielectric measurements of C. P. Smyth and his collaborators[4,5] established the existence of a substantial degree of rotational freedom in some of these materials. This chapter is concerned with a review of dielectric and acoustic relaxation properties of polar rotator-phase molecular solids.

4.2 CHARACTERISTICS OF DIPOLAR MOTION IN ROTATOR-PHASE SOLIDS

The existence of orientational freedom in a dipolar material will influence the static value of the dielectric permittivity and the position and shape of the dielectric dispersion. The idealized, general variations expected as one proceeds from the liquid state to the static solid via a rotator phase[6] are depicted in Figure 4.1.

In the high temperature rotator-phase (Figure 4.1(a)) the static value of the permittivity ϵ_0 has a high value characteristic of a system of freely rotating dipoles. The value is similar but slightly greater than that in the liquid state implying a somewhat greater degree of molecular freedom. At the rotator phase transition, ϵ_0 falls to n^2 (n = refractive index) a value characteristic of a static lattice.

Within the rotator phase a change in the frequency of observation, ω at constant temperature leads to the variations noted in Figure 4.1 (b). At low frequencies the real part, ϵ', of the complex permittivity $\epsilon^* (=\epsilon' + i\epsilon'')$ should have

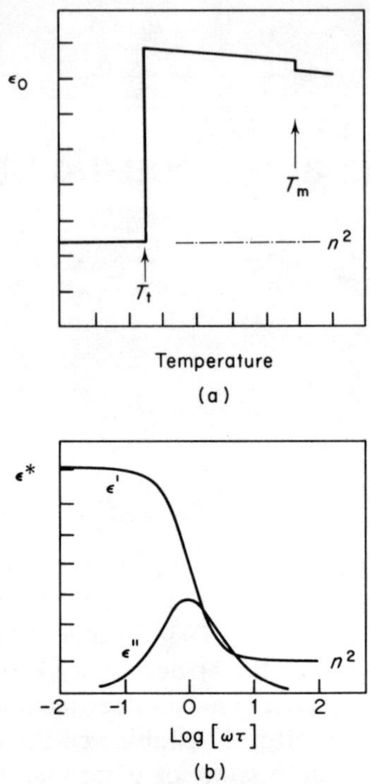

Figure 4.1 Ideal static and dynamic dielectric behaviour for a rotator phase solid. (a) Temperature dependence of the static permittivity; (b) Dynamic response of the dielectric permittivity as a function of frequency at constant temperature

a value approximately equal to the dipole moment of the gas phase. As the frequency is increased to a value greater than that for dipole rotation τ^{-1} (where τ is the relaxation time of the dipole), ϵ' falls and approaches n^2. In the region of τ^{-1} (i.e. $\omega\tau = 1$) a proportion of the reorienting dipoles will be unable to keep in phase with the oscillating electric field. Consequently, there will be a phase lag between the applied voltage and the current. ϵ'' is proportional to the phase lag. It reflects the dissipation of energy used in aligning the dipoles and which appears in the solid as heat. This and hence ϵ'' reaches a maximum at a frequency equal to the reciprocal relaxation time. Hence τ can be evaluated.

For a system of molecular dipoles in which rotations occur by a Brownian type motion, the observed dispersion is usually ~1.5 decades in width and is said to be of the Debye type.[7,8] This is the type of behaviour usually found in low viscosity liquids. If, on the other hand, interaction or correlation exists between molecular dipoles, the dispersion can be broadened to cover several decades of frequency.[9–12] The breadth of the dispersion is then indicative of the degree of correlation.

Within experimental error, the temperature dependence of τ can be described by an equation of the classical type

$$1/\tau = \nu_0 \exp(-E/RT). \qquad (4.2.1)$$

This can be interpreted on the basis of an Eyring type rate process, i.e.

$$1/\tau = k = \frac{kT}{h} \exp\left(\frac{\Delta S_E^*}{R}\right) \exp\left(\frac{-\Delta H_E^*}{RT}\right) \qquad (4.2.2)$$

k is the rate constant for the dipole reorientation process of characteristic relaxation time. ΔS_E^* and ΔH_E^* are the activation entropy and enthalpy for the process. It is generally accepted that the *ad hoc* use of the rate theory approach is inappropriate, particularly if the motion is cooperative. Its use is usually justified on the basis that the rotational motion is free. In support of this assumption, for plastic crystals, the factor kT/h is found to provide a good approximation to ν_0. ΔH_E^* is assumed to be proportional to the potential energy barrier hindering free rotation (i.e. the process is independent of pressure at least at low pressures). Comparison can then be made (at least relatively) with theoretically derived values of the latter quantity.

The degree to which the above phenomena will be observed in any particular solid system will depend on the symmetry or asymmetry of the molecules of which it is comprised. By far the greater number of molecules are asymmetric and their general behaviour can be divided into three basic types:

(i) The majority, undergo a first order transition on freezing, from a 'normal' polar liquid to a solid in which there is restricted or no molecular motion. In the liquid state the molecules reorientate with relaxation times in the range 10^{-10}–10^{-12} s. The sudden change in volume on freezing inhibits the dipolar motion. This is accompanied by a parallel substantial decrease in ϵ_0.

(ii) Highly asymmetric molecules sometimes undergo a partial degree of ordering above the freezing point—*liquid crystals*. The parallel alignment of the dipoles leads to the existence of a so-called persistence dipole. This is often markedly greater than that of a freely rotating dipole thus leading to a significant increase in ϵ_0.

Many liquid crystalline systems pass through a succession of phases in which the degree of order, perfection and packing increases as the temperature is lowered. Finally the system freezes to a static lattice with the expected decrease in ϵ_0.

(iii) Some asymmetric molecules exhibit a marked supercooling. There is little or no change in entropy, volume and hence ϵ_0 on passing through the normal crystallization temperature.[9–12] Dielectrically the super-cooled state is characterized by a markedly temperature-dependent relaxation frequency and a distribution of relaxation times. A small decrease in temperature can lead to the relaxation frequency dropping by several decades with a parallel increase in viscosity. The extent to which the motion of individual molecules may be considered cooperative is demonstrated by the magnitude of the activation energy which may be of the order of 200 kJ mol^{-1}.[12] The only plausible explanation for such a high value is that the transition state must involve several molecules moving in a cooperative manner.

In contrast, with spherically symmetric structures molecular rotation is more likely and the formation of a rotator meso-phase below the freezing point becomes possible. If this occurs, the dielectric behaviour would be roughly as predicted in Figure 4.1.

The volume and entropy change on freezing can be small. Hence, only minor changes may occur in ϵ_0 and τ. These systems can be distinguished from supercooled liquids however, since the relaxation frequency will not be so markedly temperature dependent and the dispersion will be narrower. The dipole motion will not necessarily be so completely free as envisaged above but could be subject to weak coupling effects with the surrounding lattice.

4.3 EXPERIMENTAL STUDIES[6]

In spite of the period since this type of work was initiated the number of detailed studies are few. This is primarily due to experimental limitations. Experiment requires examinations in the frequency range, typically 10^8–10^{12} Hz. Until recently only specific frequencies in the middle to upper parts of this range were available. The details of the reorientation process had to be derived from the rather tedious examination of the temperature dependence of the relaxation at these frequencies. Considerable credit is due to the early workers who produced such excellent results under these conditions.

With the advent of far infrared spectroscopy a higher frequency range ($> 10^{11}$ Hz) has become available. This has allowed a better definition of the relaxation profile and hence the relaxation process. The results for hydrogen chloride (see below) more than adequately demonstrate the improvement.

More recent experimental developments: time domain spectroscopy and improved, tunable, microwave sources, have made the lower frequency range more accessible. (10^6–10^9 Hz and 10^9–10^{11} Hz respectively). Potentially the whole frequency range can now be covered. With the wider availability of these techniques we can look forward to an expansion of interest in this area.

4.4 EXPERIMENTAL RESULTS

4.4.1 Hydrogen Halides–a Model Rotator Phase System

The hydrogen halides form an interesting group of molecules exhibiting simple molecular structures and a complex variety of phase transitions. Hydrogen fluoride forms a molecularly rigid solid lattice of small permittivity. This is in marked contrast to the other hydrogen halides where the reorientational motion in the solid is found to be no more restricted than in the liquid state.[13] This difference in behaviour can be attributed to the magnitude of the dipole–dipole interactions (μ^2/d^3, where μ is the dipole moment and d is the mean separation distance). In the case of hydrogen fluoride at the melting point, the dipole–dipole interaction exceeds kT leading to a structure of interlocking dipoles. In the remaining hydrogen halides the much smaller values of μ^2/d^3 allow reorientational freedom in the solid state just below the melting point.[13–15] Figure 4.2 summarizes the data for these solids.

Solid hydrogen chloride has a permittivity of ~20 over the range from the melting point (159 K) to approximately 100 K. At 99 K the crystal exhibits a

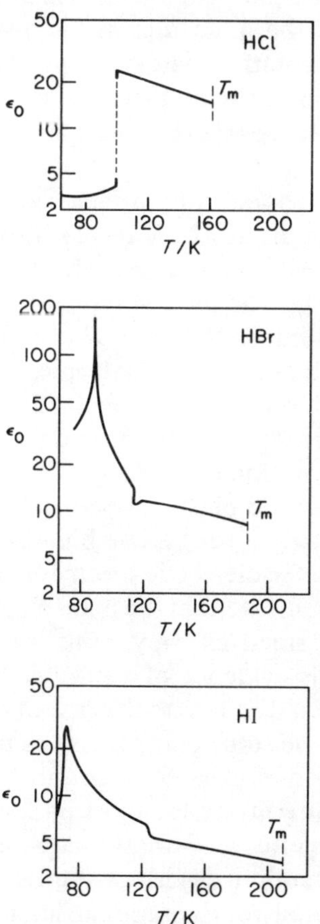

Figure 4.2 Temperature dependence of the static permittivity for the hydrogen halides

first-order solid-phase transition and the permittivity falls discontinuously to a value of 0.44 nm. A further small decrement in the permittivity typical of pre-transitional behaviour and which can be associated with oscillations (librations) in a solid[16] is observed on cooling from 99 K to 80 K. This behaviour is almost classical for a rotator-phase material.

The rotator phase of HBr is stable from the freezing point at 185 K to 120 K. At the latter temperature, the permittivity changes as a consequence of a phase change at 114 K.[17] Phase transformations in these systems are associated with either changes in the symmetry of the unit cell or the gross re-organization of the system. Both would lead to an increase in ϵ_0 and it is not possible to speculate which cause would be dominant in the present case. A further change is marked by a distinct maximum in the permittivity near 89 K. This is more likely to be associated with a change in the molecular organization since around this temperature the value of ϵ_0 rises and falls in a manner typical of a lambda-type transition. The motion of the molecules in the higher temperature phase approximates to that of free rotation. The subsequent phase changes which occur on cooling are associated with an increasing asymmetry in the rotational potential and an increased degree of rotational coupling in the system respectively.

Similar behaviour is observed in hydrogen iodide. Here, the rotator phase exists from the freezing point at 222 K to 123 K, where a discontinuous transition occurs with the appearance of a new phase. A further lambda-type transition is observed near 75 K. The highest temperature phase is isotropic and is based on a face-centred cubic lattice.[18] The symmetry is lost in the lower temperature phases which are both anisotropic. This and the fact that ϵ_0 remains high compared with $n^2(\sim 2.5\text{–}3)$ suggests that the greater polar character acquired by the system leads to a coupled dipolar reorientation which still retains a high degree of rotational freedom.

Further examinations of the plastically crystalline phase of hydrogen chloride have been made recently[19] using a combination of far infrared and microwave techniques to extend the dielectric spectrum to cover the whole relaxation region.[20,21] These studies indicate that the main relaxation lies at approximately 250 cm^{-1} at 100 K. The small entropy gain[22] of 10.87 J mol^{-1} on fusion at 158.9 K has been taken as evidence of a plastically crystalline phase existing in this material. There is no doubt as to the ease of rotation of the dipoles, however there does appear to be some controversy as to whether or not the dispersion arises from the motion of isolated or coupled dipoles. The appearance of a distinct relaxation feature in the far infrared spectrum at low temperatures has been attributed to the effects of librational motion or collisional broadening.[20,21] The librational motion describes the oscillatory perturbations of a dipole prior to its activated reorientational jump. Collisional broadening is associated with the creation of a transitory dipole as a result of an activated collision. In the latter case subsequent collisions destroy the induced dipole.

This type of relaxation behaviour is known to occur in the far infrared. Consideration of the dispersion observed in hydrogen chloride indicates that collisional processes add an insignificant contribution to the observed loss. Various models have been applied to the description of the rotational motion in hydrogen chloride.[19] A good fit of the experimental data can be obtained with a model which describes the rotational motion of the dipole in terms of activated jumps and includes the possibility of librational motion. The best agreement is obtained when the potential wells are narrowed.

Cooling the rotator phase leads to very significant changes in the absorption spectrum with the appearance of a lattice band, in addition to the broader relaxation feature discussed above, Figure 4.3. Assignment of the peak to a lattice vibration has been made on the basis of a characteristic shift on deuteration. It has been proposed[23] for rotator phase of DCl that the 12 potential wells of the face-centred cubic lattice are formed by temporary hydrogen bonds with neighbouring molecules. The extended hydrogen bonded structures so formed are of irregular length and have lifetimes comparable with the rotational relaxation time. These structures control the librational motion and are assumed to break upon reorientational relaxation of the molecular dipole. Many of the features found in the hydrogen halides are observed in the dielectric relaxation behaviour of more complex molecules.

4.4.2 Spherical Methane Derivatives

A survey of the known organic rotator phase materials[24,25] suggests that a possible criterion for the existence of essentially free reorientational behaviour

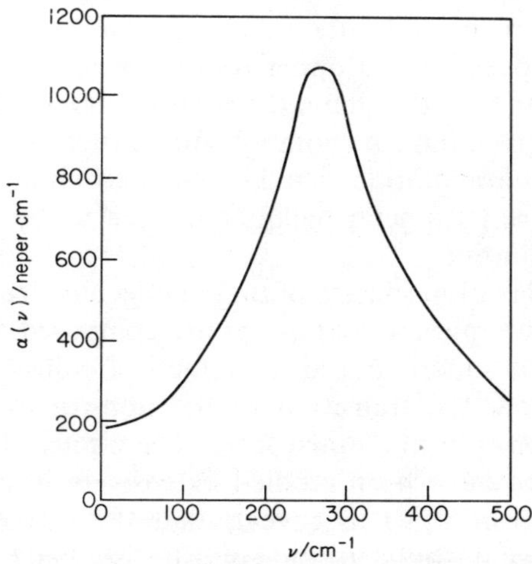

Figure 4.3 Submillimetre dielectric absorption in hydrogen chloride.

is a nearly spherical or globular molecular form. It is therefore not surprising that spherical methane derivatives exhibit rotator phase behaviour.[26] In practice it is found that this behaviour is confined to molecules where the hydrogens are substituted in part by combinations of CH_3, NO_2 or Cl.[27] Introduction of CN or Br into the methane in certain combinations can also lead to rotator phase behaviour, but this is not always the case. The polar substituted methane derivatives form an ideal group of molecules for dielectric study since they have the advantage of readily accessible liquid and solid phase.[5]

The variation with temperature of the permittivity of certain of the 'spherical' methane derivatives is summarized in Figure 4.4(a) and Table 4.1 The general behaviour is very similar to that found in the hydrogen halides. Studies of the frequency dependence of the permittivity indicate that the relaxation has an approximately Debye form in the liquid phase and is not significantly modified by cooling below the melting point of the solid. These observations are in contrast to the behaviour observed in irregularly shaped organic systems. The microwave measurements on this type of molecule performed in the sixties[27,28] have been recently complemented by far infrared studies.[29] The 2,2-disubstituted propanes have been the subject of detailed study and are worthy of discussion.

The spectra of the liquid and rotator phases are virtually identical, Figure 4.4(b). In passing from the rotator to non-rotor phase there is a significant development of structure in the far infrared region associated with the development of a well-defined phonon spectrum. The production of a rigid lattice structure modifies the low frequency far infrared region in a manner consistent with the development of domain and defect structure in an amorphous solid.[30] Although all 2,2-disubstituted propanes have similar molecular structures they exhibit a variety of types of molecular organization in the rotator phase. The 2-methyl-2-nitropropane molecule is very similar to *t*-butyl chloride and *t*-nitrobutane[100] and possesses an upper rotator phase which has a face-centred cubic structure.[30] On the other hand the rotator phases in 2,2-dichloropropane and 2,2–dinitropropane have a rhombohedral structure.[31,32] It is of interest to note that carbon tetrachloride can be obtained with either structure, the rhombohedral lattice parameter being equivalent to the body diagonal of the face-centred cubic lattice.

The dielectric relaxation spectra of these molecules are all basically similar. The far infrared and microwave regions are composed of a background absorption on which is superimposed the effects of collisional polarization and dipolar reorientation. On transition to the non-rotator phase the phonon background develops a well defined form. The effects of modification due to isotopic substitution have been studied extensively in both *t*-butyl chloride and methylchloroform.[33–35] This reveals that the strongest band in the far infrared region of solid *t*-butylchloride may be ascribed to librational motions about the C_{3v} axis of the molecule and merges into the dipolar rotational band

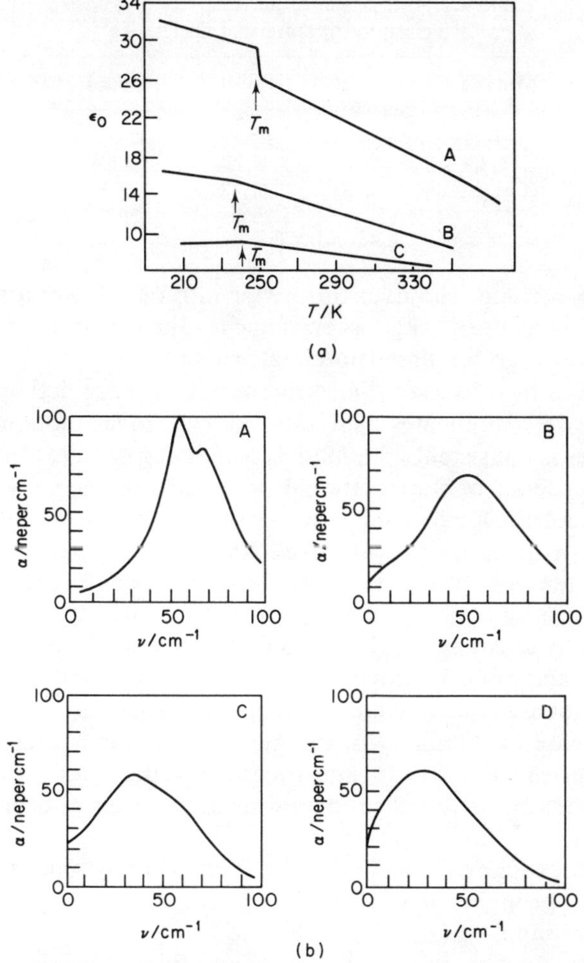

Figure 4.4 (a) Temperature dependence of the static permittivity and frequency variation of the dielectric loss for spherical methane derivatives. (A) 2 chloro-2-nitro-propane; (B) 2,2 dichloro-propane; (C) 1,1,1 trichloro-propane.
(b) Temperature dependence of the submillimetre attenuation in 2,2 dichloro-propane. (A) non-rotator-phase (182 K); (B) rotator phase (192 K); (C) and D. liquid phase (241 K and 295 K)

on transition to the rotator phase solid. A similar effect is observed in 2-methyl-2-nitropropane. Studies of the permittivity[36] indicate that in the lower temperature phase, rotation only occurs about the symmetry axis as indicated by the observed value of ϵ_0 for the plastic phase. On passing to the higher temperature phase, isotropic reorientational motion commences and the spectrum changes negligibly on passing from the rotator phase to the liquid. The

Table 4.1 Variation of permittivity (ϵ_0) with phase change for spherical molecules[27]

Molecule	Solid	Liquid
$(CH_3)_2CCl_2$	15.37	15.27
CH_3CCl_3	9.22	8.89
$(CH_3)_3CNO_2$	24.33	23.3
$(CH_3)_2CClNO_2$	29.0	29.0

observation of a slight 'shoulder' in the far infrared absorption spectrum of *t*-butyl chloride has been taken as evidence for the existence of a distribution of energies in the reorientational potential profile.[29]

Comparison of the observed dielectric spectra with predictions from simple theoretical models[33] indicates that the reorientational motion executed by these molecules is consistent with models containing definite wells and angular apertures. The detail of the rotational potential is principally a function of the lattice structure. Due to a small degree of cooperative motion in these phases the spectrum appears to be sensitive to the thermal expansion of the lattice. Lowering the temperature of the solid leads to a contraction of the lattice parameters and a 'narrowing' of the potential well whilst leaving the barrier height essentially unchanged. The barrier heights arise from a combination of van der Waals and polar interactions.

The potential barrier can be increased by the introduction of a dipolar group and/or a decrease in symmetry of the molecular structure. This trend is observed in the series CMe_nCl_{4-n}[37], and also found in the following series of nitro, chloro and methyl substituted propanes, listed in order of decreasing barrier heights:

2-chloro 2-nitropropane ($\mu = 11.7 \times 10^{-30}$ cm) (low symmetry)
2-methyl 2-nitropropane ($\mu = 12.4 \times 10^{-30}$ cm)
2,2′-dichloropropane ($\mu = 7.6 \times 10^{-30}$ c,)
t-butyl chloride ($\mu = 7.2 \times 10^{-30}$ cm) (high symmetry)

The close similarity between the shape and position of the dielectric dispersion in the liquid and the rotator phase solid is consistent with the order being only slightly lower in the former than in the latter. These observations are consistent with the solid having a low entropy of fusion and it is not surprising to find that the rotational correlation time τ_R approximates to the relationship usually used for liquids.[38]

$$\tau_R = \mathscr{G}/2kT \tag{4.3.1}$$

where $\mathscr{G}$ is the damping coefficient ascribed to viscous forces. The success of the above formula in the prediction of the relaxation time for reorientation in the rotator phase has been taken as evidence that dipole–dipole forces are of minor importance in defining the force field.

Studies of the temperature dependence of the rotator phase relaxation in the substituted methanes (Table 4.2) has revealed that in certain instances the activation enthalpies, ΔH_E^*, are lower in the solid than in the liquid.[27,28]

This fact at first appears rather surprising but can be explained in the following manner. If the compressibilities of the liquid and the rotator phase solid are assumed to be very close, the difference in the activation enthalpy values quoted at constant pressure are likely to reflect corresponding changes in the activation energy at constant volume. The observation of ΔH_E^* (solid) $\leqq \Delta H_E^*$ (liquid) is believed to result from the effects of the intermolecular potential following the formation of the highly symmetric face-centred cubic lattice found in these solids. Melting of the solid does not significantly modify the local force field of any particular molecule but it does destroy the precise geometric regularity. Cooling below the rotor phase transition temperature introduces irregularities in the rotational potential with change in azimuthal angle which are not present in the rotator phase solid.

In a system of spherical molecules reorientation is possible without the necessary introduction of translational motion. When a rotator phase is formed it is usual to find that ΔH_E^* (dielectric) $\leqq 0.5\,\Delta H_E^*$ (viscosity). This rule has been considered in terms of the Botsihinski relation for the prediction of the temperature dependence of liquid viscosities.[39] The relationship has the form

$$A/\eta = (v - b) \tag{4.3.2}$$

where A and b are constants for the liquid, the value of b may be regarded as an 'effective' molecular volume and has a value which lies between the specific volume of the liquid (v) and that of the solid. A similar relationship to equation (4.3.2) may be used to describe the Debye internal friction constant δ' ($= 2kT$) for molecular reorientation in a liquid

$$A'/\varsigma' = (v - b') \tag{4.3.3}$$

For a normal liquid $b' = b$ and reorientation ceases at the freezing point (when

Table 4.2 Activation enthalpies for reorientational motion in spherical methane molecules[27]

	$\Delta H_E^*(\tau)^a$ Liquid	/kJ mol^{-1} Solid	$\Delta H^*(\eta)^b$ Liquid
$(CH_3)_2CCl_2$	5.23	5.85	10.45
CH_3CCl_3	4.68	4.60	10.66
$(CH_3)_3CNO_2$	3.39	2.09	14.04
$(CH_3)_2CClNO_2$	6.10	5.48	13.59

[a]from dielectric studies.
[b]from viscosity studies.

$b \approx v$). For some liquids $b' < v$ (solid) which implies that rotational motion may persist in the solid state. These relationships appear to apply to the substituted methanes where $b' < v$ (solid) $< b$ is found to apply.

4.4.3 Camphene and Related Molecules

Whilst not readily allowing comparison with the liquid phase, the examination of camphene derivatives, Figure 4.5 have provided an extensive contribution to the knowledge of the dielectric properties of rotator phase solids.[40,41] Examination of the temperature dependence of the relaxation parameters, Table 4.3 obtained from dielectric measurements provides an interesting contrast with the behaviour of the substituted methanes. Camphor exhibits an activation enthalpy which varies from 5.3 to 12.0 kJ mol^{-1} within a temperature range of 373 K.[42] The barrier to reorientation appears to increase with decreasing temperature and a corresponding increase is observed in the distribution of the relaxation times. Commensurate with these changes is a decrease in the effective dipole moment as estimated from the magnitude of the value of ϵ_0. All these features suggest an increasing degree of molecular

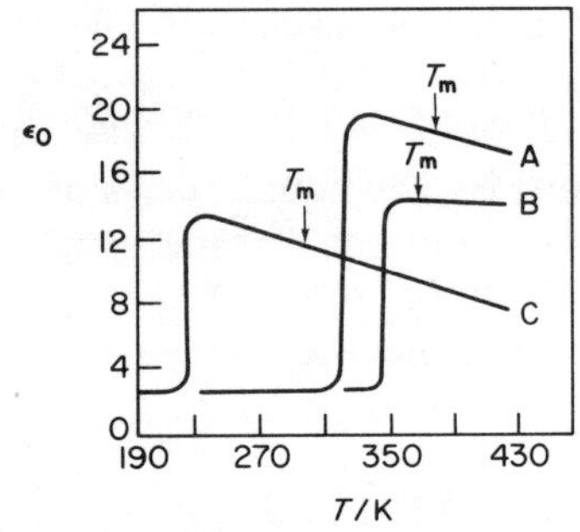

Figure 4.5 Variation of its static permittivity as a function of temperature for camphor and its derivatives. (A) nitro camphor (T_m = 373 K); (B) chloro camphor (T_m = 368 K); (C) camphor (T_m = 289 K). The arrows indicate T_m for each sample

Table 4.3 Activation parameters for camphene derivatives[28]

Molecule	ΔH^*_E /kJ mol^{-1}	ΔS^*_E /J mol^{-1} K^{-1}
Camphene	9.20 ± 3.0	−7.94 ± 8.36
Camphor	7.52 ± 3.0	+12.12 ± 11.70
Bornyl Chloride	10.45 ± 1.5	−7.94 ± 4.2
Isoborneol	22.99 ± 0.5	+25.5 ± 1.69

interaction in the solid as the temperature falls. The shape of the relaxation can be fitted to the formulism for a simple cooperative process, but critical consideration of the dielectric data indicates that it is in itself inadequate to establish the quantitative validity of the cooperative model.

In an attempt to investigate the extent to which interaction occurs, further studies have been performed on solid solutions of camphor in camphene. As expected the effective dipole moment increases on dilution towards the value found for solutions in benzene, whilst the relaxation reverts to a simple Debye type and exhibits an activation enthalpy which approaches a lower value of 5.4 kJ mol^{-1}. This latter value is very small for the reorientational process in a solid.

In isoborneol, the hydroxyl group raises the potential barrier hindering rotation to a value of 26.4 kJ mol^{-1} and is apparently constant over the temperature range 323 K to 363 K. Below 323 K the breadth of the dielectric absorption increases and is not adequately described in terms of a distribution of relaxation times. In fact the dispersion resembles that expected for a number of overlapping absorptions. The change of shape of the dielectric dispersion is partly associated with the appearance of a low frequency absorption in the MHz region characteristic of the synchronous motion of extended hydrogen bonded entities. The effect of hydroxyl group interaction increases as the temperature is lowered until at approximately 253 K the molecules become strongly associated and produce a rigid lattice structure, manifested by ϵ_0 (= 2.58) becoming equivalent to the optical value (n_D^2 = 2.55). This gradual transformation from complete reorientation of the polar molecules to an essentially rigid molecular lattice arises from increasing inter-molecular interactions which introduce a marked asymmetry in the potential and lead to order in the solid.

4.4.4 Hexasubstituted Benzenes

4.4.4(a) Methyl-containing Compounds

The behaviour observed in the polysubstituted aromatics having at least four substituents of halogen, nitro or methyl groups is typical of a molecular system which approximates to a circular disc structure and executes easy rotation about one axis. The hexasubstituted chloromethyl benzenes in particular have been subjected to systematic dielectric study, Table 4.4 below.[43–50] Over the temperature range 150 K–260 K these molecules *gradually* change from a rigid state to one in which easy rotation occurs. As indicated in Chapter 3, x-ray studies of 1,2,3-trichloro-4,5,6-trimethylbenzene and 1,2,4,5-tetramethylbenzene throughout the transition region show replacement of the six fold disorder, in which there is equal probability of the Cl or Me groups being found at each of the sites, by an ordered, orientated structure and the progressive development of super-structure.[51–60] It is clear that as the temperature is lowered the degree of cooperative motion increases. For some of these molecules, this is marked by the usual lambda type transition in the dielectric spectrum. The barrier to reorientational motion appears to be governed by electrostatic interactions. The van der Waals energy between Cl–Cl, Me–Cl, Me–Me pairs in adjacent molecules is less than 250 J mol^{-1} at the equilibrium distance whereas for Cl–Cl and Me–Me pairs the electrostatic energy difference reaches 7.3 kJ mol^{-1}. Detailed calculations[51] of the rotational potential have showed that a point dipole model for the molecule was unable to discriminate between several alternative near-neighbouring configurations possible for the low temperature state. Refinement of the model with the introduction of mesomeric and inductive effects yields a precise prediction of the energy of the rotator phase. This type of approach provided a means of estimating the transition temperature from the difference in energy (ΔU) between the randomly disordered and the ordered state, $\Delta U/R \ln 6 = 205$ K. This value is consistent with experiment.

An attempt has been made to assess whether the system was completely ordered or disordered using Monte Carlo techniques with an assembly of 240 molecules and Boltzman statistics.[61]

A good correlation between the experimentally observed Kirkwood '*g*' factor and the prediction of the Monte Carlo computation was observed for 1, 2, 3 trichloro 4, 5, 6 trimethylbenzene. The 'g' factor is a measure of the degree of interaction of a dipole and its environment and provides a means of assessing the effects of configurational change on the molecular motion of the probe. Differences between experiment and theory were attributed to the assumption that the lattice spacing is constant. The important result which appears from this correlation between experiment and theory is the possibility of superstructured domains occurring in the region of the lambda transition. These domains will appear to have a rapidly changing size and may be expected to have a significant effect on certain experiments and virtually none on others. This explanation has been put forward to explain the fact that the NMR line in certain of these molecules narrow appreciably at 200 K, whereas the corresponding dielectric change occurs at 205 K.[62]

An alternative approach to the description of these systems has been developed[63] in which prominence has been given to the detail of the methyl group interactions. A consequence of these considerations is the hypothesis that cogwheel motion rather than electrostatic interactions are responsible for the correlation in the reorientational motion of neighbouring molecules. The cogwheel effect is known to play an important role in determining the intramolecular rotational energy in substituted propanes.[64,65] It is apparent that both the effects of electrostatic and van der Waals interactions may be expected to have significant effects on the dynamic spectra of these systems; van der Waals forces acting over relatively short distances will play a definitive role in determining the detailed rotational potential, whereas the longer range electrostatic interactions lead to cooperative effects and longer range domain effects.

4.4.4(b) *Non-Methyl-containing Compounds*

As with the methyl containing compounds discussed above this group of materials can also exhibit rotator phase behaviour.[66] The relaxations occur at significantly lower frequencies and are often broader than those found in the methyl containing materials, Figure 4.6. A study of pentachlorobenzene derivatives containing amine, hydroxyl, methyl ether or *N*-ethylamine[67] as the sixth substituent were studied in an attempt to establish whether the trends found in the methyl hexa-substituted benzenes were general or not. The activation enthalpies Table 4.4 do not show the usual correlation of the activation energy with the magnitude of the dipole moment. The low values of the relaxation frequencies would suggest that there is in these systems a significant

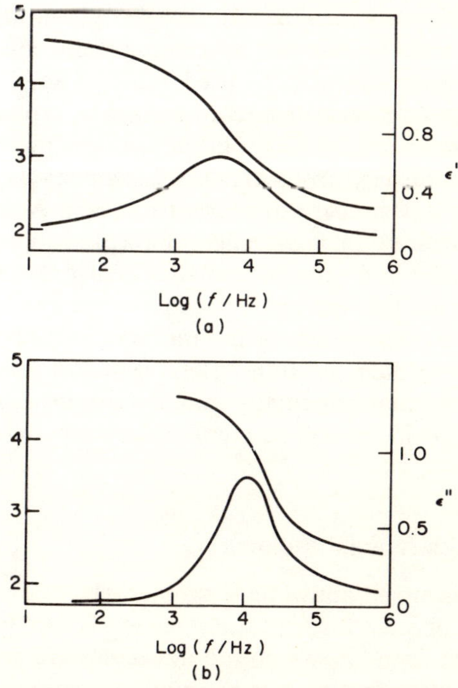

Figure 4.6 Dielectric behaviour of pressed discs of (a) pentachloronitrobenzene (disc pressed at 1.55 Mg cm^{-2}) and; (b) pentachloroaniline (disc pressed at 1.24 Mg cm^{-2}) at 296 K. The larger size of the nitro groups hinders the reorientational motion of the pentachloronitribenzene molecule. The relaxation in the case of pentachloroaniline is almost ideal

Table 4.4 Activation parameters for hexasubstituted benzenes

	ΔH_E^* /kJ mol^{-1}	M^a /g mol^{-1}	μ /D	Reference
Pentachloronitrobenzene	67.4	295.5	2.33	66
Pentachloraniline	51.5	265.5	2.63	67
Pentachlorotoluene	48.5, 50.6	264.5	1.55	57
3,5,6-Trichloro-1,2,4-trimethylbenzene 43.1	43.1	223.5	1.85	67
1,2,3-Trichloro-4,5,6-trimethylbenzene	42.7	223.5	3.15	51, 67
1,2-Dichloro-3,4,5,6-tetramethylbenzene	32.2	203.0	2.94	51, 67

[a] Molecular weight.

degree of coupling of neighbouring groups. Consequently the observed moment will not correlate closely with the bond moment for the constituent molecule. It will be noted that the value of the activation energy decreases with molar mass, probably reflecting increasing steric effects. These observations are in fact in agreement with the predictions of Smyth[68] who pointed out that with solid penta and hexasubstituted benzenes, replacement of a methyl by an ethyl group apparently blocks rotation. In agreement with this, replacement of hydrogen in the head group of pentachlorophenol and pentachloroaniline by a methyl group appears to prevent rotation. A study of mixtures of pentachloronitrobenzene[67] in naphthalene indicated that the rotation of the polar molecule is easier in the nonpolar matrix than in the pure solid and points to the role of dipolar interactions on the rotational potential.

It is clear that the replacement of the methyl groups by other substituents leads to a system in which rotation is less easy and indicates that the perturbation of the rotational potential by methyl groups is infact an important factor in determining the ease of rotation of the former class of rotator phase solids.

4.4.5 Molecules with Internal Rotation

The molecules discussed above have dipoles which are rigidly attached to the molecular axis about which rotation occurs. A number of rotator phase materials have been found in which the molecules are capable of rotational isomerism and this adds a further complication to the discussion of the dynamic disorder in these systems. An example of this type of system is succinonitrile,[69] which can exist in both *trans* and *gauche* isomeric states.[70,71] Internal rotation of succinonitrile can transform the highly polar *gauche* isomer into the nonpolar *trans* structure. It will be readily apparent that such a process provides a mechanism for dipolar annihiliation and hence an associated 'relaxational' effect. Succinonitrile exhibits one rotator phases. The transformation to this phase occurs at 233 K with an associated entropy change of 28.54 J mol^{-1}, the solid melting transition occuring at 331.3 K with an entropy of 11.20 J mol^{-1}. In the plastic phase the molecule exists in three isomeric structures, two isoenergetic *gauche* forms and one trans form, the concentration of the latter being about 20%.[70] The isomerism will itself induce extrinsic disorder in the plastic phase. An unusually high plasticity in this molecular system is explained by the occurrance of regions of high defect density which consequently give rise to a high intensity of light scattering. Like the substituted methanes, succinonitrile yields an activation enthalpy which is lower in the solid state than in the liquid: ΔH_E^* (liquid)—2.8 kJ mol^{-1}; ΔH_E^*(solid) = 8.4 kJ mol^{-1}. A major point of contention in the discussion of relaxation in rotator phases is the extent to which conformational change influences the rotational potential. No definitive discussion of this point exists, however certain general guide lines can be drawn. If the intramolecular potential is of the order of kT then there is a high probability of conformational change within the time scale of molecular reorientation. The effect of these additional relaxation effects is a perturbation of the detailed shape of the reorientational potential. In the other extreme, if the intramolecular potential is several times that of the reorientational potential then the molecule will appear rigid. The only effect is one of disorder as a consequence of a mixture of two molecularly 'dissimilar' species; one of which is capable of long range dipolar interactions. The effect of both conditions is to lead to an uncertainty in the detailed shape of the reorientational potential and consequently a high probability of co-operative interactions. The dielectric relaxational behaviour of a number of di- and tri-substituted ethanes has recently been published.[76,77] These demonstrate the problems which arise in interpretation of the dynamic data as a consequence of rapid conformational change. It is however clear that those molecules which possess intrinsic 'cigarrillo' shapes may be expected to exhibit rotator phase behaviour.

4.5 ACOUSTIC RELAXATION IN MOLECULAR CRYSTALS

The reorientational motion in most of the systems discussed in the previous section occurs on a time scale which is significantly shorter than that accessible to acoustic experiments, (5 to 100 MHz). This fact, coupled with the hypothesis that the reorientational motion occurs between molecular orientations which are virtually isoenergetic and are similar in volume, would suggest that the acoustic absorption in these systems would be very small. In fact, the few acoustic studies of molecular crystals[78] yield absorption coefficients which are significantly higher than those observed in either glasses[79] or other amorphous solids.[80] The absorption is similar to that found for those liquids (e.g.

CCl_4) in which vibrational/translational interactions make a significant contribution to the total absorption. This discrepancy between hypothesis and experiment must be associated with an absorption mechanism in which the pressure–temperature wave in its perturbation of the thermal distribution of the lattice energy achieves a dynamic storage of energy.

Possible sources of this excessive acoustic absorption are the influence of structural defects, thermal conductance or the interaction of acoustic and optical modes. The experimental observation that acoustic absorption is not significantly modified by the density of structural defects[81,82] precludes the first explanation. The second will be negligible in the present materials and theoretical calculations related to the third process in naphthalene[83] shows the contribution from this source to be less than 2% of the total.

It must therefore be suggested that the large absorption in many molecular crystals originates from the storage of energy in collisionally induced vibrational polarization. The mechanism would be as follows: two suitably activated molecules in the process of collision convert part of their momentum (thermal lattice energy) into an internal distortion of a bond or group of atoms.

This process is analogous to an increase in the vibrational specific heat of that mode. The energy so stored is returned to the thermal bath by further interaction with the phonon distribution of the lattice and appears as heat. The probability function associated with this process is a measure of the strength of the intermolecular potential, the symmetry of the molecule and its surroundings and the effective masses of the molecular species involved.

A molecular crystal can be thought to possess two classes of thermal vibration; displacements about a centre of gravity of the molecule and the vibration of the lattice as a whole. The acoustic wave will perturb the translational and rotational specific heats of the lattice and through phonon–phonon interactions modify the internal vibrational distribution. The lattice rotational-modes do not constitute a mechanism for energy storage but the interaction of these low-lying modes with the lowest vibrational modes provides the mechanism for efficient conversion of momentum to vibrational energy.

Calculations carried out for naphthalene agree well with this hypothesis.[83] For this solid it is possible to lower the temperature sufficiently for the relaxation frequency associated with the exchange to become similar to the frequency of the perturbation and dispersion is observed. The maximum in the absorption coefficient, Figure 4.7 shifts to higher temperatures with increasing frequency, which is in line with the theoretical predictions. The fact that the probability does not agree exactly with that predicted by the theory has been interpretated as a consequence of the non-colinearity of molecular and atomic coordinates of the crystal. It is clear that the intermolecular interactions which govern the probability of vibrational energy exchange also determines whether or not a particular system will be a rotator phase solid. This area of research is therefore one which should prove productive in the future.

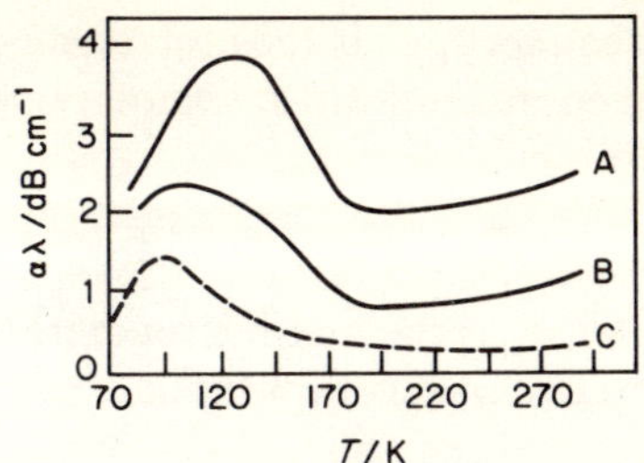

Figure 4.7 Acoustic loss as a function of temperature and frequency in naphthalene. *A* and *B* experimental values at 45 MHz and 25 MHz respectively; *C* is the theoretical derived variation corresponding to 25 MHz

4.5.1 Calculation of the Rate of Energy Transfer

The current theory used to describe acoustic energy transfer in molecular crystals is the same as that used for liquids.[84] The observation of a high loss is regarded as a consequence of the high intrinsic symmetry of the molecule and its environment and also of a weak coupling between its motion and that of neighbouring molecules. In the simple theory, the lattice is assumed to be strictly harmonic and to obey a Hookian type law. The exchange probability is obtained as a consequence of a perturbation of the motion from its ideal form under the effects of elastic deformation. In the theory due to Liebermann[85] the intramolecular potential is assumed to have a Lennard–Jones form and the lattice a cubic structure. The transition probability may be calculated using quantum mechanical perturbation theory[86] and leads to an equation of the form

$$K = \frac{9A^2(kT)^3N^4}{8\,h\nu_{01}\,m_{01}\theta_{\mathrm{m}}(4\pi^2M)} \tag{4.5.1}$$

where K is the transition probability, A is the coupling parameter evaluated from a Lennard–Jones potential and is a measure of the strength of the intermolecular binding energy, k is Boltzmann's constant, T is the temperature, N is Avogardo's number, h is Planck's constant, θ_{m} is the Debye frequency available from specific heat data, M is the molecular weight and ν_{01} and m_{01} are respectively, the frequency and effective mass of the internal vibrational motion. Although the lattice vibrational frequencies are usually lower than those of the internal vibrations, the Liebermann theory predicts the possible coupling between the higher lattice-rotational modes and the lowest internal vibrational motions. Atomic vibrations higher than the first are usually neglected at room temperature and below.[89]

The acoustic absorption coefficient is given by

$$\alpha = \tfrac{1}{2}\nu(1 - C_{\mathrm{v}}/C_{\mathrm{p}})\, C'/(C_{\mathrm{p}} - C_{\mathrm{v}})\, \omega^2K/\omega^2 + K^2 \tag{4.5.2}$$

where ω is the angular frequency, C' is the molar heat capacity associated with the internal degree of freedom, and v is the velocity of sound. The molar heat capacities may be related by

$$C_p - C_v = A_0 C_p^2 T \tag{4.5.3}$$

where $A_0 = 0.0214/T_m$ and T_m is the melting temperature. Using this relationship, the absorption coefficient α' may be written

$$\alpha' = \alpha v = \frac{A_0 C_p C' T \omega^2 K}{2\,(C_p - C')\,\omega^2 + K^2} \tag{4.5.4}$$

The intermolecular molar heat capacity, C' is associated with the lowest internal vibrational frequency and is evaluated from the Einstein specific heat formula

$$C' = RE(x_i) \tag{4.5.5}$$

$$\text{where } x_i = h\omega_i/kT \text{ and } E(x_i) = x_i^2 e^{x_i}/\,(e^{x_i} - 1)^2 \tag{4.5.6}$$

4.5.2 Experimental Observations on molecular Solids

Other than for naphthalene[82,88] acoustic vibrational relaxation has been observed in a number of rotator phase systems and has been studied in more detail in carbon tetrachloride,[87] benzene[90] and more hindered systems such as *p*-dichlorobenzene[99] and cyclohexane. These studies have been principally concerned with the observation of the frequency dependence of the absorption coefficient below the relaxation frequency. In all samples studied, α' varies with ω^2. This is in contrast to the behaviour of most amorphous materials where the absorption follows a first power relationship.[88] This square law type of behaviour is typical of systems in which a simple molecular relaxation process occurs.[90] These systems have not been studied in as much detail as naphthalene. Within the limitations of the measurements, the results are inadequate agreement with the theory discussed in the previous sections.

4.5.3 Acoustic Velocity Measurements

The propogation of an acoustic wave is controlled by the elastic constants of the material which it moves. The study of acoustic velocities has been used for a number of years to provide data on the form of intermolecular potentials in solids. A number of studies have been reported on molecular crystals,[78] certain of which are rotator phase solids. Materials such as adamantane,[91] succinonitrile,[92] pivalic acid,[92] and benzene[93] have been studied in detail. The elastic modulii of succinonitrile and pivalic acid are one or two orders of magnitude smaller than those associated with the 'harder' covalent, ionic, or metallic crystals. This difference reflects the relatively weak nature of the intermolecular van der Waals interactions in these systems. In theory, agreement should be found between the values of acoustic velocity obtained from Brillouin scattering measurements (see Chapter 6) and the ultrasonic observations. In practice, differences are reported both in the absolute values and the ratio, A, of the acoustic velocity in different crystal directions, observed by the two techniques. For an isotropic solid the factor A should equal unity: the value of A deviates from unity with the development of anisotropy in the elastic constants.

In adamantane, the ratios of the elastic constants obtained from the acoustic measurements would suggest that anisotropy A of the system (~ 2) is significantly different from that expected for a spherical molecule. The origin of this behaviour is at present obscure. The A factor is expected to tend to unity as the melting point is approached.

In the case of succinonitrile (body-centred cubic) the A factor has a value of 0.858[92] whereas in hexamethyltetramine (also b.c.c.) it has a value of 0.851. The close correspondence between these values of A infers that similar intermolecular interactions exist in these crystals. This is surprising since succinonitrile exhibits plastic properties whereas hexamethylenetetramine is essentially a normal molecular material. It may be suggested that the principle difference between these materials is that succinonitrile can exhibit rotational isomerism. Anisotropy in this case could arise from a local ordering of conformational states which are, on the average, orientationally disordered.

Comparison of the velocity in single crystals and polycrystalline samples of benzene, naphthalene, cyclohexane, *p*-xylene, diphenyl and *n*-tricosane ($C_{23}H_{48}$)[93] indicates that in polycrystalline samples the frequency dependence of the velocity exists over a much larger range than in the single crystal. The velocity of sound in fine-grained structures is smaller than in the larger-grained samples. Both these observations are consistent with discontinuities in the elastic constants of the material being associated with the grain structure of the material.

In the case of tricosane, the velocities at all frequencies pass through a minimum at about 312.5 K. A similar effect has been observed for the velocity of shear waves.[94] The occurrence of this sound velocity minimum is attributed to the fact that at this temperature a transition takes place from one modification to another. In the second form, rotation about internal bonds of the hydrocarbon backbone is assumed to become possible leading to an orientationally disordered phase.

The elastic constants of organic crystals are, in general one or two orders of magnitude less than those of covalent and ionic crystals. Comparison of the values of the compressibility tensors in stilbene,[95] dibenzyl,[95] toluene[95] with those of diphenyl[96] indicate that the maximum value of the tensor in all the crystals studied corresponds to the direction virtually normal to the long axis of the molecules. Similarily, comparison of the velocity data for naphthalene[97] and anthracene[98] indicates that the tensor is again normal to the molecular long axis. The temperature dependence of the elastic constants show similar behaviour, the average value of $d(\ln c_{ij})/dT$ for anthracene was found to be half that for naphthalene. This relationship may be predicted theoretically by considering the mass ratio of naphthalene to anthracene.[97,98] A detailed discussion of the elastic constants and the significance of the observed anisotropy in terms of molecular orientation in the sample has been reported.[99] It is evident that the extension of such measurements in closely related series of organic molecules and especially orientationally disordered phases would provide very useful data on the nature of intermolecular forces in such systems.

REFERENCES

1. L. Pauling, *Phys. Rev.*, **36**, 430 (1930).
2. J. Frenkel, *Acta. Physiochem.*, USSR **3**, 23 (1955).
3. J. Timmermans, *J Chim. Phys.*, **35**, 331 (1938).
4. C. P. Smyth, *Trans. Faraday Soc.*, **42**, 175 (1946).
5. C. P. Smyth and F. H. Brannin, *J. Chem. Phys.*, **20**, 1121 (1952).
6. N. E. Hill, *Proc. Phys. Soc.*, **72**, 1532 (1958); N. E. Hill, M. Davis, W. E. Vaughan and A. Price, *Dielectric Behaviour and Molecular Structure*, Van Nostrand, New York, 1969.
7. J. G. Kirkwood, *J. Chem. Phys.*, **7**, 911 (1938).
8. L. Onsager, *J. Amer. Chem. Soc.*, **58**, 1486 (1936).
9. G. P. Johari and C. P. Smyth, *J. Amer. Chem. Soc.*, **91**, 5168 (1969).
10. R. M. Foss, '*The Chemistry of Large Molecules*', Ed. K. E. Burk and O. Grummitt, Interscience, New York, 1943.
11. G. P. Johari and M. Goldstein, *J. Chem. Phys.*, **53**, 2372 (1970).
12. M. Goldstein, *J. Chem. Phys.*, **51**, 3728 (1969).
13. R. H. Cole and S. Havriliak, *Disc. Faraday Soc.*, **23**, 31 (1951).
14. N. L. Brown and R. H. Cole, *J. Chem. Phys.*, **21**, 1920 (1953).
15. R. W. Swenson and R. H. Cole, *J. Chem. Phys.*, **22**, 284 (1954).
16. F. I. Mopsik and R. H. Cole, *J. Chem. Phys.*, **44**, 1015 (1965).
17. G. Adam, *J. Chem. Phys.*, **43**, 662 (1965).
18. E. Sander and R. C. Farrow, *Nature*, **213**, 171 (1967).
19. I. W. Larkin, *J. C. S., Faraday II*, **70**, 1457 (1974).
20. G. M. Arnold and R. Hastie, *Chem. Phys. Letters*, **1**, 51 (1967).
21. S. H. Walmsley and H. A. Gebbie, *Mol. Phys.*, **7**, 401 (1963).
22. D. Iacocca, A. L. De Iacocca, and O. Cerceau, *Acta Cientif. Venezulana*, **15**, 189 (1974).
23. E. Sandor and R. F. C. Farrow, *Disc. Faraday Soc.*, **48**, 82 (1969).
24. C. P. Smyth, *Physics and Chemistry of the Organic Solid State*, ed. D. Fox, M. M. Labes and A. Weissberger, Interscience, New York, 723 (1963).
25. M. Davis, *J. Chem. Phys.*, **63**, 67 (1966).
26. C. P. Smyth, *J. Phys. Chem. Solids*, **18**, 40 (1961).
27. C. Clemett and M. Davis, *Trans. Faraday Soc.*, **58**, 1705 (1962).
28. C. Clemett and M. Davis, *J. Chem. Phys.*, **29**, 1347 (1960).

29. R. Hoffmans and I. W. Larkin, *J. C. S., Faraday II,* **68**, 1729 (1972).
30. R. Rudman and B. Post, *Mol. Cryst.,* **5**, 95 (1968).
31. S. C. Abrahams, *J. Chem. Phys.*, **21**, 1218 (1953).
32. J. J. Villings and A. W. Nolle, *J. Chem. Phys.*, **29**, 214 (1958).
33. I. W. Larkin, *J. C. S., Faraday II,* **69**, 1278 (1973).
34. J. R. Durig, S. M. Craven, and J. Bragin, *J. Chem. Phys.,* **51**, 5663 (1969).
35. J. R. Durig, S. M. Craven, K. K. Lau, and J. Bragin, *J. Chem. Phys.,* **54**, 474 (1971).
36. R. W. Crowe and C. P. Smyth, *J. Amer. Chem. Soc.,* **72**, 4009 (1950).
37. B. Lassier and C. Brot, *J. Chim. Phys.,* **65**, 1723, (1968).
38. W. G. Rothschild, *J. Chem. Phys.,* **53**, 990, 3265 (1970).
39. N. E. Hill, *Trans. Faraday Soc.,* **55**, 2000 (1959).
40. J. G. Powles, *J. Chem. Phys.,* **20**, 1048 (1952).
41. C. Clemett and M. Davis, *Trans. Faraday Soc.,* **58**, 1718 (1962).
42. C. Clemett, *Ph. D. Thesis, University of Wales*, (1961).
43. A. H. White and W. S. Bishop, *J. Amer. Chem. Soc.,* **62**, 16 (1940).
44. W. A. Yager and S. O. Morgan, *Ind. Eng. Chem.,* **32**. 1519 (1940).
45. S. O. Morgan, *Ann. N. Y. Acad. Sci.,* **40** (**5**), 357 (1940).
46. C. P. Smyth and G. L. Lewis, *J. Amer. Chem. Soc.*, **62**, 949 (1940).
47. C. Brot and I. J. Darmon, *J. Chim. Phys.,* **63**, 100 (1960).
48. Y. Balcou and J. Meinnel, *J. Chim. Phys.,* **63**, 114 (1966).
49. I. J. Darmon and C. Brot, *Mol. Crys.,* **2**, 301 (1961).
50. C. Brot, I. J. Darmon, and N. Dat-Xuong, *J. Chim. Phys.,* **64**, 1061 (1967).
51. C. Brot and I. J. Darmon, *J. Chem. Phys.,* **53**, 2271 (1970).
52. A. Tulinsky and J. G. White, *Acta. Cryst.,* **11**, 7 (1958).
53. R. Fourme, M. Renaud, and D. Andre, *Mol. Cryst. Liq. Cryst.,* **17**, 209 (1972).
54. C. P. Charbonneau and J. Totter, *J. Chem. Soc.*, 2032 (1967).
55. R. Fourme and M. Renaud, *Mol. Cryst. Liq. Cryst.,* **17**, 223 (1972).
56. J. C. Messager and M. Sanquer, *Mol. Cryst. Liq. Cryst.,* **26**, 373 (1974).
57. M. Sanquer and J. C. Messager, *Mol. Cryst.,* **20**, 107 (1973).
58. E. G. Cox, D. W. J. Cruickshank, and J. A. S. Smith, *Proc. Roy. Soc.,* **247A**, 1 (1958).
59. J. E. Anderson, *J. Chem. Phys.,* **43**, 3575 (1965).
60. D. W. Cruickshank, *J. Chem. Phys.,* **45**, 5971 (1970).
61. J. G. Kirkwood, *J. Chem. Phys.,* **7**, 911 (1939).
62. M. Davis, *Annual Reports Chemical Society,* **67**, 86 (1970).
63. J. C. Messager and M. Sanquer, *Mol. Cryst. Liq. Cryst.,* **26**, 373 (1974).
64. G. J. Szasz, *J. Chem. Phys.,* **23**, 2449 (1955).
65. A. B. Dempster, K. Price, and N. Sheppard, *Spectrochim. Acta.,* **A25**, 1381 (1969).
66. A. Aihara, C. Kitazawa, and A. Nohara, *Bull. Chem. Soc. Japan,* **43**, 3750 (1970).
67. P. G. Hall and G. S. Horsfall, *J. C. S., Faraday II,* **69**, 1071 (1974).
68. C. P. Smyth, *Dielectric Behaviour and Structure*, McGraw Hill, New York, 163 (1955).
69. S. Kondo and T. Oda, *Bull. Chem. Soc., Japan,* **27**, 567 (1954).
70. W. E. Fitzgerald and J. G. Janz, *J. Mol. Spectry.,* **1**, 49 (1957).
71. G. J. Janz and W. E. Fitzgerald, *J. Chem. Phys.,* **23**, 1973 (1955).
72. C. Clemett and M. Davis, *J. Chem. Phys.,* **32**, 316 (1960).
73. D. E. Williams and C. P. Smyth, *J. Amer. Chem. Soc.,* **84**, 1808, (1962).
74. H. M. Hawthorne, *Ph.D. Thesis, University of Strathclyde*, (1966).
75. A. J. Hyde, J. Kevorkian, and J. N. Sherwood, *Disc. Faraday Soc.,* **48**, 19 (1969).

76. I. Larkin and M. Evans, *J. C. S., Faraday II,* **70**, 477 (1974).
77. C. P. Smyth, *Internal Rotation in Molecules*, ed. W. J. Orville Thomas, Wiley, New York (1974). p. 29.
78. R. A. Pethrick and A. M. North, *Transfer and Storage of Energy by Molecules*, ed. G. Burnett, A. M. North and J. N. Sherwood, Wiley. N. Y. (1974) Chapter 8.
79. M. F. Lewis, *Cryogenics,* **10**, 38 (1970).
80. R. J. Bell and P. Dean, *Amorphous Materials, Proc. Int. Conf., on the Physics of Non-Crystalline Solids*, Sheffield 1970, Wiley Interscience, New York (1972) p. 443.
81. L. Liebermann, *J. Acoust. Soc. Amer.*, **31**, 1073 (1959).
82. R. Rasmussen, *J. Chem. Phys.*, **36**, 1821 (1962).
83. H. F. Danielmeyer, *Accustica*, **17**, 102 (1966).
84. T. A. Litovitz, *J. Chem. Phys.,* **26**, 469 (1957).
85. L. Liebermann, *Phys. Rev.*, **113**, 1052 (1959).
86. L. Schiff, *Quantum Mechanics*, McGraw-Hill, New York, (1949).
87. R. A. Rasmussen, *J. Chem. Phys.*, **46**, 211 (1967).
88. J. D. Wilson and S. S. Yun, *J. Acoust. Soc. Amer.*, **50**, 164 (1971).
89. S. S. Yun and R. T. Beyer, *J. Chem. Phys.*, **40**, 2538 (1964).
90. A. J. Matheson, *Molecular Acoustics*, Wiley, New York (1971).
91. J. C. Damien, *Solid State Comm.*, **16**, 1271 (1975).
92. H. Fontaine and C. Moriamez, *J. Chim. Phys.*, **65**, 969 (1968).
93. N. A. Dobromyslov and N. I. Koshkin, *Soviet Physics Acoustics,* **15**, 386 (1970).
94. W. Pechnold, W. Dollhopf, and A. Engel, *Acoustica,* **17**, 60 (1966).
95. V. F. Teslenko, A. P. Ryzhenkov, T. V. Ryzhkova, K. S. Aleksandrov, and A. T. Kitaigorodskii, *Soviet Physics Crystallography*, **10**, 744 (1966).
96. A. I. Krupnyi, K. S. Aleksandrov, and G. S. Belikova, *Soviet Physics. Crystallography,* **15**, 507 (1970).
97. G. K. Afanaseva, *Soviet Physics Crystallography*, **13**, 892 (1969).
98. G. K. Afanaseva and R. M. Myasnikova, *Soviet Physics Crystallography,* **15**, 156 (1970).
99. K. S. Aleksandrov and O. V. Nosikov, *Soviet Physics, Acoustics,* **2**, 256 (1956).
100. P. Freundlich, J. Kalenik, E. Narewski, and L. Sobozyk, *Acta. Physica Polonica.,* **48**. 701 (1975).

5

NMR Studies of Plastic Crystals

N. Boden

5.1 INTRODUCTION

NMR provides a simple, widely applicable, laboratory technique for characterizing and measuring the rates of rotational and translational motions in molecular crystals. In the case of plastic crystals, the detailed changes in the reorientational motion at the ordered solid–plastic phase transition (rotational melting) and the translational motion at fusion (translational melting) may be investigated. In the ordered solid, the translational motion is too slow to be detected by NMR and only the rotational motion can be studied, while in the plastic phase both motions are accessible.

One of the most attractive features of the NMR experiment is its selectivity. First, it is selective in the sense that the resonances of different nuclear species may be independently monitored and studied; this is particularly useful in studying organic solids where deuterium atoms may be substituted for hydrogen at selected sites in the molecule. Secondly, NMR relaxation measurements detect those frequencies in the motional spectrum which correspond to the precession or Larmor frequency of the spins which is an experimental variable. This is very important because it enables motions occurring at different rates to be independently studied; for example, the rotational and translational diffusion of molecules in plastic phases are easily distinguished.

The NMR methods employed in the study of plastic crystals are well established. The basic theories and techniques are described in the texts by Abragam,[1] Slichter,[2] and Goldman,[3] the book by Farrar and Becker[4] offers a simple introduction to pulse experiments. The reviews by Ailion[5] and Allen[6] are specifically concerned with the application of NMR to the study of molecular motion in solids and should, therefore, be particularly useful to the reader. An introduction to the basic principles of NMR spectroscopy would clearly be superfluous here. The object of this chapter is an attempt to answer the question: What can NMR tell us about the properties of plastic crystals? The approach will be to indicate those NMR techniques which have been, or will be, useful in this direction, to explain the relationships between the experimental measurements and these properties and, most importantly, to assess the significance of the results.

5.2 FREQUENCY SPECTRUM

5.2.1 Experimental Measurement

The frequency spectrum may be measured either directly or indirectly. In the direct method, the energy absorbed by the spin system is measured as a function of the frequency of an applied weak cw radiofrequency field; the result is the 'frequency domain spectrum' $F(\omega)$. In the indirect method, the time response function of the spin system is observed following the application of an intense radiofrequency pulse; the result is the 'time domain spectrum' $G(t)$, often referred to as the free induction decay (FID) signal. $F(\omega)$ and $G(t)$ are Fourier transforms of each other[1]

$$F(\omega) = \int_{-\infty}^{+\infty} G(t) \exp(-\mathrm{i}\omega t)\, \mathrm{d}t;$$

$$G(t) = \frac{1}{2\pi}\int_{-\infty}^{+\infty} F(\omega) \exp(\mathrm{i}\omega t)\, \mathrm{d}\omega. \tag{5.2.1}$$

so that, at least in principle, we get exactly the same information from each experiment. The pulse method has many attractive features and in many applications it is gradually superceding the cw method. For symmetrical frequency spectra, such as dipolar broadened ^{1}H resonances in solids, resonant 90° radiofrequency pulses are usually employed, while for asymmetric spectra, off-resonance pulses are necessary. The recent development of quadrature detection methods[7] obviates the need for off-resonance excitation in the better cases. A serious technical difficulty in the pulse method is that the initial part of the FID signal is obscured by the finite time it takes the spectrometer receiver to recover from the powerful radiofrequency pulses. This so-called 'dead-time' is typically of the order 5 to 10 μs which is significant compared with the duration, 50–100 μs, of the FID signal in solid samples. It is essential to correct for this loss of signal, since the Fourier transformation is very sensitive to the signal at short times. Analytical techniques may be used; these rely on fitting the FID signal to an appropriate analytical function at long times and obtaining the signal at short times by extrapolation.[1,3,8] A variety of experimental techniques have been developed[9–12] to provide for 'zero-time' resolution of the signal. It is also necessary to correct for distortion of the signal due to the finite bandwidth of the receiver system, but this is straightforward.[13]

5.2.2 Spectra Dominated by Dipolar Interactions

In molecular crystals, the shape $F(\omega)$ of a ^1H or ^{19}F resonance spectrum is invariably determined by the direct dipole–dipole interactions between the spins. These spectra are broad, typically 10–100 kHz, symmetrical and structureless due to the multiplicity of these interactions. $F(\omega)$ often approximates to a simple Gaussian function. Unfortunately, it is not possible to calculate the explicit form of $F(\omega)$ in a solid due to the 'many body' nature of the spin interactions. It is, however, possible to relate $F(\omega)$ to the properties of the spin system by use of the van Vleck method of moments.[14]

5.2.2(a) van Vleck Moments

The nth moment M_n of $F(\omega)$ is given by

$$M_n = \int_0^\infty (\omega - \omega_0)^n F(\omega)\,\mathrm{d}\omega \bigg/ \int_0^\infty F(\omega)\,\mathrm{d}\omega \tag{5.2.2}$$

where ω_0 is the centre of the resonance line. Alternatively, in terms of $G(t)$, the moments are given by[1,3]

$$M_n = (-1)^n \left[\frac{\mathrm{d}^n G(t)}{\mathrm{d}t^n}\right]_{t=0}. \tag{5.2.3}$$

van Vleck[14] has shown how these moments may be related to the properties of the spin system. The second moment M_2 is of particular practical importance because it is directly related to the geometric arrangement of the spins in the crystal. For a single crystal, the second moment, due to interactions between like spins, is given by

$$M_2^{\mathrm{II}} = \left(\frac{\mu_0}{4\pi}\right)^2 \frac{3}{4}\gamma^4 h^2 I(I+1)\frac{1}{n}\sum_{j,k}\frac{(1-3\cos^2\theta_{jk})^2}{\mathbf{r}_{jk}^6} \tag{5.2.4}$$

where θ_{jk} is the orientation of the vector $\mathbf{r}_{jk}$ from spin j to k with respect to the direction of the applied static magnetic field $\mathbf{B}$. For each of the n distinct spins j in the unit cell, the sum must be taken over all its neighbours k within the crystal. γ is the gyromagnetic ratio and I the spin quantum number for the nucleus in question. In solids containing two dipolar coupled species I and S, the second moment of the resonance line will have contributions from both I–I and I–S interactions and is given by

$$M_2^{\mathrm{I}} = M_2^{\mathrm{II}} + M_2^{\mathrm{IS}} \tag{5.2.5}$$

where

$$M_2^{\mathrm{IS}} = \left(\frac{\mu_0}{4\pi}\right)^2 \frac{1}{3}\gamma_{\mathrm{I}}^2\gamma_{\mathrm{S}}^2\hbar^2 S(I+1)\frac{1}{n}\sum_{j,k}\frac{(1-3\cos^2\theta_{jk})^2}{\mathbf{r}_{jk}^6}. \tag{5.2.5a}$$

$M_2^{\rm II}$ and $M_2^{\rm IS}$ may be selectively measured by spin–echo techniques[15]. Most studies of molecular solids have been made on powdered samples containing crystallites of random orientations for which

$$M_2^{\rm II} = \left(\frac{\mu_0}{4\pi}\right)^2 \frac{3}{5} \gamma^4 \hbar^2 I(I+1) \frac{1}{n} \sum_{j,k} r_{jk}^{-6}. \tag{5.2.6}$$

In the presence of thermal motions of the spins the experimentally observed second moment is reduced to [16]

$$M_2^{\rm expt} = \left(\frac{\mu_0}{4\pi}\right)^2 \frac{3}{4} \gamma^4 \hbar^2 I(I+1) \frac{1}{n} \sum_{j,k} \overline{\left|\left\langle \frac{(1 - 3\cos^2\theta_{jk})}{r_{jk}^3} \right\rangle_{\rm Av}\right|^2} \tag{5.2.7}$$

when

$$M_2^{1/2}\tau \ll 1,$$

where τ is the correlation time[1] for the motion of the spin-spin vector. The $\langle\ \rangle_{\rm Av}$ indicates an average over the motion and the bar an average over crystallite orientations in powder samples. Figure 5.1 depicts the temperature dependence of M_2 expected for a solid in which the molecules undergo thermally activated reorientation. The sharp drop in the value of $M_2^{\rm expt}$ corresponds to the 'motional narrowing' of the resonance line when $M_2^{1/2}\tau \cong 1$. M_2 for ^{1}H in organic solids is typically 10 to 20 $\times$ 10^{-8}T^2 {1 T(Tesla) $\equiv$ 10^4 G(Gauss)}, so that motional narrowing is observed when $\tau \cong 10^{-5}$ s. Thus, provided that the crystal structure is known, the mechanism of the rotational process causing line narrowing may be identified from the 'rigid lattice' and 'motionally averaged' values of M_2. Furthermore, values for τ are easily esti-

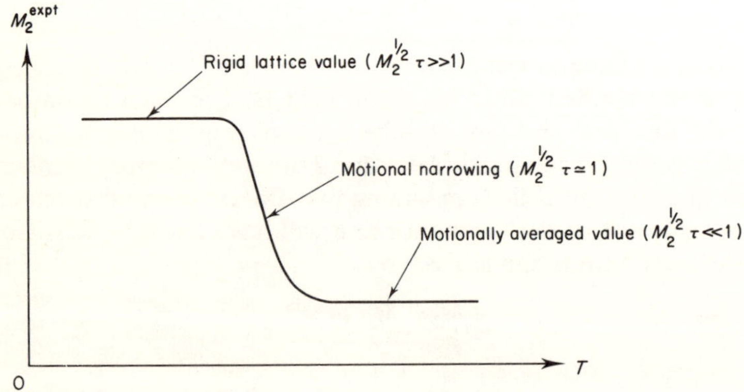

Figure 5.1 Second moment $M_2^{\rm expt}$ as a function of temperature in a solid in the presence of molecular reorientation

mated from M_2^{expt} in the transition region using the relationship[16,17]

$$M_2^{\text{expt}} = \overline{M_2} + (M_2 - \overline{M_2}) \frac{2}{\pi} \tan^{-1} (\alpha\, M_2^{1/2}\, \tau) \tag{5.2.8}$$

where $\alpha \cong 1$ and M_2 and $\overline{M_2}$ are, respectively, the rigid lattice and rotationally averaged second moments. The kinetic parameters obtained in this way are not, in general, very reliable. This is attributable to the fact that the second moment is not uniquely defined in the transition region and that, the transition in M_2^{expt} corresponds to a change of one to two orders of magnitude in τ. At structural phase transitions, M_2 usually changes abruptly due to changes in both structure and motion, enabling the transition to be characterized.

5.2.2(b) Second Moment Studies of Molecular Motion

Figures 5.2 to 5.6 show the ^{1}H second moments measured as a function of temperature, by the cw experiment, in a selection of plastic crystal forming solids; they are representative of the entire spectrum of observed behaviour.

Figure 5.2 illustrates the behaviour of adamantane, a tetrahedral molecule with composition $C_{10}H_{16}$.[18,19] The large reduction in M_2 at about 143 K corresponds to the averaging of the entire intramolecular and part of the intermolecular contributions to M_2 by quasi-isotrope reorientation of the molecule. It is interesting that there is no detectable change in M_2 at the solid–plastic transition (209 K) suggesting the molecular packing is very similar in the two phases. The measurements stop well below the melting point 542 K.

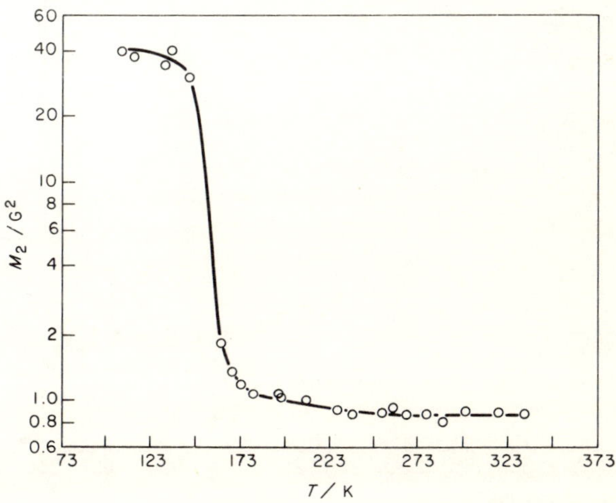

Figure 5.2 The ^{1}H second moment measured as a function of temperature in adamantane.[18] (Reproduced by permiss of *J. Chem. Phys.*)

Figures 5.3 and 5.4 give the ^{1}H second moments as a function of temperature measured in, respectively, triethylenediamine[20] and hexamethylethane[21,22]. The two plots are very similar and are typical of solids where the motion in the ordered solid phase is restricted to reorientation about one molecular symmetry axis. In triethylenediamine, the reduction in M_2 at 190 K is due to hindered reorientation about the N–N axis; note, the reduction in M_2 is only about one-half of that observed for adamantane. At the transition to the plastic phase at 351 K, M_2 drops sharply, yet continuously, to a value of approximately 1 G^2 characteristic of isotropic reorientation. Hexamethylethane, Figure 5.4, is seen to behave similarly. The transition in M_2 centred on 90 K is due to C_3' reorientation of the *t*-butyl group, i.e. simultaneous reorientation of the $(CH_3)_3C$-group about the C–C direction and the methyl groups about their C_3 axes. Note here too that M_2 appears to change continuously from the solid to the plastic phase indicative of molecular disordering as a precursor to the transition.

In cyclohexane, Figure 5.5, there is a motional averaging of M_2 centred on 150 K due to hindered reorientation of the molecules about their triad axes, but note how in this solid the motionally averaged value is not properly established before there is a sudden transition to the plastic phase[23]. In perfluorocyclohexane the motional averaging is inextricably tied up with the phase transition[24] while in *d*-camphor, Figure 5.6, there is no motional averaging whatso-

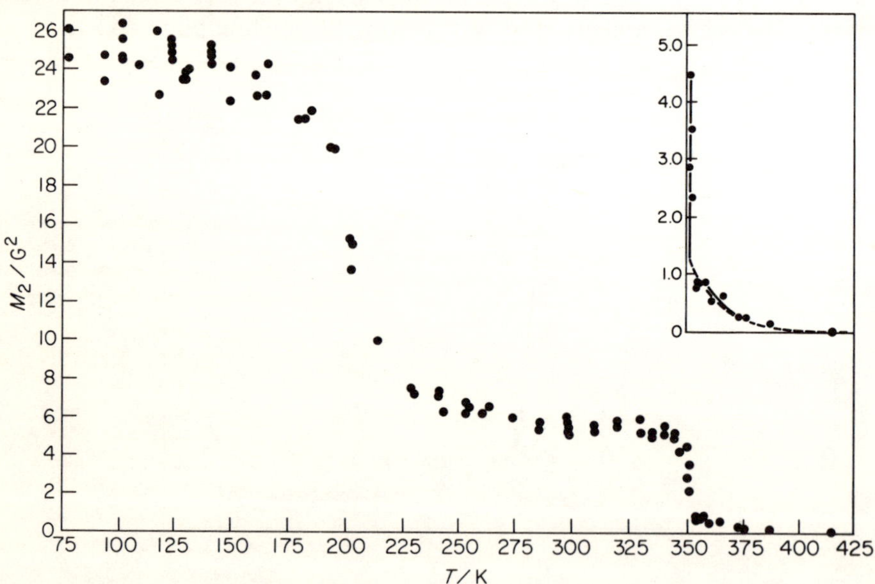

Figure 5.3 The ^{1}H second moment measured as a function of temperature in triethylenediamine.[20] (Reproduced by permission of *J. Chem. Phys.*)

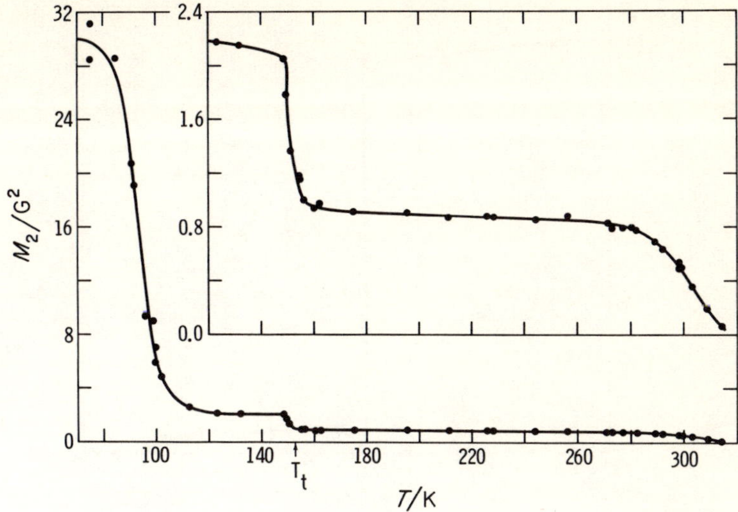

Figure 5.4 The ^{1}H second moment measured as a function of temperature in hexamethylethane.[21] (Reproduced by permission of *J. Chem. Phys.*)

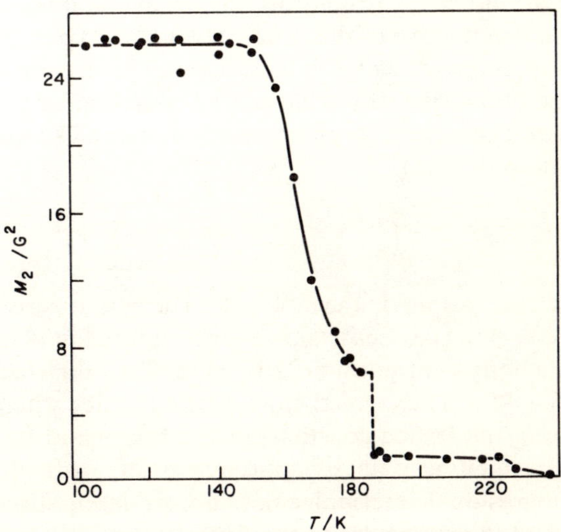

Figure 5.5 The ^{1}H second moment measured as a function of temperature in cyclohexane.[23] (Reproduced by permission of The Royal Society, London)

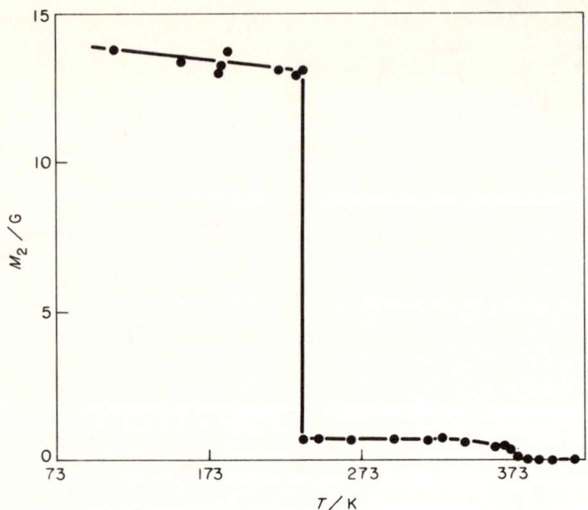

Figure 5.6 The ^{1}H second moment measured as a function of temperature in *d*-camphor.[25] (Reproduced by permission of *J. Chem. Phys.*)

ever below the phase transition;[25] actually, M_2 is partially averaged by the rotation of the methyl groups about the C_3 axes.

The above results show that the nature of the rotational motion in the solid phase and how it changes through the phase transition are clearly linked to the symmetry of the molecule. In contrast, the behaviour observed in the plastic phase is the same irrespective of the molecular shape. Typically, $M_2 \cong 1\ G^2$ and is entirely due to intermolecular interactions. These rotationally averaged values M_{2r} correspond to the values calculated by assuming all spins in a molecule are located at its centre[26]; for interactions between like spins I in a powder this is, from equation (5.2.7),

$$M_{2r} = c\left(\frac{\mu_0}{4\pi}\right)^2 \frac{3}{5}\gamma^4\hbar^2 I(I+1)\sum_k r_{jk}^{-6} \tag{5.2.9}$$

where c is the number of spins in the molecule. The lattice sums $\Sigma_k r_{jk}^{-6}$ are easily evaluated for the various lattice structures encountered in plastic phases; they are 29.05 a_0^{-6} for a body-centred cubic lattice 115.63 a_0^{-6} for a face-centred cubic lattice and 14.45 a^{-6} for a hexagonal close-packed lattice where a is the side of the hexagon and a_0 the lattice constant for the body- and face-centred cubic structures. As the melting point is approached, $M_2 \rightarrow 0$; this is due to the averaging of the residual intermolecular dipolar interactions by the translational diffusion of the molecules on the crystal lattice. When $M_2^{1/2}\tau \ll 1$, however, the resonance line has a Lorentzian shape and M_2 is no longer defined[1].

In this region.

$$(\Delta\omega)_{1/2} = 2/T_2 \text{ or } G(t) = G(0) \exp(-t/T_2) \qquad (5.2.10)$$

where $(\Delta\omega)_{1/2}$ is the full linewidth at half-maximum intensity. T_2 is the spin–spin relaxation time and is related to the mean time interval τ between successive molecular jumps by

$$\frac{1}{T_2} = M_{2r}\,\tau. \qquad (5.2.11)$$

Linewidth measurements have been employed to study molecular diffusion in many plastic crystals. These measurements are simple, but do not give very reliable kinetic parameters. It is preferable to study the kinetics of this process through spin–lattice relaxation measurements. The results of both kinds of study will be described in Section 5.3.5.

5.2.3 Spectra Dominated by Nuclear Electric Quadrupole Interaction

The solid state NMR spectra of nuclei with $I > \frac{1}{2}$ are often dominated by the nuclear electric quadrupole interaction[1]. This interaction is usually much greater than the dipolar one, making these spectra simpler to interpret than those observed for protons. For studying motion in organic solids, deuteron ($I = 1$) spectra are particularly useful because deuterium may be selectively substituted for hydrogen. Experimentally the resonance is much weaker, making measurements on single crystals advantageous though not essential. The spectrum for a deuteron interacting with an axially symmetric electric field gradient is, for a single crystal, a simple doublet with separation[1,2]

$$\Delta\nu = \frac{3}{4}\left(\frac{e^2qQ}{h}\right)\langle 3\cos^2\theta - 1\rangle_{av}. \qquad (5.2.12)$$

e^2qQ/h is the electric quadrupole coupling constant and θ, the angle between the direction z' of the principle component, V_{zz}, of the electric field gradient tensor and the direction z of magnetic field **B**. The angular brackets denote an average over the motion necessary when $2\pi\Delta\nu \ll 1/\tau$;

$$\langle 3\cos^2\theta - 1\rangle_{av} = (3\cos^2\phi - 1)\frac{(3\cos^2\gamma - 1)}{2} \qquad (5.2.13)$$

where γ is the angle between z' and the axis of rotation, while ϕ is the orientation of the latter with respect to the z-axis. By studying the shape of the spectrum in the region of $2\pi\Delta\nu \cong 1/\tau$, values of τ may be extracted[1,27,28]. The value of e^2qQ/h is dependent on the nature of the chemical bonding and is typically of the order 200 kHz, e.g. for $>CD_2$ it is 175 kHz while for C_6D_6 it is 182 kHz. The dipolar interactions are only of the order of a few kHz for deuterons and cause either a resolvable splitting or simply a broadening of each line of the doublet depending on the complexity of the system. Broadening of the resonance lines due to dipolar interactions with non-resonant protons may be eliminated by double resonance techniques[29] to give an improvement in the signal sensitivity.

In a powder sample, equation (5.2.12) must be averaged over the random orientations of the crystallites. The resulting spectrum for a three-dimensional

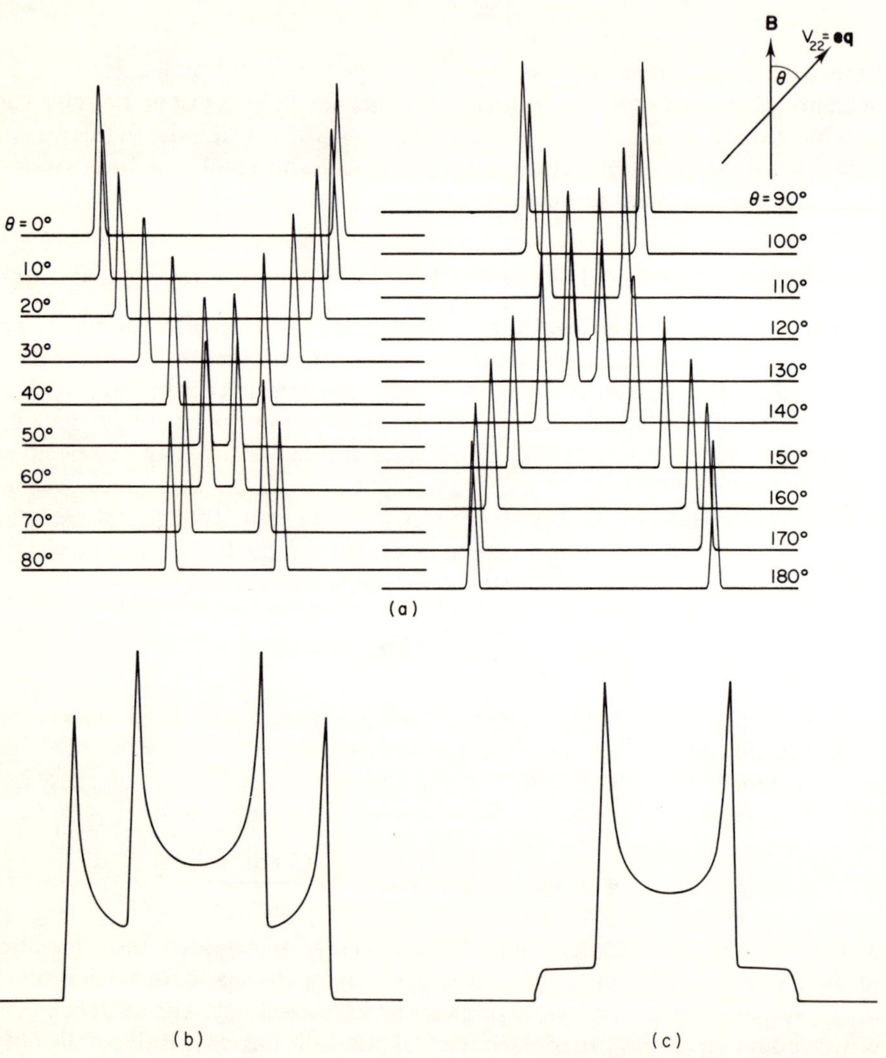

Figure 5.7 ^{2}H spectra, in arbitrary frequency units, for a single crystal, (a), and two- and three dimensional powder samples, (b) and (c), respectively for an axially symmetric electric field gradient tensor

powder is shown in Figure 5.7(c) and is seen to be dominated by a doublet (pair of singularities) whose separation is, for a 'rigid' lattice, (i.e. $2\pi\Delta\nu\tau \gg 1$), $\frac{3}{4}(e^2qQ/h)$ or, for a 'motionally averaged' lattice (i.e. $2\pi\Delta\nu t \ll 1$) $\frac{3}{4}(e^2qQ/h)$ $(3\cos^2\gamma - 1)/2$. For non-axially symmetric electric field gradient tensors, the spectra are more difficult to interpret.

The study of molecular motion in $(CD_3)_3CCl$ by O'Reilly *et al.* is one of the few examples to date of the use of ^2H resonance spectroscopy to study molecular motion in an organic solid. The doublet separation in the cw powder spectrum is plotted as a function of temperature in Figure 5.8. Below 100 K, the splitting corresponds to a value of 173 kHz for (e^2qQ/h), the value expected for a 'rigid' methyl group. There are successive reductions by 1/3 in the separation first at 100 K and then at 150 K. The first is due to rotation of the methyl group about its triad axes ($\gamma = 70.5°$) and the second to the reorientation of the molecule about the C–Cl axis ($\gamma = 70.5°$, the angle between the triad axis of the methyl group and the C–Cl bond). The mechanism of the motion is unchanged in going from the monclinic to the tetragonal phase, whilst at the transition to the face-centred cubic plastic phase the spectrum changes abruptly to a single line characteristic of fast isotropic reorientation.

5.2.4 Spectra Dominated by Anisotropic Chemical Shielding Interaction

Frequency shifts due to the anisotropy of the shielding tensor are only resolved when either the strong dipole–dipole interactions are removed by the use of multiple-pulse line-narrowing techniques or the spectrum is measured in

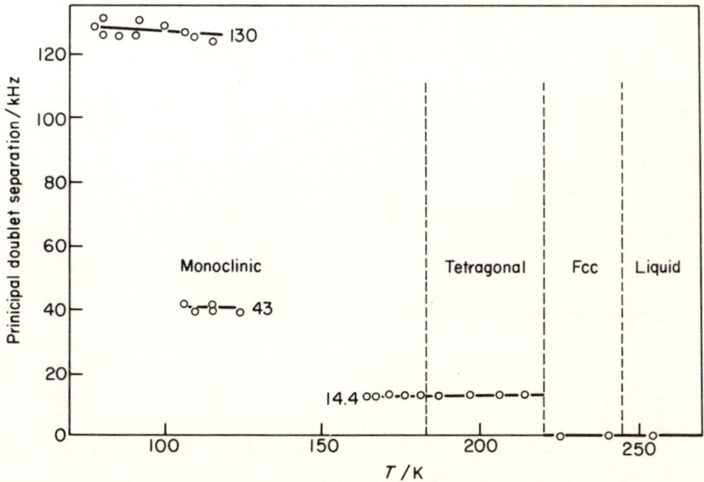

Figure 5.8 Separation of the principal doublet in the ^2H powder spectrum of $(CD_3)_3CCl$.[30] (Reproduced by permission of *J. Chem. Phys.*)

a magnetic field large enough to make the chemical shielding interaction larger than the dipolar one. The recently published text by Haeberlen[31] gives an excellent account of these experiments and the information obtainable from studies of chemical shielding in solids. There is only one example to date of such a study in a plastic crystal forming solid; it is of solid white phosphorus, P_4, which has been studied at 92 MHz using a superconducting magnet[32]. Figure 5.9 shows the ^{31}P powder spectrum in the 'rigid' lattice at 25 K and in the plastic phase at 293 K. In the plastic phase, the frequency shifts due to the anisotropy of σ are completely averaged by the rapid reorientation of the P_4 molecules. The spectrum in the ordered solid phase at 25 K is typical of an axially symmetric shielding tensor σ: the frequency shifts due to the anisotropy of σ give the spectrum a width of about 37 kHz which is much larger than the estimate of 2.3 kHz for

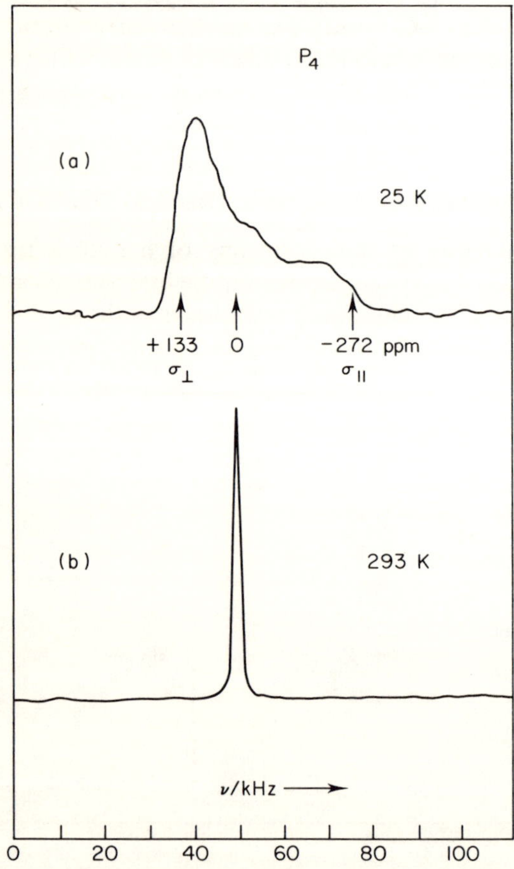

Figure 5.9 ^{31}P powder spectra of solid white phosphorus at 92 MHz: (a) ordered solid at 25 K, (b) plastic phase at 293 K.[32] (Reproduced by permission of *Chem. Phys.*)

the dipolar interaction. From the spectrum it is found that $\sigma_{\parallel} = -272$ p.p.m. and $\sigma_{\perp} = +133$ p.p.m. relative to the isotropic shift measured in the plastic phase. The transition from the 'rigid' lattice spectrum, Figure 5.9(a), to the motionally averaged one, Figure 5.9(b), takes place in the temperature interval between 80–120 K where $2\pi\Delta\nu \cong 1/\tau$. Values for τ may be derived from the lineshape in this region, but it is simpler and more accurate to determine such data from spin–lattice relaxation measurements as described later.

Measurements of ^{13}C chemical shielding anisotropies in organic solids using the technique of proton-enhanced nuclear induction[33] would be a useful source of information about the nature of the reorientational motion in the ordered solid just below the transition to the plastic phase. The calculation of powder lineshapes for axially symmetric shielding tensors in the presence of molecular motion has been described by Spiess.[34]

5.2.5 General Properties of Spectra in Plastic Phases

The frequency shifts due to the intramolecular dipolar, nuclear electric quadrupole and anisotropic chemical shielding tensors are always averaged to zero by the rapid quasi-isotropic molecular rotation in the plastic phase. Figure 5.10 shows the ^{13}C spectra measured by proton-enhanced nuclear induction spectroscopy in the plastic phases of adamantane and camphor;[35] the spectra are identical to those expected for the liquid giving direct confirmation of the isotropy of motion.

The intermolecular dipolar interaction is only partially averaged by the rotational motion and always dominates the shape of the ^1H or ^{19}F frequency spectrum in this phase. Measurements of the second moment of the rotationally averaged spectrum can provide an accurate value for the lattice parameter through equation (5.2.9). Near to the melting point the intermolecular interaction is averaged by the translational motion and linewidth measurements through the line narrowing region can provide an estimate of the diffusion rate, though this parameter is better measured by spin relaxation measurements.

5.3 NUCLEAR SPIN RELAXATION

5.3.1 Experimental Measurements

Nuclear spin relaxation measurements are particularly suited to studying the rotational and translational diffusion of molecules in both ordered and plastic phases. These experiments yield information on both the rates and mechanisms of such processes. Values of τ may be measured in the range 1 to 10^{-12} s.

The experimentally measurable quantities are the laboratory and rotating frame spin–lattice relaxation times T_1 and $T_{1\rho}$, respectively, the dipolar relaxation time T_{1D} and the spin–spin relaxation time T_2. T_1 is the time constant describing the evolution of the nuclear magnetization **M** along the applied static

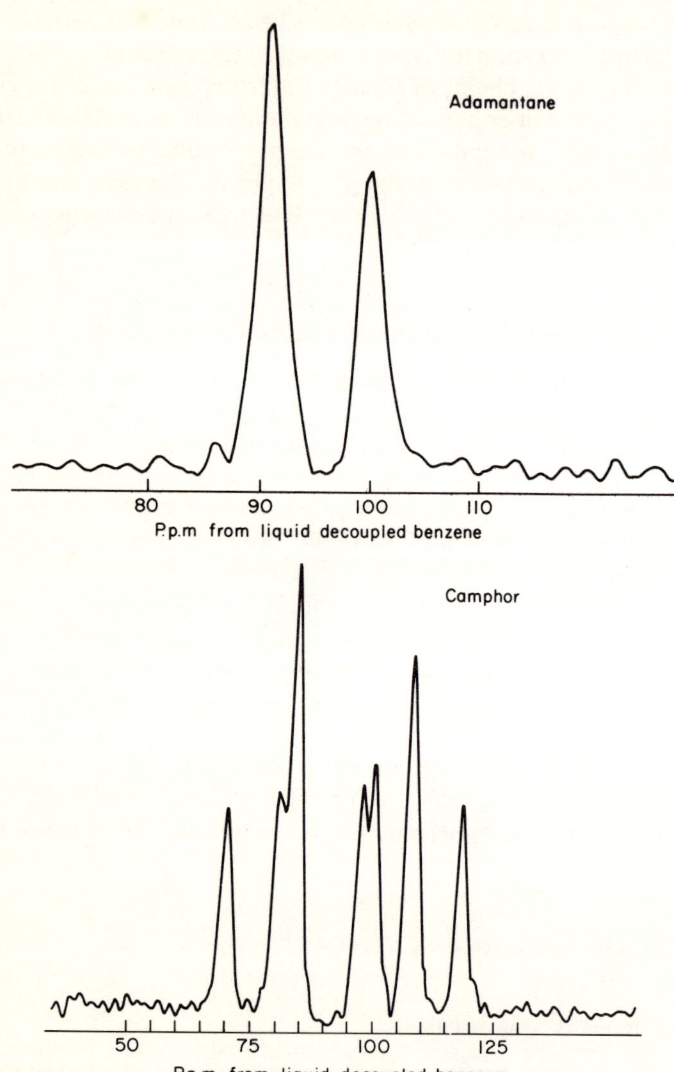

Figure 5.10 ^{13}C spectra of adamantane and camphor in their plastic phases at room temperature. Note that the number of lines in each spectrum corresponds to the number of magnetically inequivalent carbon atoms in the corresponding molecules. (By courtesy of S. Kaplan, Bruker.)

magnetic field **B** due to the relaxation of Zeeman (spin system) to thermal (lattice) energy. $T_{1\rho}$ denotes the spin–lattice relaxation time measured along the field $\mathbf{B}_1$ rotating perpendicularly about **B** at the Larmor precession frequency of the spins. T_2 here describes the decay of **M** in a direction perpendicular to **B** under conditions where the frequency spectrum is Lorentzian; the half-height linewidth and T_2 are thus related by $\Delta\omega_{1/2} = 2/T_2$; equations (5.2.10). T_D^{-1} is a measure of the rate at which dipolar as opposed to Zeeman energy relaxes to the lattice. These relaxation times are measured by pulse techniques described together with the theory of spin relaxation in the standard texts[1,2,3] various reviews[5,6,36–39] and early papers[40,41] on this subject.

5.3.2 Relationship between Measurements and Motional Variables

We now consider the relationship between T_1, $T_{1\rho}$ and T_2 (T_{1D} will be considered later) and the variables describing the molecular motion. Theory[1,2,3] expresses these relaxation times in terms of the Fourier transforms of the correlation functions for the spin-dependent interactions which couple the nuclear spins to the lattice. In plastic crystals, the nuclei which have been most widely studied are ^{1}H and ^{19}F. The relaxation of these is dominated by either rotational or translational modulation of the dipolar interactions. For ^{31}P nuclei there may also be significant contributions to the relaxation from anisotropic chemical shielding and spin rotation interactions whilst the relaxation of ^{2}H and ^{35}Cl nuclei is invariably dominated by their nuclear quadrupole interactions. These interactions are entirely intramolecular in origin and are therefore only affected by molecular rotation. First, relaxation by dipolar interactions will be more or less exclusively considered, the discussion of relaxation by these other interactions will be left until the section on investigations of molecular rotation.

Relaxation measurements are usually carried out in fields **B** and $\mathbf{B}_1$ large compared with the magnitude of the local dipolar fields $\mathbf{B}_L$. The dipolar interaction may therefore be treated as a small time-dependent perturbation on the Zeeman energy; this is the so-called 'high-field' or 'weak-collision' theory first developed by Bloembergen, Purcell, and Pound (BPP)[40] and subsequently, revised and generalized by Kubo and Tomita[41] and Redfield[36]. For a pair of dipolar coupled, like nuclei, of spin I[1,3]

$$\frac{1}{T_1} = \frac{3}{2}\gamma^4\hbar^2 I(I+1)\,\{J^{(1)}(\omega_0) + J^{(2)}(2\omega_0)\} \tag{5.3.1}$$

$$\frac{1}{T_2} = \frac{3}{8}\gamma^4\hbar^2 I(I+1)\,\{J^{(0)}(0) + 10J^{(1)}(\omega_0) + J^{(2)}(2\omega_0)\} \tag{5.3.2}$$

$$\frac{1}{T_{1\rho}} = \frac{3}{8}\gamma^4\hbar^2 I(I+1)\,\{J^{(0)}(2\omega_1) + 10J^{(1)}(\omega_0) + J^{(2)}(2\omega_0)\} \tag{5.3.3}$$

where $\omega_0 = \gamma B_0$ and $\omega_1 = \gamma B_1$ ($\mathbf{B} = \mathbf{k}B_0$; $\mathbf{B}_1 = \mathbf{k}B_1$). Equations (5.3.1) and (5.3.2) were first derived by BPP and require $B_0 \gg B_L$ (in the laboratory frame $B_L^2 = \frac{5}{2} M_2$); while equation (5.3.3) was calculated by Look and Lowe[42] and requires $B_0 \gg B_1 \gg B_L$ (in the rotating frame $B_L^2 = \frac{1}{3} M_2$). The functions $J^{(q)}(\omega)$ are the spectral densities of the correlation functions $G^{(q)}(\tau)$ for the dipolar interactions, i.e.

$$J^{(q)}(\omega) = \int_{-\infty}^{\infty} G^{(q)}(\tau) \exp(\mathrm{i}\omega t)\, \mathrm{d}\tau. \tag{5.3.4}$$

The $G^{(q)}(\tau)$ are, for uncorrelated pair interactions,

$$G^{(q)}(\tau) = \sum_j \langle F_{ij}^{(q)}(t)\, F_{ij}^{(q)*}(t + \tau)\rangle_t \tag{5.3.5}$$

where

$$F_{ij}^{(0)}(t) = r_{ij}^{-3}(1 - 3\cos^2\theta_{ij}) \tag{5.3.6}$$

$$F_{ij}^{(1)}(t) = r_{ij}^{-3}\sin\theta_{ij}\cos\theta_{ij}\exp(\mathrm{i}\phi_{ij}) \tag{5.3.7}$$

$$F_{ij}^{(2)}(t) = r_{ij}^{-3}\sin^2\theta_{ij}\exp(2\mathrm{i}\phi_{ij}). \tag{5.3.8}$$

Here r_{ij}, θ_{ij} and ϕ_{ij} are the spherical coordinates of the vector $\mathbf{r}_{ij}$ from spin i to j in a laboratory coordinate system in which the magnetic field $\mathbf{B}$ is along the z axis. Spin relaxation is determined by the time dependence of these coordinates.

Before any information about molecular motion can be obtained from the relaxation time measurements, $G^{(q)}(\tau)$ must be calculated for a physically plausible model for the motion; this is by far the most difficult problem in any relaxation study. For isotropic small-step rotational diffusion of the spin vector $\mathbf{r}_{ij}$, $G^{(q)}(\tau)$ is a simple exponential function of time and it is easily shown that

$$J^{(1)}(\omega) = \frac{1}{6}J^{(0)}(\omega) = \frac{1}{4}J^{(2)}(\omega) = \frac{4}{15}\frac{1}{r_{ij}^6}\frac{\tau_2}{1 + \omega^2\tau_2^2} \tag{5.3.9}$$

which is the well-known Debye-type spectral density function. τ_2 is the autocorrelation time for the second order spherical harmonics $F_{ij}^{(q)}$ and is usually referred to as the reorientational correlation time. Substituting equations (5.3.9) into equations (5.3.1), (5.3.2) and (5.3.3) and summing over all interacting pairs of spins within the same molecule, assuming they behave independently, i.e. they are uncorrelated, gives

$$\frac{1}{T_1} = \frac{2}{3}M_2\left\{\frac{\tau_2}{1 + \omega_0^2\tau_2^2} + \frac{4\tau_2}{1 + 4\omega_0^2\tau_2^2}\right\} \tag{5.3.10}$$

$$\frac{1}{T_2} = \frac{1}{6}M_2\left\{6\tau_2 + \frac{10\tau_2}{1 + \omega_0^2\tau_2^2} + \frac{4\tau_2}{1 + 4\omega_0^2\tau_2^2}\right\} \tag{5.3.11}$$

$$\frac{1}{T_{1\rho}} = \frac{1}{6} M_2 \left\{ \frac{6\tau_2}{1 + 4\omega_1^2\tau_2^2} + \frac{10\tau_2}{1 + \omega_0^2\tau_2^2} + \frac{4\tau_2}{1 + 4\omega_0^2\tau_2^2} \right\} \tag{5.3.12}$$

where M_2 is the intramolecular second moment, equation (5.2.6), averaged by the rotation of the molecule. Intensity functions similar to equations (5.3.9) are also obtained for a molecule jumping randomly between geometrically equivalent orientations in a solid; the correlation time τ_2 now corresponds to the average lifetime of a molecule in any of its equilibrium orientations. The intramolecular relaxation rates for polycrystalline samples are given, in this case, by equations (5.3.10), (5.3.11), and (5.3.12), but with coefficients fM_2 instead of M_2, where f is the fractional reduction of M_2 effected by molecular reorientation.

T_1, $T_{1\rho}$ and T_2 as given by equations (5.3.10–5.3.12) are plotted as functions of $\omega_0\tau$ in Figure 5.11. The minimum in the value of T_1 corresponds to 14.892 ω_0/M_2 and occurs when $\omega_0\tau = 0.6158$. That for $T_{1\rho}$ is $\cong 4.0\ \omega_1/M_2$ at $\omega_1\tau = 0.5$. Thus, T_1 and $T_{1\rho}$ measurements detect motions with frequencies in the

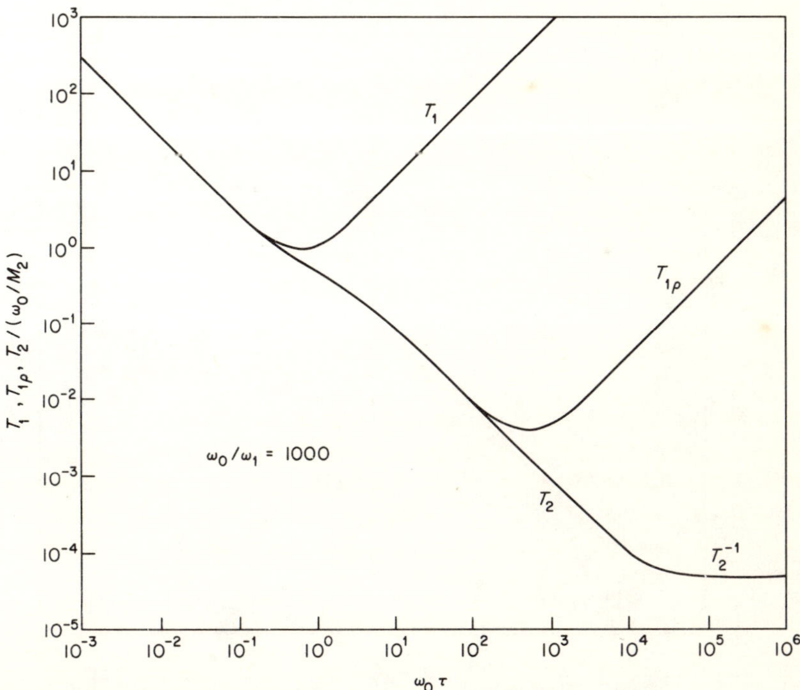

Figure 5.11 Relaxation times T_1, $T_{1\rho}$ and T_2 versus $\omega_0\tau$ calculated for a simple exponential correlation function. The relaxation times are in units of M_2/ω_0 and $\omega_0/\omega_1 = 1000$. The limiting value of $T_2 \equiv T_2^{r1} = 1/M_2^{1/2}$ is taken as 5×10^{-5} s

neighbourhood of ω_0 (10^7 to $10^9\,s^{-1}$) and ω_1 (10^4 to $10^6\,s^{-1}$), respectively. When the temperature dependence of τ_2 is of the form $\tau_2 = \tau_0 \exp(E_a/RT)$, the plot of log T_1 ($T_{1\rho}$ or T_2) versus T^{-1} at a given ω_0 (and ω_1) will be similar in appearance to that in Figure 5.11. Comparison of the theoretical and experimental minima provides a check for the motional model, but should be noted that it is not possible to distinguish between continuous and discrete jump rotational diffusion. Equation (5.3.10) predicts that when $\omega_0\tau_2 \gg 1$, i.e. slow reorientation at low temperatures,

$$1/T_1 \propto 1/\omega_0^2\tau_2$$

and when $\omega_0\tau_2 \ll 1$, i.e., fast reorientation at high temperatures,

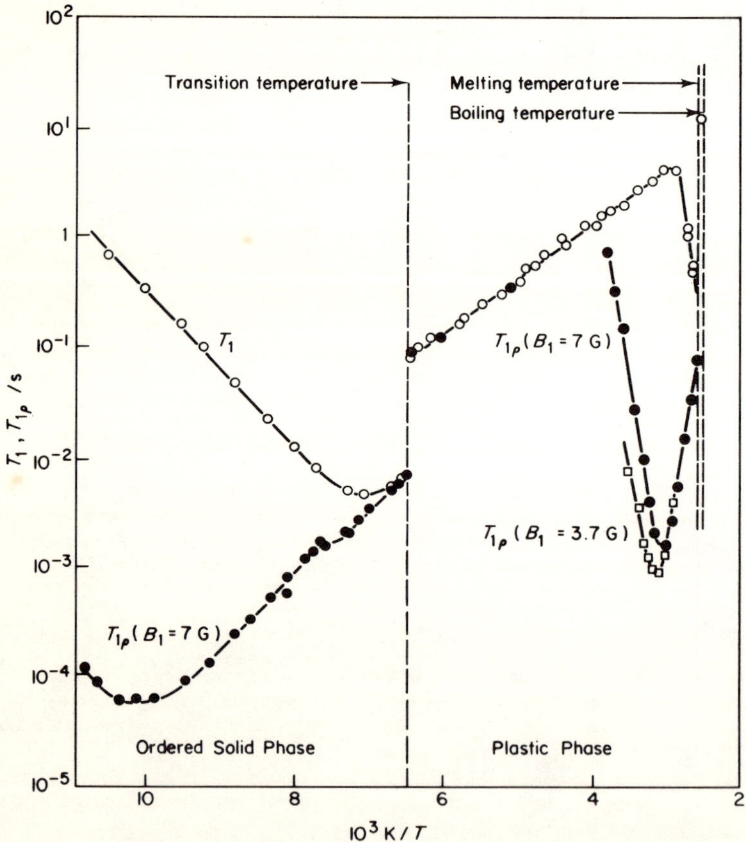

Figure 5.12 Proton relaxation times T_1 and $T_{1\rho}$ as a function of reciprocal temperature in the ordered solid and plastic phases of hexamethylethane.[22] (Reproduced by permission of *Mol. Phys.*)

$$1/T_1 \propto \tau_2.$$

Thus, on the low temperature side of the T_1 minimum, $T_1 \propto \omega_0^2$, but is frequency independent on the high-temperature side. We also note that the slope of the T_1 plot on either side of the minimum corresponds to E_a. The same arguments apply to $T_{1\rho}$ measurements. These features are illustrated by the proton T_1 and $T_{1\rho}$ measurements of Chezeau *et al.*[22] in hexamethylethane (Figure 5.12) which are typical of plastic crystalline materials. These measurements have been confirmed by Albert *et al.*[43] In the low temperature ordered solid phase the measurements are determined by the reorientation of the molecule about its three-fold axis. The minima in both T_1 and $T_{1\rho}$ are realized and the slopes of the plots are the same. In the high temperature plastic phase the T_1 measurements are determined principally by the rapid quasi-isotropic reorientation of the molecules which satisfies the condition $\omega_0\tau_2 \ll 1$. At temperatures just below the melting point, T_1 starts to decrease with increasing temperature. This is due to an increasing contribution from the modulation of the intermolecular dipolar interactions by the relatively slower translational motions of the molecules for which $\omega_0\tau \gg 1$. This holds even at the melting point where τ is typically 10^{-7} to 10^{-6} s. In contrast, $T_{1\rho}$, which is sensitive to spin motions with frequencies of the order $10^5\ \mathrm{s}^{-1}$, is dominated by the translational motions of the molecules near to the melting point, but by the reorientational motions at lower temperatures where $T_{1\rho} = T_1$ (as predicted by equations (5.3.10) and (5.3.12)). Thus, the sensitivity of relaxation measurements to frequencies of the order of ω_0 or ω_1 is very important since it enables processes occurring at different rates to be independently studied.

In the application of equations (5.3.10), (5.3.11) and (5.3.12) to the study of molecular rotation the following points must be considered. First, the Debye-type spectrum, equation (5.3.9), is a consequence of the choice of an exponential correlation function and pertains to the reorientation of a molecule between geometrically equivalent orientations in a solid or isotropic small-step rotational diffusion which is more appropriate to viscous liquids than solids. On the other hand, for jumps between inequivalent sites, i.e. sites of differing potential energies such as might be experienced for an asymmetric molecule in a disordered solid, the correlation function will be non-exponential and equations (5.3.10)–(5.3.12) are inappropriate. Non-exponential correlation functions are also encountered for the translational motion of spins on a crystal lattice. In these cases simple analytic expressions for T_1, $T_{1\rho}$ and T_2 as a function of τ are not usually obtainable; the relaxation times must be calculated numerically as a function of $\omega\tau$ using correlation functions computed for physically plausible models. Comparison of the measurements of T_1 as a function of temperature and frequency with the predictions of the model can in some cases lead to a characterization of the motional process and the determination of τ.

Second, we must examine the assumption that all interacting spin pairs may be treated independently. This is fundamentally incorrect as the relative motions of the various pairs of interacting spins attached to a rigid molecular frame will be correlated. Such cross-correlations in the dipole–dipole interactions give rise, at least in principle, to non-exponential relaxation[44]. In practice, such effects have only been observed for H^1 relaxation in relatively isolated methyl groups; isolation is necessary because intra- and inter-group or intermolecular interactions are uncorrelated. It is, therefore, an adequate approximation to treat the spin interactions idenpendently as assumed in the derivation of equations (5.3.10), (5.3.11) and (5.3.12).

Third, so far only intramolecular interactions have been considered. The calculation of the relaxation rate due to intermolecular interactions is far more difficult; the crystal structure as well as the rotation mechanism must be taken into account. The calculation is not often attempted for organic solids. A procedure which is frequently employed is to assume that only a single correlation time τ_2 is involved and that the total T_1 is given by

$$\left(\frac{1}{T_1}\right)_{\text{obs}} = C\left(\frac{\tau_2}{1 + \omega_0^2\tau_2^2} + \frac{4\tau_2}{1 + 4\omega_0^2\tau_2^2}\right). \qquad (5.3.13)$$

A value for the proportionality constant C is obtained from the experimental T_1 minimum. If the value obtained for C corresponds to $\frac{2}{3} f M_{2,\text{intra}}$ as calculated from equation (5.2.7) then equation (5.3.10) is exactly applicable. A value of $C > \frac{2}{3} f M_{2,\text{intra}}$ is indicative of significant contributions from intermolecular dipolar interactions. For correlated molecular reorientation, only a single correlation time τ_2 is involved and these equations are still valid. On the other hand, for random independent reorientational jumps—single molecule rotation—the intermolecular interaction fluctuates at $\tau_2/2$ in addition to τ_2; this is due to the fact that the interaction between a pair of spins on neighbouring molecules changes in magnitude twice during the time τ_2 since both molecules may jump. In this case the position of the T_1 minimum will occur at a value between 0.62 and 1.24 depending upon the relative importance of the two contributions. The contribution which fluctuates at $\tau_2/2$ is more important when $\omega_0\tau \gg 1$ than when $\omega_0\tau \ll 1$. It should, therefore, be possible, at least in principle, to use T_1 measurement to distinguish between these two rotational models. These effects are, however, very small and it is usually a good approximation to assume that only a single correlation time is involved and to use equation (5.3.13) to obtain values for τ_2 as a function of temperature. Noack *et al.*[45] have recently shown using T_1 measurements that correlated molecular rotation occurs in solid benzene.

In some crystals, molecules are located at sites of different symmetries and as a consequence may rotate at quite different frequencies and have distinctly different relaxation times. In most solids, the Zeeman energy is transmitted throughout the entire (proton) spin system by the process of 'spin-diffusion' at a rate much greater than the relaxation rates of these different spin subsystems. The total spin system is therefore maintained in internal equilibrium and relaxes exponential to the lattice with a time constant given by

$$\left(\frac{1}{T_1}\right)_{\text{obs}} = \sum_j a_j (1/T_1)_j \tag{5.3.14}$$

where a_j is the fractional population of the jth spin subsystem. Thus, when the relaxation rate of one of the spin subsystems is much greater than any of the others, it effectively acts as a relaxation sink for the entire spin systems and determines the observed T_1. The reorientation of the molecules in hexafluorobenzene[46] provide a good example of this behaviour. An analogous situation is encountered in organic solids containing methyl groups. At low temperatures the rotating methyl groups invariably control the observed relaxation times. In the ordered phase of *d*-camphor the log T_1 versus T^{-1} plot exhibits a minimum due to the internal rotation of the methyl groups, yet overall molecular rotation is quenched in this phase[25]. Other examples are the normal alkanes[47] and polymers.[48]

In disordered molecular crystals continuous distributions of rotational correlation times may be encountered. This will be indicated by the slope of the log T_1 (or $T_{1\rho}$) versus T^{-1}plot. The minimum will be shallower, the ratio $T_1 : T_{1\rho}$ at their respective minima smaller, and the width broader than predicted by equation (5.3.10) or (5.3.12); moreover, the slope of the plot will reflect the shape of the distribution. Information about the nature of the distribution may be obtained by generalizing equation (5.3.10) or (5.3.12) as follows

$$\frac{1}{T_1} = \frac{2}{3} M_2 \left(\int_0^\infty \frac{\tau D(\tau) d\tau}{1 + \omega_0^2 \tau^2} + \int_0^\infty \frac{4\tau D(\tau) d\tau}{1 + 4\omega_0^2 \tau^2} \right) \tag{5.3.15}$$

where $D(\tau)$ is defined

$$\int_0^\infty D(\tau) d\tau = 1.$$

The choice of $D(\tau)$ is entirely empirical; expressions for $1/T_1$ for a wide variety of distribution functions are given in the review by Noack[37]. Discrete distributions of correlation times are frequently encountered in molecular crystals; for example in hexafluorobenzene there are two distinct lattice sites at which the molecules re-orientate with characteristic frequencies.[46]

5.3.3 Spin Relaxation in Low Fields and Dipolar Relaxation

$T_{1\rho}$ measurements detect molecular motions with frequencies of the order ω_1. By reducing the amplitude of B_1, slower motions may be detected, but when $B_1 \cong B_L$ the high-field theory breaks down. Slichter and Ailion[49] have developed a theory for spin relaxation in the strong-collision region $B_1 \leqq B_L$, $T_2^{rl} \ll \tau \ll T_1$ which gives

$$\frac{1}{T_{1\rho}} = \frac{1}{T_{1D}} \frac{B_L^2}{B_1^2 + B_L^2} \tag{5.3.16}$$

where

$$\frac{1}{T_{1D}} = \frac{2(1-p)}{\tau} \tag{5.3.17}$$

is the dipolar relaxation rate which may be directly measured.[10,38] The function $(1 - p)$ is a measure of the fractional change in the dipolar energy due to a single rotational or translational jump; it is easily calculated for a given type of motion and crystal structure. Typically, $(1 - p) \cong 1$ so that $T_{1D} \cong \tau$. The above theory breaks down for $\tau \leqq T_2^{rl}$, here, T_2 is the rigid lattice T_2. When $\omega_0\tau \ll 1$

$$\frac{1}{T_{1D}} = \frac{2}{T_{1\rho}} = \frac{2}{T_1}$$

for uncorrelated motion of the interaction spins and

$$\frac{1}{T_{1D}} = \frac{3}{T_{1\rho}} = \frac{3}{T_1}$$

for correlated motion [3]. Thus, we see that T_{1D} passes through a minimum when $\tau \cong T_2^{rl}$ though, in this region, there is no simple relationship between T_{1D} and τ.[50] The theory and experimental techniques used to study very slow motions in low fields are described in the review by Ailion.[5] Moreover, the typical ranges of motion accessible by the various relaxation measurements are: T_{1D}, 10 to 10^{-4} s; $T_{1\rho}$, 10^{-2} to 10^{-7} s; T_1, 10^{-6} to 10^{-12} s. The application of these techniques to the study of the rotational and translational motions in the ordered and plastic phases of molecular crystals is reviewed in the following sections.

5.3.4 Studies of Molecular Reorientation

5.3.4(a) Ordered Phases

(i) Molecules with Tetrahedral or Octahedral Symmetry. Studies of frequency spectra have shown that molecules with tetrahedral (e.g. adamantane[18] and P_4[32]) or octahedral (e.g. MoF_6[51]) symmetry undergo quasi-isotropic rotation in the solid below the phase transition. Their log T_1 versus T^{-1} plots are very similar in appearance to those for hexamethylethane (Figure 5.12). In white phosphorus there are significant contributions from the anisotropic chemical

shielding interaction in addition to the dipolar interaction and this complicates the observed frequency dependence of T_1[52]. The values of τ_2 obtained for these spherically symmetric molecules by T_1 measurements are summarized in Table 5.1. The most notable feature is that at the solid–plastic transition $\tau_2 \cong 5 \times 10^{-9}$ s. In going to the plastic phase τ_2 decreases by roughly an order of magnitude. It is interesting to note that hexamethylenetetramine, which is an ordered body-centred cubic solid, has the value 4×10^{-9} s for τ_2 at its melting point.

(ii) Molecules with an Axis of Symmetry. In hexamethylethane the proton T_1 and $T_{1\rho}$ measurements (Figure 5.12) are determined by correlated reorientation of the methyl and *t*-butyl groups[43,53]. In contrast, in hexamethyldisilane the methyl and $(CH_3)_3Si$ groups reorientate independently with respective activation energies of 6.5 and 31.4 kJ mol^{-1}.[53] The latter motion dominates $T_{1\rho}$ while the fast methyl group rotation dominates T_1, thus enabling the two motions to be independently studied. The highly hindered rotation of the methyl groups in hexamethylethane is somewhat anomalous; in most organic solids the barriers hindering methyl group rotation are small and quantum mechanical tunnelling affects the spin relaxation measurements at low temperatures.[6] Triethylenediamine and *t*-butylchloride also undergo axial rotation and the values of τ_2, given in Table 5.1, have been obtained, respectively, from measurements of the ^{14}N pure NQR T_1[54] and from the NMR T_1 for ^2H and ^{35}Cl[30]. At the transition temperature, τ_2 for axial reorentation is of the same magnitude as observed for spherically symmetric molecules.

Somewhat limited proton T_1[23], $T_{1\rho}$[55] and T_{1D}[56] measurements have been reported for cyclohexane. The phase transition now occurs before the T_1 minimum is reached and the $T_{1\rho}$ minimum is asymmetric and indicative of a distribution of correlation times. The ^{19}F T_1[24] and $T_{1\rho}$[55] measurements for perfluorocyclohexane exhibit similar, though more extreme, behaviour to cyclo-

Table 5.1 Kinetic parameters for molecular reorientation in ordered solid phases

	τ_0/s	E_a/kJ mol^{-1}	$\tau_{2,t}$/s^a	Reference
Adamantane	1.1×10^{-15}	27.2	7.1×10^{-9}	58
White phosphorus	1.5×10^{-13}	15.7	2.3×10^{-9}	32, 52, 126
Molybdenum hexafluoride	1.3×10^{-18}	48.2	5.0×10^{-9}	51, 127
Hexamethylene-tetramine	1.2×10^{-16}	80.7	4.3×10^{-9}	58
Hexamethylethane	1.4×10^{-13}	13.4	5.6×10^{-9}	22, 43
Hexamethyldisilane	1.2×10^{-15}	31.4	2.9×10^{-8}	43, 101
Triethylendiamine	7.4×10^{-15}	34.2	9.1×10^{-10}	54
t-butylchloride	7.5×10^{-12}	11.7	4.7×10^{-9}	30

^{a}Values of τ_2 at the solid–plastic transition temperature calculated from $\tau_2 = \tau_0 \exp(E_a/RT)$.

hexane. Clearly, the orientational disordering process sets in well below the transition temperature in these two materials. More extensive studies might reveal motional precursors to the phase transition in other solids too.

(iii) Asymmetric Molecules. The rotational motion of these molecules is generally quenched below the disordering transition. The internal rotation of methyl groups, when present, will dominate the proton spin relaxation in these materials, as for example in *d*-camphor[25].

5.3.4(b) Plastic Phase

(i) Spin Relaxation by Dipolar Interactions. In the plastic phase, the molecules undergo fast quasi-isotropic rotation such that $\omega_0\tau \ll 1$, irrespective of their structure. The notable exception is pivalic acid where intermolecular hydrogen-bonding restricts the rotational, but, interestingly, not the translational motion[57]. The information obtainable from nuclear spin relaxation measurements is, therefore, severely restricted as for liquids, but investigations are simpler in plastic solids where the reorientational and translational diffusion rates differ substantially enabling their contributions to the observed relaxation rates to be unambiguously separated.

In the limit $\omega_0\tau \ll 1$, equation (5.3.10) reduces to

$$\frac{1}{T_1} = C\tau_2. \tag{5.3.18}$$

We see that the relaxation measurements are now directly related to τ_2, i.e., they depend on the area under the rotational correlation function[28]

$$\tau_2 = \int_0^\infty G(\tau)\mathrm{d}\tau, \tag{5.3.19}$$

provided that $\lim_{\tau\to\infty} G(\tau) = 0$, but are insensitive to its actual shape. Thus, τ_2 may be determined provided we know C.

Resing[58] has studied the reorientational motion in both the solid (tetragonal) and plastic face-centred cubic phase of adamantane. Using a value for C derived from the T_1 minimum in the solid phase, values for τ_2 in the plastic phase have been deduced; this is a reasonable assumption since, as we have already seen, the value of M_2 is unchanged at the transition. In both phases, τ_2 exhibits an Arrhenius temperature dependence; in the solid phase, $\tau_2 = 1.1 \times 10^{-15}$ s exp(27.2 kJ mol$^{-1}/RT$), while in the plastic phase, between 208 and 500 K $\tau_2 = 9.4 \times 10^{-14}$ s exp(12.9 kJ mol$^{-1}/RT$). At the phase transition, there is an order of magnitude decrease in the value of τ_2 and a considerable change in the reorientational activation parameters, yet the value of M_2 is unchanged. This latter observation implies that the packing density is similar in the two phases; indeed, in going from the tetragonal to the f.c.c. phase, the arrangement of the molecules is unchanged except for a 90° tilt about the C axis.[59] The change in the rotational kinetics must reflect the accompanying change in the interaction of the molecule with its environment. At the melting point, 542 K, τ_2 is estimated as 1.6×10^{-12} s and is getting close to the value of the correlation time for classical free rotation $(I/kT)^{1/2}$, which is 0.8×10^{-12} s. This means that, on average, one half of the nearest-neighbour molecules must be simultaneously rotating. It is not, however, appropriate to consider this as evidence for collision modulated free rotation since in plastic crystals the nearest neighbour separation is always smaller than a molecular diameter[60]. Correlated molecular reorientation is a more plausible explanation. Values of τ_2 have not been measured for many other organic plastic solids, though the reorientational contribution to T_1 for protons is typically 10–20 s at the melting point as compared to 16 s for adamantane. A value of $\tau_2 \cong 2 \times 10^{-12}$ s at the melting point, would therefore seem to be a characteristic property of plastic organic solids. The notable exception is pivalic acid where at the melting point $\tau_2 \simeq 5 \times 10^{-9}$ s; in this case molecular reorientation involves the breaking of hydrogen-bonded dimers with an activation energy of 36 kJ mol^{-1}.[57] Table 5.2 summarizes values for the activation energies for molecular rotation in plastic solids as obtained from the slopes of the log T_1 versus T^{-1} plots. This procedure

Table 5.2 Activation energies for molecular reorientation in plastic crystals

Material	T_t/K	T_m/K	$\Delta S_f/R$	E_a/kJ mol^{-1}	Reference
Adamantane	209	542	2.51	12.9	58
Hexamethylethane	153	374	2.41	9.2	22, 43
Perfluorocyclo-hexane	168	336	2.30	5.0 (168–270) 1.8 (270–336)	91
Methylchloroform	224	243	2.26	18.8	128
3-Azabicyclononane	298	467	1.78	28.5	96
Norbornane	131	360	1.53	non.Arrhenius	113
Neopentane	140	257	1.46	4.2	128
d-Camphor	238	449	1.40	11.7	25
Succinonitrile	236	331	1.35	25.1	129
Hexamethyldisilane	222	288	1.25	6.3	43
Norbornene	129	320	1.22	11.2 (129–146) 7.1 (146–250)	61
dl-Camphene	175	323	1.15	4.9 (175–220) 3.5 (220–323)	61
Cyclohexane	186	280	1.10	6.40	76
t-Butylchloride	183	245	0.95	3.0	30
White phosphorus	196	317	0.95	5.7	70
Phosphine	88	139	0.90	1.30	71
Norbornadiene	202	254	0.79	7.9	113
Pivalic acid	280	310	0.78	36	51

is possible even when absolute values of τ_2 cannot be confidently determined since $1/T_1 \propto \tau_2$.

There are some solids for which the log T_1 versus T^{-1} plots are distinctly non-linear. For *dl*-camphene (b.c.c.), norbornene (h.c.p.) and perfluorocyclohexane (f.c.c.), these plots exhibit two linear regions with an abrupt change from one to the other as shown in Figure 5.13 for *dl*-camphene where this transition occurs at 220 K.[61] This behaviour must be due to an orientational disordering phase transition as there is no change in either the crystal structure or density at these temperatures. In *dl*-camphene, there is a significant change in molecular scattering factor between 208 and 297 K showing an increase in the spherical symmetry of the molecular packing.[62] A fundamentally different behaviour is observed in the h.c.p. phase of norbornene: the apparent activation energy decreases continuously with temperature indicating a progressive increase in

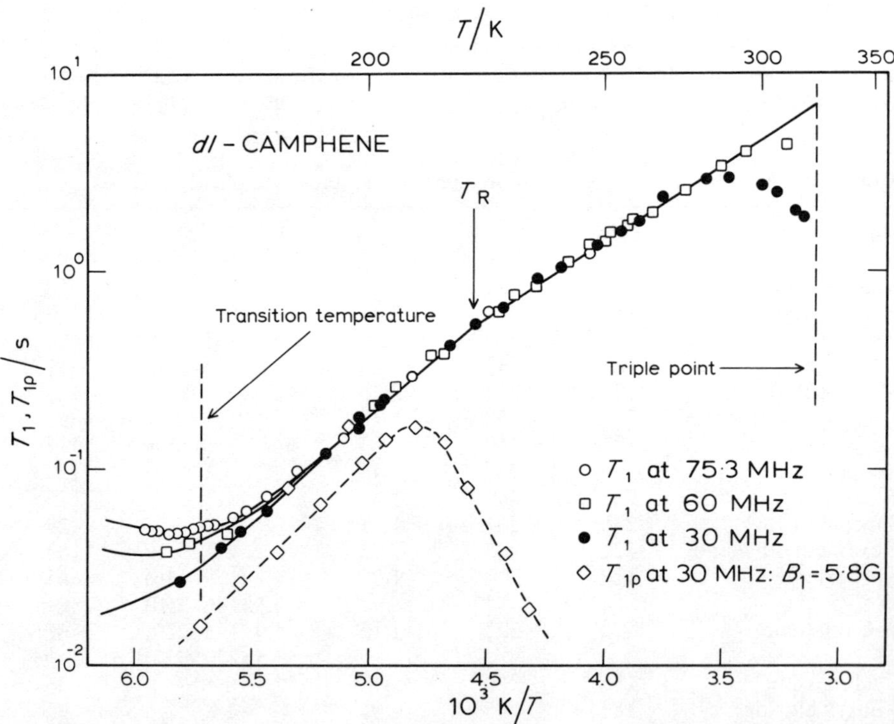

Figure 5.13 Proton relaxation times T_1 and $T_{1\rho}$ as a function of reciprocal temperature in the plastic phase of *dl*-camphene. (N. Boden, S. Hanlon, M. Mortimer, and S. Ross, unpublished results.)

the orientational disorder. The T_1 measurements fit

$$\frac{1}{T_1} = A \exp\{E_a/(T - T_0)\}$$

with $A = 1.62 \times 10^{-2}\,\mathrm{s}^{-1}$, $E_a = 2.86\,\mathrm{kJ\,mol}^{-1}$ and $T_0 = 65.5$ K. This behaviour is reminiscent of viscous liquids and it is interesting to note that $T_0 = T_t/2$ the anticipated ideal glass transition temperature for the plastic phase[63]. It is no doubt significant that *dl*-camphene, norbornane and norbornene are all camphor type molecules. The log T_1 versus T^{-1} plots for *d*-camphor and *dl*-camphor[25] are indistinguishable in the plastic phase and appear to be non-linear though the measurements are not sufficiently defined to draw any definite conclusions. Dielectric measurements[64] do indicate a non-Arrhenius temperature dependence for the rotational motion in camphor.

The activation volumes for molecular reorientation have been determined from measurements of proton T_1 as a function of pressure and are typically 0.05–0.09 V_m in wide variety of plastic organic solids[65–67]. These values are small, but consistent with the relatively unhindered reorientational motion of the molecules. In the high ΔS_f materials adamantane and hexamethylethane, the activation volumes increase linearly with temperature while in the low ΔS_f solids norbornane, cyclohexane, and hexamethyldisilane it decreases. Thus, there appears to be a fundamental difference in the activation parameters describing molecular reorientation in the high and low ΔS_f solids; similar differences in behaviour are observed for translational diffusion as will be described in the following section.

(ii) Relaxation by Nuclear Electric Quadrupole, Anisotropic Chemical Shielding and Spin Rotation Interactions. Neither ^{1}H or ^{19}F relaxation measurements are ideally suited to studies of molecular rotation in plastic crystals since unambiguous values for τ_2 are not easily determined due to the difficulties of separating the intra- and inter-molecular dipole–dipole interactions and the multiplicity of spin–spin vectors involved in complex organic molecules. It is preferable, where possible, to study relaxation due to either nuclear electric quadrupole or anisotropic chemical shielding interactions.

The spin–lattice relaxation rate due to modulation of the electric quadrupole interaction of a spin $I > \frac{1}{2}$ is given, in the limit $\omega_0\tau \ll 1$, by[1]

$$\frac{1}{T_1} = \frac{3}{40}\,\frac{2I + 3}{I^2(2I - 1)}\left(1 + \frac{1}{3}\eta^2\right)\left(\frac{e^2qQ}{\hbar}\right)^2 \tau_2 \tag{5.3.20}$$

where $e^2qQ/\hbar$ is the electric quadrupole coupling constant and η is the asymmetry parameter.

The spin–lattice relaxation rate due to the anisotropic chemical shielding interaction for a spin in a molecule with tetrahedral or octahedral symmetry is, for $\omega_0\tau_2 \ll 1$ and assuming the chemical shielding tensor to be diagonal in the the molecular frame, given by[1]

$$\frac{1}{T_1} = \frac{2}{15}(\sigma_\parallel - \sigma_\perp)^2(\gamma B_0)^2\tau_2 \tag{5.3.21}$$

where $\sigma_\parallel$ and $\sigma_\perp$ are the principal components of the chemical shielding tensor. The values of τ_2 in equations (5.3.20) and (5.3.21) are not necessarily the same, they will depend on the orientation of the principal axes systems for the two tensors with respect to the molecular axes. For molecules with tetrahedral or octahedral symmetry these axes systems may coincide, then the two correlation times will be identical. The value of τ_2 determined from dipolar relaxation measurements will also depend on the orientation of the spin–spin vector relative to the molecular axes system; for this reason it is desirable to study relaxation due to selected pairs of spins. For organic molecules it would be useful to measure relaxation times of ^{13}C nuclei since their relaxation is usually dominated by dipolar interaction with the directly bonded protons.

The spin–rotation interaction has also been utilized in studies of the rotation of small molecules in plastic phases. For a spin $I = \frac{1}{2}$ in a spherical top molecule, when $\omega_0\tau_2 \ll 1$[68]

$$\frac{1}{T_1} = \frac{2\overline{I}kT}{\hbar^2}\overline{C}^2\tau_J \tag{5.3.22}$$

where the spin-rotation constant $\overline{C}$ is given by

$$\overline{C} = \tfrac{1}{3}(C_\parallel + 2C_\perp) \tag{5.3.23}$$

and the moment of inertia $\overline{I}$ by

$$\overline{I} = \tfrac{1}{3}(I_\parallel + 2I_\perp). \tag{5.3.24}$$

τ_J is the angular momentum correlation time which characterizes the fluctuations in the angular momentum of the molecule. Nuclear spin relaxation measurements are the only source of values for τ_J. Their importance lies in their use in conjunction with values of τ_2 to test the applicability of theoretical models for molecular rotation[69].

When a combination of the above mechanisms are operative, it is usually possible to separate their contributions to the measured relaxation rates, so that, provided independent values for the spin interaction constants are available, unambiguous values for τ_2 or τ_J can be obtained. Studies of this kind have mainly been restricted to plastic phases of inorganic materials; these include, P_4,[70] PH_3/PD_3,[71] HCl/DCl,[72] $(CD_3)_3CCl$.[30]

The ^{31}P spin–lattice relaxation times measured in the plastic phase of white phosphorus[70] exhibit a complex frequency and temperature dependence, as illustrated in Figure 5.14. Compare the behaviour of the proton T_1 measure-

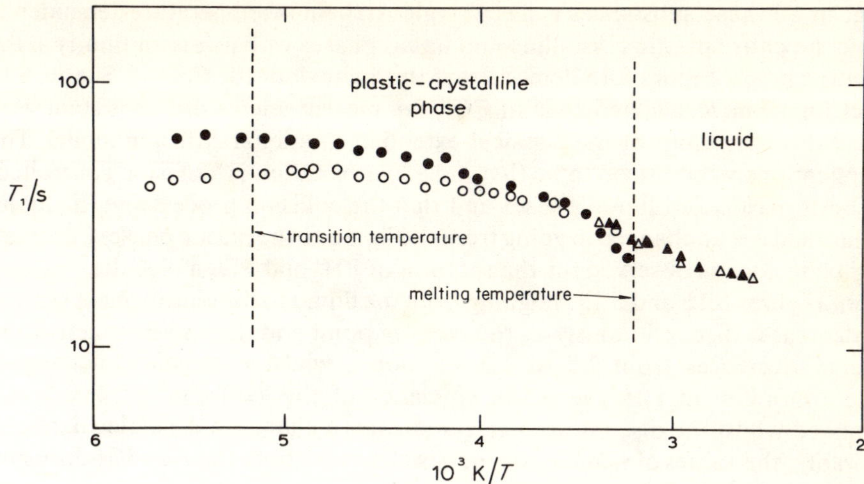

Figure 5.14 ^{31}P spin–lattice relaxation times T_1 as a function of reciprocal temperature in the plastic crystalline and liquid phases of white phosphorous: ●, 0 solid at 10 and 30 MHz respectively; △, ▲ liquid at 10 and 30 MHz, respectively.[70] (Reproduced by permission of *Chem. Phys. Letters*)

ments for hexamethylethane as shown in Figure 5.12. The frequency dependence, except that just below the melting point where translational diffusion affects T_1, is due to the anisotropic chemical shielding interaction: using for $\sigma_\parallel - \sigma_\perp$ in equation (5.3.21) the value 405 p.p.m.,[32] gives $\tau_2 = 2.4 \times 10^{-12}$ s $\exp(5.7 \text{ kJ mol}^{-1}/RT)$. At the melting point, 317.4 K, $\tau_2 = 2.0 \times 10^{-12}$ s which is only a factor of three greater than the value of 7.6×10^{-13} s for $(\bar{I}/kT)^{1/2}$, behaviour identical to adamantane. The decrease of T_1 with temperature is characteristic of spin–rotation relaxation. The values of τ_J obtained from the spin-rotation relaxation rates are not consistent with classical rotational diffusion models[69]. For fast, uncorrelated jumps between discrete potential wells[70]

$$\frac{1}{T_1}_{\text{SR}} = \frac{1}{3}\left(\frac{\bar{I}\bar{C}\langle\theta^2\rangle^{1/2}}{\hbar}\right)^2 \frac{1}{\tau_2} \tag{5.3.25}$$

where $\langle\theta^2\rangle^{1/2}$ is the root-mean-square jump angle. Using for $\bar{C}$, the value of 1.99 kHz, calculated from the chemical shift, gives a value of 93° for $\langle\theta^2\rangle^{1/2}$ suggesting the P_4 molecules undergo large angle reorientational jumps over 90° about their D_{2d} axes.

Figure 5.14 also shows that there is no discontinuity in T_1 in going from the plastic to the liquid phase, indicative of a similar mechanism for rotation in both phases. Similar behaviour is observed for other spherical or approximately spherical molecules. Studies of Raman band shapes in CCl_4[73,74] and C_6H_{12}[73] and ^2H spin–lattice relaxation in CD_4[75], DCl[72], C_6D_{12}[76], and $(CD_3)_3CCl$[30] show that in all these substances τ_2 has a single Arrhenius temperature dependence over the entire plastic crystalline and liquid phases with no discontinuity at the melting point. Figure 5.15 demonstrate this behaviour for C_6D_{12}[76]. Sunder and McClung[74] have claimed that in CCl_4 the measurements are consistent with the J-diffusion limit of the classical extended rotational diffusion model. The implications would seem to be that molecular rotation is 'collision controlled' in both plastic and liquid phases and that the collision process and the molecular field are unchanged in going from the liquid to the plastic phase. Contrasting behaviour is observed for the rotation of PH_3 and PD_3 molecules[71] as seen from Figures 5.16 and 5.17. In going from the liquid to the plastic phase (f.c.c.), τ_2 decreases discontinuously at the melting point and the apparent activation energy decreases from 2.5 to 1.3 kJ mol^{-1}, whilst τ_J remains unchanged. The continuity in τ_J suggests the efficiency of the 'collision' process is unchanged whilst the discontinuity in τ_2 indicates a change in the molecular field. Notably, the values of τ_2 and τ_J are inconsistent with both the *J*- and *M*-diffusion models.

Inspection of the available data for τ_2 shows, that for molecules such as adamantane, P_4, C_6H_{12}, CCl_4 and *t*-butyl chloride $\tau_2 \cong 2 \times 10^{-12}$ s at T_m, whilst for the much lighter molecules CH_4 and HCl the corresponding values are an order of magnitude shorter. It is probably more significant, however, to compare the values for τ_2/τ_f, where $\tau_f = (\bar{I}/kT)^{1/2}$; these are approximately 2 for the former group and 1 for the latter. Thus, the difference between the two groups of molecules is not as marked as suggested by the absolute values of τ_2.

Brot and Lassier–Govers have recently reviewed[77] the various types of

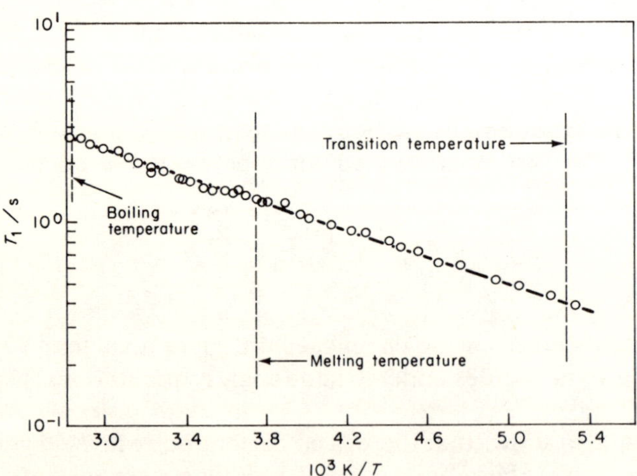

Figure 5.15 ^{2}H spin–lattice relaxation times in the plastic crystalline and liquid phases of C_6D_{12}. In the temperature interval 188–357 K; $\tau_2 = 9.8 \times 10^{-14}$ s exp(6.40 kJ mol^{-1}/*RT*). (Reproduced by permission of *J. Chem. Phys.*)

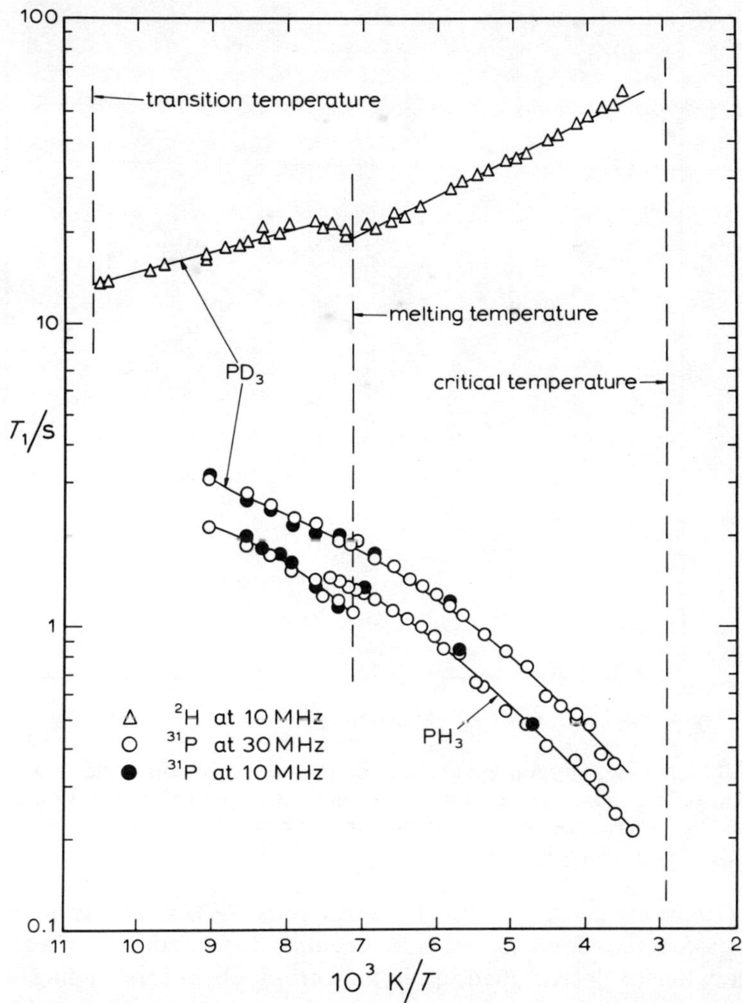

Figure 5.16 ^{2}H and ^{31}P spin–lattice relaxation times T_1 as a function of reciprocal temperature in the plastic crystalline and liquid phases of PH_3 and PD_3. The plastic crystalline phase transition temperature for PD_3, 93.5 K, is significantly higher than for PH_3, 88.2 K.[70] (Reproduced by permission of *Chem. Phys. Letters*)

models which might be applicable to rotational motion in plastic crystals, including those developed for liquids.[69] They forcefully argue that collision models are physically unacceptable for plastic crystals and that discrete site jump models are more appropriate. It is difficult to envisage how such models could be applicable near the melting point where $\tau_2 \cong \tau_f$ except by postulating that the rotation of neighbouring molecules is correlated.

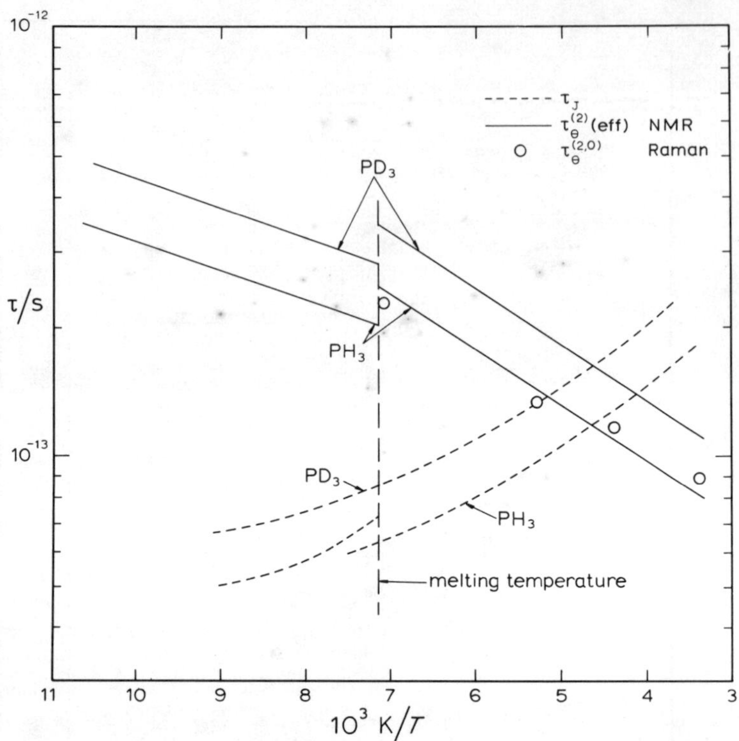

Figure 5.17 Reorientational and angular momentum correlation times as a function of reciprocal temperature in the plastic crystalline and liquid phases of PH_3 and PD_3.[70] (Reproduced by permission of *Chem. Phys. Letters*)

(iii) Comments on the Utility of Nuclear Spin Relaxation Measurements for Studying Molecular Rotation in Plastic Crystals. Measurements of nuclear spin relaxation due to electric quadrupole, anisotropic chemical shielding or dipolar interaction between well defined pairs of spins are particularly suited to studies of the reorientational motion of molecules in crystals. Specifically, they provide values for the correlation times $\tau_{2,m}$ which characterize the fluctuations in the spherical harmonics $Y_{2,m}(\Omega)$ where Ω describes the orientation of a vector fixed in a molecule with respect to a space fixed coordinate system. Relaxation measurements of selective spins, e.g. deuterium atoms substituted at selected sites in organic molecules, may be used to investigate the anisotropy of the rotation. Furthermore, measurement of relaxation due to the spin-rotation interaction for spins such as ^{31}P provide values for τ_J which cannot be obtained by any other technique. The interpretation of spin relaxation due to rotational modulation of intermolecular dipolar interactions is far more complex. It could, in principle, be used to distinguish between correlated and uncorrelated

molecular rotation, an aspect of the motion we know very little about. This would require measurements about the T_1 minimum which could become possible, with the development of magnets with higher field strength. A more promising attack on this problem is, however, by comparison of the NMR τ_2, which refers to the average single molecule, with values of τ_2 as measured by Rayleigh light scattering which is determined by a collective reorientational correlation function. Indeed, the unambiguous characterization of the rotational motion in the plastic phase will require the comparison of the correlation times as measured by as many different techniques as possible; these include dielectric and infra-red absorption, slow-neutron scattering, Rayleigh and Raman scattering and of course, NMR relaxation. The use of these techniques in the quest for information about molecular motion in solids has recently been described by Powles[78] and is also discussed in the article by Brot and Lassier-Govers.[77]

Any model for molecular rotation in the plastic phase must take account of the molecular disorder. NMR does not give any direct information about this property, at best inferences may be drawn from the kinetics of molecular rotation. X-ray and neutron diffraction measurements must be employed to this end.

5.3.5 Translational Diffusion

The defect structure of the plastic crystalline state is not yet well understood at least in comparison with that of metals. Self-diffusion data are an important source of information about the structure and properties of the point defects. Nuclear spin relaxation measurements provide a particularly attractive method for the investigation of self-diffusion in plastic solids because polycrystalline samples may be used hence avoiding the difficulties of handling the well-defined single crystals required for radiotracer measurements (see Chapter 2). On the other hand, the relation of the measured relaxation rates to the mean time τ between successive jumps of a molecule at a lattice site is dependent on the model chosen for the microscopic molecular motion and the development of realistic models is mathematically difficult. The value of τ obtained may, however, be checked as it is related to the self-diffusion coefficient D through the Einstein relationship

$$D = \langle r^2 \rangle / 6\tau \tag{5.3.26}$$

where $\langle r^2 \rangle$ is the mean-square-jump distance. Provided an appropriate model is employed and there is no ambiguity in the choice of value for $\langle r^2 \rangle$, the value of D thus obtained should be in agreement with the value determined directly from radiotracer measurements.

In this section we will be concerned with examining the reliability of the NMR method for the measurement of D and assessing the implications of the measurements to date for the structure of the point defects in plastic solids. Comparison with radiotracer measurements, where possible, will be made.

The difference in the rotational and translational diffusion rates, typically five orders of magnitude at the melting point, enables the contributions from these two processes to the observed spin relaxation rates to be unambiguously separated as has already been mentioned. We will assume this has been done and, henceforth, will be solely concerned with relaxation arising from modulation of the intermolecular dipolar interactions by self-diffusion of the molecules on the crystal lattice. Furthermore, since the molecules are rotating much faster than they are diffusing, all the spins in a molecule may be considered to be equivalent and to be located at its centre. This assumption is necessary: it enables us to use theories developed for the translational motion of single spins on monatomic crystal lattices to extract values for τ from the experimental measurements.

5.3.5(a) Theoretical Considerations

The calculation of the spectral density functions $J^{(q)}(\omega)$ for the fluctuations in the dipolar interactions produced by the translational motion of spins on a crystal lattice is difficult. Calculations for realistic point defect diffusion models have only recently been accomplished by Wolf.[79,80] The measurements reported for plastic crystals have, with few exceptions, been interpreted using spectral density functions first calculated in 1954 by Torrey[81,82] and in 1963 by Resing and Torrey[83] for a model which considers the spins to undergo a random-walk on an 'isotropic' crystal lattice. In this model, the jumps are not restricted to the nearest neighbour lattice sites, but are allowed to all points on the surface of a sphere of radius corresponding to the nearest-neighbour distance *e*. That this is a reasonable approximation for polycrystalline samples of b.c.c. and f.c.c. solids has been shown by the recent calculations of Sholl[84,85] which take account of the discrete lattice structure. A more serious limitation of Torrey's model is that it neglects the correlation between successive jumps of the spins associated with a point defect diffusion mechanism. Wolf[79] has recently calculated the spectral density functions for a monovacancy mechanism, considered as the most likely one to be encountered in a plastic crystal. Comparison of these results with those of Torrey (see later) shows that for polycrystalline samples the values of τ derived using Torrey's model may be corrected for the effects of correlation without any serious loss of accuracy. This is an important result, because it removes uncertainties in any consideration of the significance of the published data.

To justify the above conclusion, we will summarize the results of Torrey's isotropic random-walk model and compare them with the predictions of models for a random-walk on a discrete crystal lattice and a monovacancy diffusion mechanism. Torrey's expressions for the spectral density functions are

$$J^{(1)}(\omega) = \frac{1}{6}J^{(0)}(\omega) = \frac{1}{4}J^{(2)}(\omega) = \frac{8\pi n\tau}{15k^3l^3}G(k, y). \tag{5.3.27}$$

The function $G(k, y)$ is defined by

$$G(k, y) = \int_0^\infty J_{3/2}^2(kx)\frac{1 - \sin x/x}{(1 - \sin x/x)^2 + y^2}\frac{dx}{x} \tag{5.3.28}$$

where $J_{3/2}$ is a Bessel function of order 3/2. y is $\omega\tau/2$ and k is a normalization parameter dependent on the lattice structure and given by

$$k^3 = 4\pi n_0/3a_0{}^3l^3 \sum_j r_{ij}{}^{-6}. \tag{5.3.29}$$

n_0 is the number of lattice sites per unit cell of edge a_0. The lattice sums $\sum_k r_{jk}{}^{-6}$ have the values 115.631 $a_0{}^{-6}$ and 29.045 $a_0{}^{-6}$ for face-centred and body-centred cubic lattices, respectively, giving corresponding values of 0.74280 and 0.76293 for k. n is the spin density and is given by $n_0c/a_0{}^3$ where c is the number of spins per molecule. Values for $G(k, y)$ as a function of y have been calculated and tabulated for face-centred[82] and body-centred cubic[83] lattices.

Substitution of equation (5.3.27) in equations (5.3.10) (5.3.11), and (5.3.12) gives

$$\frac{1}{T_1} = M_2\tau\{G(k, \omega_0\tau/2) + 4G(k, 2\omega_0\tau/2)\}, \tag{5.3.30}$$

$$\frac{1}{T_{1\rho}} = M_2\tau\left\{\frac{3}{2}G(k, 2\omega_1\tau/2) + \frac{5}{2}G(k, \omega_0\tau/2) + G(k, 2\omega_0\tau/2)\right\} \tag{5.3.31}$$

and

$$\frac{1}{T_2} = M_2\tau\left\{\frac{3}{2}G(k, 0) + \frac{5}{2}G(k, \omega_0\tau/2) + G(k, 2\omega_0\tau/2)\right\} \tag{5.3.32}$$

where

$$M_2 = c\left(\frac{\mu_0}{4\pi}\right)^2\frac{4}{5}\gamma^4\hbar^2 I(I + 1)\sum_k r_{jk}{}^{-6}$$

$$= c\left(\frac{u_0}{4\pi}\right)^2\frac{4}{5}\gamma^4\hbar^2 I(I + 1)\pi n_0/a_0{}^3k^3l^3 \tag{5.3.33}$$

which for plastic crystals corresponds to M_{2r}, equation (5.2.9), the apparent second moment of the rotationally averaged frequency spectrum prior to diffusional narrowing. Furthermore, since in the plastic phase $\omega_0\tau \gg 1$, only the first terms in equations (5.5.31) and (5.3.32) need to be considered; note, too, that equation (5.3.32) is only applicable in the region $M_2^{1/2}\tau \ll 1$ where the frequency spectrum has a Lorentzian shape and T_2 is defined.

Sholl[84,85] and Wolf[86] have recently reported rigorous treatments of uncorrelated random-walk lattice diffusion, taking into account the discrete structure of the lattice. To obtain expressions for the relaxation rates in polycrystalline samples, Sholl calculates the averages of the correlation functions over all orientations of the crystal with respect to the applied static field **B**, while Wolf takes the average of the relaxation rates which is the more plausible procedure. Their results are, however, very similar and comparison with Torrey's isotropic model shows that errors due to the assumption of isotropy are not experimentally significant. For f.c.c. lattices, the Torrey model gives values for $1/T_1$ or $1/T_{1\rho}$ that are 2.5% too small for $\omega\tau \ll 1$ (high-temperature side of minimum) and 1.4% too large for $\omega\tau \gg 1$ (low-temperature side of minimum); for b.c.c. lattices, the deviations are larger, 7.5% for $\omega\tau \ll 1$ and 3.3% for $\omega\tau \gg 1$. Since $1/T_1 \propto \tau$ for $\omega\tau \ll 1$, while $1/T_1 \propto \tau^{-1}$ for $\omega\tau \gg 1$, the net effect will be to shift the minima to shorter values of $\omega\tau$ and to reduce the activation energies deduced from τ by about 1%.

In the random-walk model the spins are allowed to make random independent jumps to nearest-neighbour sites, while in the more realistic point defect model the defect itself is considered to walk randomly. It is important to realize that spin relaxation will depend only on the relative displacement of a pair of interacting spins as a result of an encounter of one of them with a defect and, of course, on the frequency of these encounters. In the calculation of relaxation rates for a point defect mechanism it is therefore necessary to take into account that in a given encounter the displacement of a spin is not restricted to a nearest neighbour site. Also that more than one jump may occur *albeit* they are too fast to affect the relaxation process, i.e., both 'spatial' and 'temporal' correlations must be considered in contrast to radiotracer measurements where only spatial correlation effects are involved. Wolf[79,80] has very recently calculated T_1, T_2, and $T_{1\rho}$ as a function of $\omega\tau$ for both monovacancy and divacancy diffusion mechanisms in single and polycrystalline samples of face- and body-centred cubic solids as a function of their orientation in **B**. His results for a polycrystalline f.c.c. solid are given in Figure 5.18 where they are compared with those for uncorrelated random-walk diffusion[86]. The main effect is to displace the log T^{-1} versus $\omega_0\tau$ plots to smaller values of $\omega_0\tau$, the displacement being greater for the divacancy mechanism. Note, however, that experimental measurements of T_1 or $T_{1\rho}$ as a function of temperature will only reflect the shape of these plots and not their absolute position. The shapes for the monovacancy and divacancy processes are identical and slightly narrower than for random-walk diffusion. This narrowing is of the order 2% one decade down from the maxima in $1/T_1$ and $1/T_{1\rho}$ and it is doubtful if it could be detected experimentally. Thus, the shapes of the log T_1 versus T^{-1} plots are very similar for both correlated and uncorrelated diffusion models.

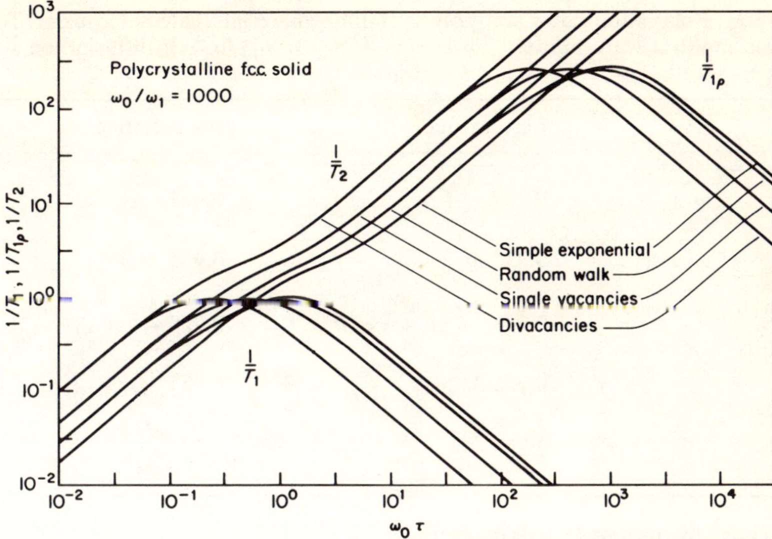

Figure 5.18 Comparison of the relaxation rates T_1^{-1}, $T_{1\rho}^{-1}$ and T_2^{-1} calculated for monovacancy, divacancy and uncorrelated random-walk diffusion mechanisms and for a simple exponential correlation function in polycrystalline fcc solids ($\omega_0/\omega_1 = 1000$).[80] (Reproduced by permission of Proc. 18th Ampere Congress, Nottingham)

Monovacancy and divacancy mechanisms may be distinguished in practice due to the different activation energies associated with the two processes.

We conclude that the Torrey model may be used, without serious error, to derive values for τ from the relaxation measurements provided that the values so obtained are corrected for correlation effects, i.e. $\tau = f\tau_T$, although it would be more appropriate to use Wolf's monovacancy model where such a mechanism is operative. Values for the correlation factor f for a monovacancy mechanism may be obtained by comparison of the limiting expressions for the relaxation rates as given by the two models in Tables 5.3 and 5.4; the results are summarized in Table 5.5. The values are similar to those previously estimated by Eisenstadt and Redfield[87] and Stoebe *et al.*[88] for monovacancy diffusion on an f.c.c. lattice: $f = 0.55$ for $\omega\tau \gg 1$ and $f = 0.67$ for $\omega\tau \ll 1$.

Wolf[79] has calculated relaxation rates for single crystals (b.c.c. and f.c.c.) as a function of the crystallographic orientation in the static field **B** and it is pertinent to refer here to the type of information obtainable from such measurements. T_1, $T_{1\rho}$ and T_2 are predicted to be isotropic on the high-temperature side of the T_1 minimum, i.e., for $\omega_0\tau \ll 1$, while on the low temperature side, $\omega_0\tau \gg 1$, the conditions encountered in plastic crystals, they are angular dependent. The dominating factor is the angular dependence of

Table 5.3 Relaxation rates for polycrystalline materials calculated using Torrey's isotropic random-walk model[81–83] ($\omega_0\tau \gg 1$, $\omega_0 \gg \omega_1$) for self-diffusion on a crystal lattice

		f.c.c. lattice	b.c.c. lattice
$\frac{1}{T_{1\rho}}$	$\omega_1\tau \gg 1$	$0.429\,\frac{M_2}{\omega_1^2\tau}$	$0.421\,\frac{M_2}{\omega_1^2\tau}$
	$\omega_1\tau \ll 1$	$0.922\,M_2\tau$	$0.955\,M_2\tau$
$\frac{1}{T_{1\rho}}$ (min)		$0.228\,\frac{M_2}{\omega_1}$ at $\omega_1\tau = 0.869$	$0.228\,\frac{M_2}{\omega_1}$ at $\omega_1\tau = 0.869$
$\frac{1}{T_2}$		$0.922\,M_2\tau$	$0.955\,M_2\tau$
$\frac{1}{T_1}$		$2.29\,\frac{M_2}{\omega_0^2\tau}$	$2.24\,\frac{M_2}{\omega_0^2\tau}$

M_2 is as given by equation (5.3.33) in the text.

Table 5.4 Relaxation rates for polycrystalline materials calculated by Wolf[79] for a monovacancy diffusion mechanism ($\omega_0\tau \gg 1$, $\omega_0 \gg \omega_1$)

		f.c.c. lattice	b.c.c. lattice
$\frac{1}{T_{1\rho}}$	$\omega_1\tau \gg 1$	$0.216\,\frac{M_2}{\omega_1^2\tau}$	$0.204\,\frac{M_2}{\omega_1^2\tau}$
	$\omega_1\tau \ll 1$	$1.32\,M_2\tau$	$1.40\,M_2\tau$
$\frac{1}{T_{1\rho}}$ (min)		$0.236\,\frac{M_2}{\omega_1}$ at $\omega_1\tau = 0.447$	$0.235\,\frac{M_2}{\omega_1}$ at $\omega_1\tau = 0.427$
$\frac{1}{T_2}$		$1.32\,M_2\tau$	$1.40\,M_2\tau$
$\frac{1}{T_1}$		$1.15\,\frac{M_2}{\omega_0^2\tau}$	$1.09\,\frac{M_2}{\omega_0^2\tau}$

M_2 is as given by equation (5.3.33) in the text.

the dipolar interaction, the dependence on the mechanism is small and probably not sufficient when compared with the accuracy of experimental measurements to be used to distinguish between different mechanisms.

Table 5.5 Correlation factors f for monovacancy diffusion for NMR[a] and radiotracer[b] measurements

Measurements		f.c.c.	b.c.c.
$\frac{1}{T_1}, \frac{1}{T_{1\rho}}$	$\omega\tau \gg 1$	0.51	0.49
$\frac{1}{T_1}, \frac{1}{T_{1\rho}}$	$\omega\tau \ll 1$	0.69	0.65
$\frac{1}{T_2}$		0.69	0.65
Radiotracer		0.78	0.73

[a] $\tau = f\tau_T$ where τ_T is the correlation time obtained using Torrey's model.[81–83]
[b] $D = fD$ where D is the self-diffusion coefficient determined by radiotracer measurements.[115]

5.3.5(b) Experimental Studies

Here we will be particularly interested in looking for consistency between the Torrey model and the measurements as evidence for the expected monovacancy mechanism. Where possible comparisons with radiotracer measurements will be made. Surprisingly, the results are found to be dependent on the entropy of fusion ΔS_f of the solid. For high entropy of fusion solids the measurements are found to be consistent with both Torrey's model and the radio-tracer measurements, while for low entropy of fusion solids Torrey's model is inapplicable and the NMR and radiotracer measurements are inexplicably inconsistent. For this reason the results will be examined in order of decreasing ΔS_f. Table 5.6 summarizes the self-diffusion data as obtained from NMR measurements; the values of τ have in most cases been obtained using the Torrey model and have not been corrected for correlation effects. Table 5.7 summarizes the radiotracer measurements; the values of τ have been calculated from D using equation (5.3.26) and taking the nearest neighbour distance l for $\langle r^2 \rangle$, i.e. a simply monovacancy mechanism has been assumed.

The $T_{1\rho}$ measurements for hexamethylethane (b.c.c., $\Delta S_f = 2.41\ R$) cover the temperature interval 250–374 K (Figure 5.12) and are claimed by Chezeau *et al.*[22] to be consistent with the Torrey model, though the minima are 15% higher than predicted using the value of M_2 calculated from equation 5.3.33 or measured from the frequency spectrum[21,22]. The values of τ cover the range 3.4×10^{-2} s to 7.0×10^{-8} s and are stated to exhibit an Arrhenius temperature dependence with an activation energy of 82 kJ mol^{-1} which is a little lower than the value 86 kJ mol^{-1} obtained from T_2 measurements above 320 K by both Baughman and Turnbull[89] and Lockhart and Sherwood[90]. A closer exam-

Table 5.6 Self-diffusion parameters of cubic solids determined by NMR measurements

Material	T_m/K	$\Delta S_f/R$	ν/Hz × 10^{12}	τ_0/s	E_a/kJ mol^{-1}	τ_m/s	$\frac{\Delta G_{D,m}}{T_m}$ J mol^{-1} K^{-1}	$\frac{E_a}{T_m}$ J mol^{-1} K^{-1}	$E_a/\Delta H_f$	E_a/L_S	Reference
Face-centred cubic											
Adamantane	542	2.51	1.12	1.6×10^{-21}	153.5	9.9×10^{-7}	136	283	13.5	2.3	58
Perfluorocyclohexane	336	2.30	0.56	3.6×10^{-16}	59.2	5.7×10^{-7}	126	176	9.2	1.8	91
Triethylenediamine	433	2.06	1.15	7.6×10^{-19}	96.4	3.2×10^{-7}	127	222	13.0	1.8	96
3-Azabicyclononane	467	1.78	1.09	1.7×10^{-16}	83.6	3.8×10^{-7}	128	179	12.0	1.5	96
Xenon	161	1.71	0.96	4.6×10^{-17}	31.0	5.0×10^{-7}	130	192	13.4	2.1	130
Neon	24.5	1.64	1.31	1.4×10^{-15}	3.96	3.5×10^{-7}	130	162	11.8	1.9	131
Norbornane	360	1.53	1.19	1.1×10^{-16}	64.8	2.8×10^{-7}	126	180	14.2	1.8	96
Neopentane	257	1.46	1.14	1.2×10^{-13}	33.0	6.6×10^{-7}	133	128	10.6	1.0	55
Cyclohexane	280	1.10	1.11	9.1×10^{-15}	42.0	6.2×10^{-7}	133	150	16.4	1.0	97
Pivalic acid	310	0.78	1.04	2.7×10^{-17}	63	1.1×10^{-6}	137	203	31.5	1.0	57
Hexagonal close-packed											
Norbornene	320	1.22	1.06	2.1×10^{-14}	45.7	6.0×10^{-7}	132	142	10.9	1.4	96
Norbornadiene	254	0.79	1.14	4.6×10^{-14}	35.7	1.0×10^{-6}	136	140	17.5	1.2	96
Body-centred cubic											
Tellurium hexafluoride	235	4.07	0.68	2.4×10^{-19}	52.3	1.0×10^{-7}	110	224	6.6	2.0	127
Sulphur hexafluoride	222	2.72	0.92	8.8×10^{-20}	52.3	1.8×10^{-7}	117	235	10.4	2.2	127
Chloropentamethylethane	408	2.72	0.99	7.0×10^{-18}	80.3	1.3×10^{-7}	120	197	8.7	1.8	93
Promopentamethylethane	424	2.60	0.87	4.0×10^{-18}	84.5	1.0×10^{-7}	112	199	9.2	1.8	93
Hexamethylethane	374	2.41	1.04	2.5×10^{-19}	82	6.8×10^{-8}	110	219	10.9	2.0	22, 132, 133
Molybdenum hexafluoride	291	1.80	0.82	2.6×10^{-17}	52.3	6.4×10^{-8}	108	180	12.4	1.4	127
Succinonitrile	331	1.35	1.38	1.7×10^{-17}	62.8	1.4×10^{-7}	118	190	16.9	0.9	128
Hexamethyldisilane	288	1.25	0.72	—	43.7	2.3×10^{-7}	117	152	14.4	1.2	101
dl-Camphene	323	1.15	0.85	4.6×10^{-17}	56.5	6.3×10^{-8}	108	175	18.2	1.1	97

Table 5.7 Self-diffusion parameters of cubic solids determined by radiotracer measurements

Solid	a_0/nm	$D_0/\mathrm{m^2\,s^{-1}}$	$E_a/\mathrm{kJ\,mol^{-1}}$	τ_0/s	$\frac{\Delta G_m}{T_m}$ J mol^{-1} K^{-1}	E_a/L_s	Temperature interval/K	Reference
				f.c.c.				
Adamantane	0.945	1.6	138.5	4.6×10^{-20}	136.8	2.1	505–537	134
Cyclohexane	0.875	1.0×10^{5}	92	6.3×10^{-25}	116.5	2.2	243–279	135
Pivalic acid	0.882	5.5×10^{2}	91.2	1.3×10^{-22}	126.0	1.6	280–300	136
				b.c.c.				
dl-Camphene	0.78	2.0×10^{4}	96.2	3.8×10^{-24}	94.7	1.9	270–320	107
Hexamethylethane	0.769	2.2	85.7	3.2×10^{-20}	103.6	2.0	300–356	90
Hexamethyldisilane	0.847	1.8×10^{-3}	52.0	5.1×10^{-17}	100.3	1.4	229–277	101
Succinonitrite	0.64	4.3×10^{-4}	69.8	1.2×10^{-16}	116.5	0.8	263–326	137

ination of Chezeau's measurements reveals a similar increase in E_A at temperatures above 315 K, behaviour indicative of a transition from a monovacancy region to one where both mono- and divacancies contribution to the molecular jumps. The values of τ determined in the range 320–374 K from the T_2 measurements are in close agreement with those obtained in the corresponding temperature interval by radiotracer measurements (see Tables 5.6 and 5.7). There is a difference of 10% between the tracer and NMR values of τ, but this could originate in correlation effects.

We see from Tables 5.6 and 5.7 that there is good agreement between the NMR and tracer data for adamantane (f.c.c., $\Delta S_f = 2.51\ R$). The NMR data was obtained by Resing[58] using low-field $T_{1\rho}$ measurements in the interval 370–462 K and the Ailion-Slichter theory[49] for monovacancy diffusion on an f.c.c. lattice. Values of τ were also derived from T_2 measurements between 475–540 K; the E_A is identical to that obtained from the $T_{1\rho}$ measurements, but the absolute values are inexplicably a factor of 0.68 smaller.

Perfluorocyclohexane is another relatively high entropy of fusion solid (f.c.c. $\Delta S_f = 2.30\ R$) for which convincing agreement is obtained between the measurements and the Torrey model. Detailed T_1, $T_{1\rho}$ and T_2 measurements (Figure 5.19) have been reported by Boden *et al.*[91] for the ^{19}F spins in a sample containing less than 100 p.p.m. of impurities. The values of τ obtained using equation (5.3.31) cover five decades and exhibit an Arrhenius temperature dependence as shown in Figure 5.20. The τ values obtained from the T_1 and T_2 measurements at high temperatures are seen to be in good agreement with those obtained from the $T_{1\rho}$ measurements. There are no radiotracer data available for comparison, but the relaxation measurements appear to be consistent with the Torrey model implying that a simple monovacancy mechanism is operative in this material. Roeder and Douglas[55] have made similar, but less detailed $T_{1\rho}$ measurements, while Bladon *et al.*[92,93] have reported that the high-temperature linewidths exhibit an anomalous impurity induced broadening. Chadwick *et al.*[94] have suggested that this is an 'inhomogeneous' broadening effect due to the formation of vapour snakes, but Bladon *et al.* argue that their samples were free from such imperfections. Alternative explanations should not, therefore, be rejected at this time; indeed, Volino,[95] has proposed that the observed behaviour is due to the quenching of self-diffusion by impurities.

Folland *et al.*[96] have reported values of τ deduced from $T_{1\rho}$ measurements as a function of temperature in triethylenediamine (f.c.c., $\Delta S_f = 2.06\ R$) and 3-azabicyclononane (f.c.c., $\Delta S_f = 1.78\ R$). The $T_{1\rho}$ measurements are not given, but the values of τ quoted exhibit an Arrhenius temperature dependence in both solids, a behaviour indicative of consistency with the Torrey model. However, for reasons given below, it is difficult in the case of f.c.c. or h.c.p. solids to unambiguously confirm the relevance of a particular model from the shape of the $T_{1\rho}$ versus temperature plot alone. Indeed, in the same paper, these authors

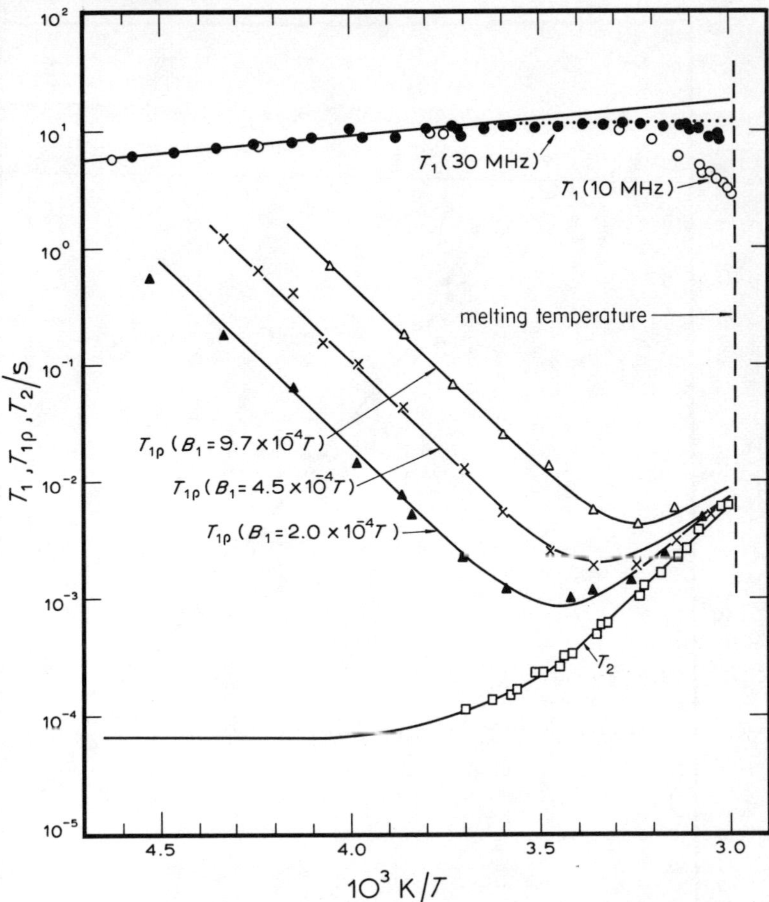

Figure 5.19 Semi-log plots of the ^{19}F relaxation times T_1, $T_{1\rho}$ and T_2 as a function of reciprocal temperature in plastic perfluorocyclohexane. The limiting low temperature value for T_2 was calculated from $T_2 = 1/(\gamma^2 M_{2r})^{1/2}$. The solid lines drawn through the $T_{1\rho}$ and high-temperature T_2 measurements were calculated from the τ in figure 5.20 using, respectively, equations (5.3.31), and (5.3.32).[91] (Reproduced by permission of *Mol. Phys.*)

also argue that the Torrey model satisfactorily accounts for the $T_{1\rho}$ measurements in norbornene (h.c.p. $\Delta S_f = 1.21\ R$), whereas a closer inspection of their measurements, given in this case, shows they are better explained by a different model as argued below.

Figure 5.21 summarizes the values of T_1 and $T_{1\rho}$ measured by Boden *et al.*[97] for the protons in *dl*-camphene (b.c.c., $\Delta S_f = 1.15\ R$) between 159 K and the melting point 323 K; also shown are the T_2 measurements made near the melting point. The values of τ calculated from the $T_{1\rho}$ measurements above

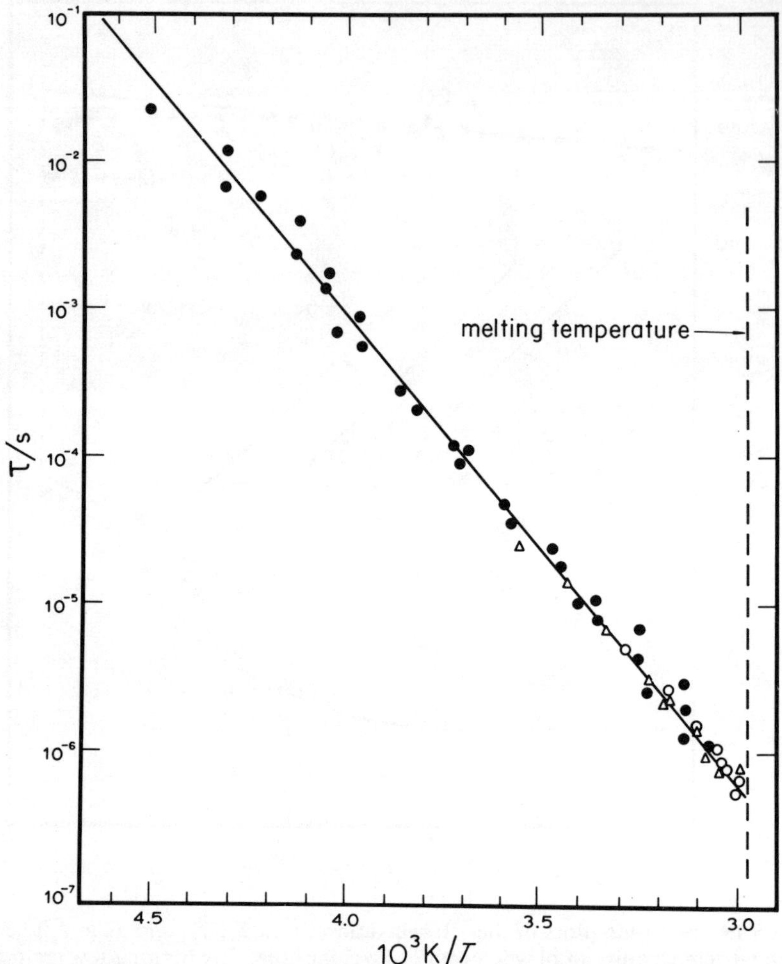

Figure 5.20 Values of τ is plastic crystalline perfluorocyclohexane as derived, using the Torrey model, from measurements of $T_{1\rho}$, ●; T_2,△; and T_1, 0. The solid line drawn through the points corresponds to $\tau = (3.6 \pm 0.8) \times 10^{-6}$ s exp(59.2 ± 1.0 kJ mol^{-1}/RT).[91] (Reproduced by permission of *Mol. Phys.*)

220 K and from the T_1 and T_2 measurements above 270 K, using the Torrey model are shown in Figure 5.22. It is immediately apparent that the fit to this model is unsatisfactory. The values obtained from the low- ($\omega_1\tau \gg 1$) and high-temperature ($\omega_1\tau \ll 1$) $T_{1\rho}$ measurements are seen to lie on two displaced yet parallel straight lines with the transition from one to the other centred on the $T_{1\rho}$ minimum. The values calculated from the T_2 measurements are in good agreement with those obtained from the high-temperature

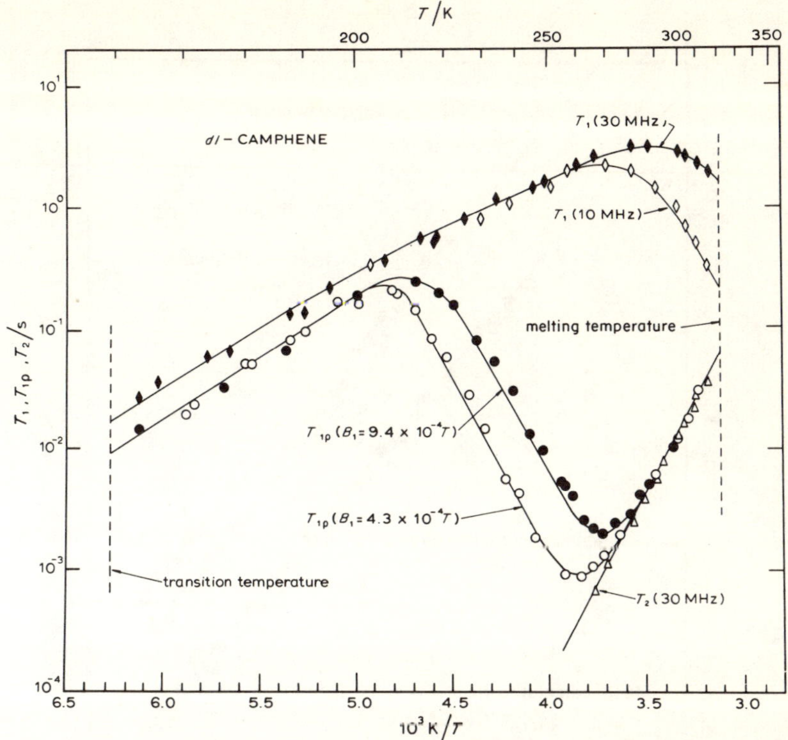

Figure 5.21 Semi-log plots of the proton relaxation times T_1, $T_{1\rho}$ and T_2 as a function of reciprocal temperature in polycrystalline plastic *dl*-camphene. The solid lines drawn, through the $T_{1\rho}$ measurements in the diffusion dominated region were calculated from the τ in Figure 5.22 using equation (5.3.36) and a value of $0.70 \times 10^{-8}\ T^2$ for the constant C.[97] (Reproduced by permission of *Mol. Phys.*)

$T_{1\rho}$ measurements, but the values calculated from the T_1 measurements are a factor of two larger. The log $T_{1\rho}$ versus T^{-1} plot is clearly significantly narrower than is predicted by either Torrey's model or Wolf's monovacancy model. The observed narrowing is 10 per cent one decade up from the minimum, compared with 2 per cent for the monovacancy model. The shape of the log $T_{1\rho}$ versus T^{-1} plot was found to be better described by the Debye-type spectral density function

$$J^{(1)}(\omega) = \frac{1}{6}J^{(0)}(\omega) = \frac{1}{4}J^{(2)}(\omega) = \frac{8\pi n}{45d^3}\frac{\tau}{1 + (\omega\tau/2)^2} \qquad (5.3.34)$$

where d is the distance of closest approach of the interacting spins. (The correlation time τ_c is here $\tau/2$, not τ, since both spins of an interacting pair may jump.) This spectrum obtains for translational diffusion when the spins under-

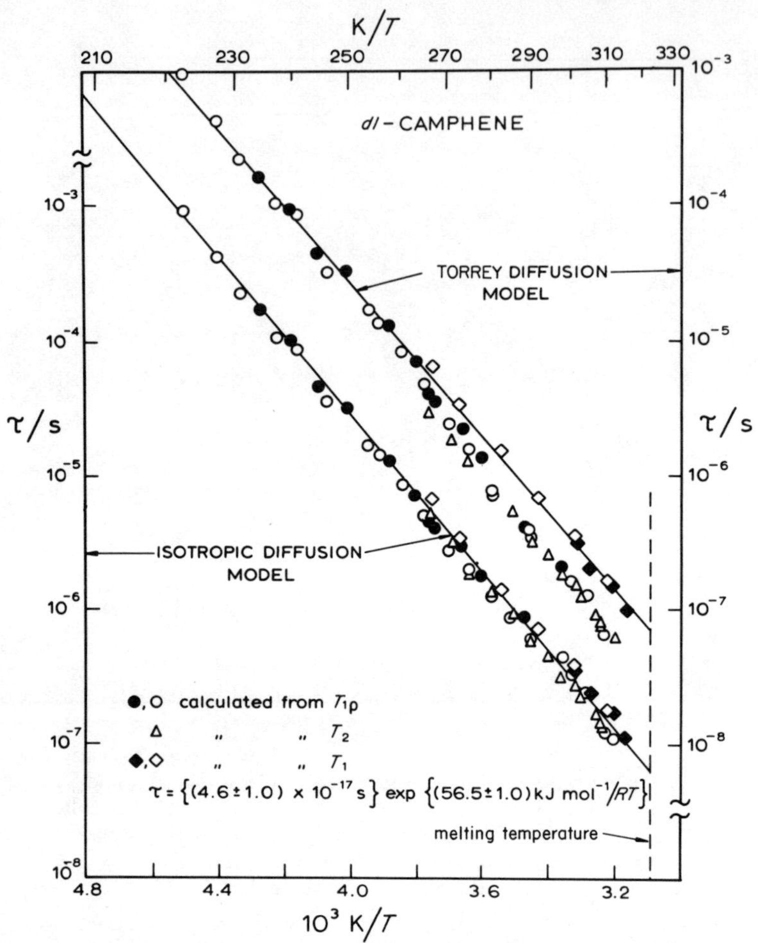

Figure 5.22 Semi-log plots of τ as a function of reciprocal temperature in *dl*-camphene.[97] (Reproduced by permission of *Mol. Phys.*)

go random isotropic jumps with a distribution of jump lengths and $\langle r^2 \rangle \gg d^2$.[81,98,99] In this limit the spectral density function is independent of the form of the distribution. This model is clearly more appropriate to diffusion in a liquid than on a crystal lattice and its applicability to the measurements for camphene is an unexpected result. Accepting this result empirically, at least for the moment, substitution of equation (5.3.34) in equations (5.3.30), (5.3.31) and (5.3.32) gives for $\omega_0\tau \gg 1$ and $\omega_0 \gg \omega_1$.

$$\frac{1}{T_1} = \frac{1}{3} C\tau \left\{ \frac{1}{1 + (\omega_0\tau/2)^2} + \frac{4}{1 + 4(\omega_0\tau/2)^2} \right\} \tag{5.3.35}$$

$$\frac{1}{T_{1\rho}} = \frac{1}{2} C\tau \frac{1}{1 + 4(\omega_1 \tau/2)^2} \tag{5.3.36}$$

and

$$\frac{1}{T_2} = \frac{1}{2} C\tau, \tag{5.3.37}$$

where

$$C = \left(\frac{\mu_0}{4\pi}\right)^2 \frac{4\pi}{5} \gamma^4 \hbar^2 I(I+1) \frac{n}{d^3} \tag{5.3.38}$$

and is equivalent to M_2 in equation (5.3.33) since $d = kl$ for diffusion on a crystal lattice. The values of τ calculated using these equations are seen, in Figure 5.22, to fall on a good straight line confirming that the isotropic diffusion model provides a more satisfactory representation of the measurements than either Torrey's model or the 'equivalent' monovacancy model.

Comparison of the expressions for the limiting relaxation rates as predicted by the Torrey (Table 5.3) and isotropic (Table 5.8) diffusion models, shows that the predictions of the two models are essentially equivalent at low temperatures ($\omega\tau \gg 1$), while at high temperatures ($\omega\tau \ll 1$) the isotropic model gives values of τ approximately double that predicted by Torrey's model, as observed for *dl*-camphene (Figure 5.22). It is surprising how spin relaxation measurements are not very sensitive to the microscopic details of self-diffusion. An important consequence is that the diffusion parameters obtained from NMR measurements are unlikely to be seriously in error due to modelling. On the other hand, detailed, accurate $T_{1\rho}$ measurements on both sides of the minimum are required if we are to have a chance of distinguishing between dif-

Table 5.8 Relaxation rates calculated for isotropic diffusion in the large step limit, i.e. $\langle r^2 \rangle \gg d^2$

$\frac{1}{T_{1\rho}}$	$\omega_1\tau \gg 1$	$0.5 \frac{M_2}{\omega_1^2 \tau}$
	$\omega_1\tau \ll 1$	$0.5\, M_2 \tau$
$\frac{1}{T_{1\rho}}$ (max)		$0.25 \frac{M_2}{\omega_1}$ at $\omega_1\tau = 1.0$
$\frac{1}{T_2}$		$0.5\, M_2$
$\frac{1}{T_1}$		$2.67 \frac{M_2}{\omega_0^2 \tau}$

ferent models. Comparison of the corresponding τ values obtained from $T_{1\rho}(\omega\tau \ll 1)$ and $T_1(\omega_0\tau \gg 1)$ are also necessary to eliminate the possibility of distortion of the shape of the $T_{1\rho}$ minimum due to a non-Arrhenius temperature dependence of τ at high temperatures due to a contribution from a second diffusion mechanism with a higher E_a e.g. a divacancy mechanism. The major limitation in the case of plastic crystals is the restricted range of accessible measurements on the high-temperature side of the $T_{1\rho}$ minimum. The most favourable materials for study are those with b.c.c. structures because their melting points are found to correspond to values of τ which are roughly an order of magnitude shorter than for either face-centred cubic or hexagonal close packed solids.[100] This phenomenon is immediately apparent on inspection of the relation of the $T_{1\rho}$ minima with respect to the melting points for perfluorocyclohexane (Figure 5.19) and *dl*-camphene (Figure 5.21). This no doubt explains why the breakdown of the Torrey model had not been detected earlier for measurements in other low entropy of fusion solids such as cyclohexane (f.c.c., $\Delta S_f = 1.10\ R$)[55] and norbornene (h.c.p., $\Delta S_f = 1.21\ R$).[96]

To determine which model was applicable to cyclohexane, Boden *et al.*,[97] made detailed proton T_1 and $T_{1\rho}$ measurements in this material. These are illustrated in Figure 5.23. Note how the $T_{1\rho}$ minimum is only 37 K below the melting point as compared to 63 K for the same B_1 value in *dl*-camphene. The values of τ obtained using the Torrey and isotropic models are given in Figure 5.24 and we see that their behaviour is identical to that observed for camphene (Figure 5.25). Boden *et al.* also examined the relevance of the two models to the $T_{1\rho}$ measurements reported by Folland *et al.*[96] for norbornene with similar results.

These results for *dl*-camphene, cyclohexane, and norbornene which have structures representative of the entire range exhibited by plastic solids, so eliminating the influence of specific structural effects, appear to typify the behaviour of the low ΔS_f materials. Hexamethyldisilane (b.c.c., $\Delta S_f = 1.25\ R$) is another relatively low entropy of fusion solid for which $T_{1\rho}$ measurements are available;[43,101] however the τ are distinctly non-Arrhenius in their temperature dependence making it impossible to distinguish between the two different models.

5.3.5(c) *Comparison of NMR and Radiotracer Self-diffusion Data*

Turning to the radiotracer data, in Table 5.7 above, we see that the values of τ_0 and E_a for the low ΔS_f solids cyclohexane, pivalic acid and *dl*-camphene differ considerably from the NMR ones, behaviour in marked contrast to that for the high ΔS_f materials adamantane and hexamethylethane for which there is reasonable agreement. The two sets of data for succinonitrile and hexamethyldisilane are reported to be in approximate agreement, but it has been shown[102] that, in the latter case, the previously published tracer measurements[101] may be influenced by impurity induced grain-brundaries. Apart from the uncertainties in these measurements, there are clearly serious discrepancies between the

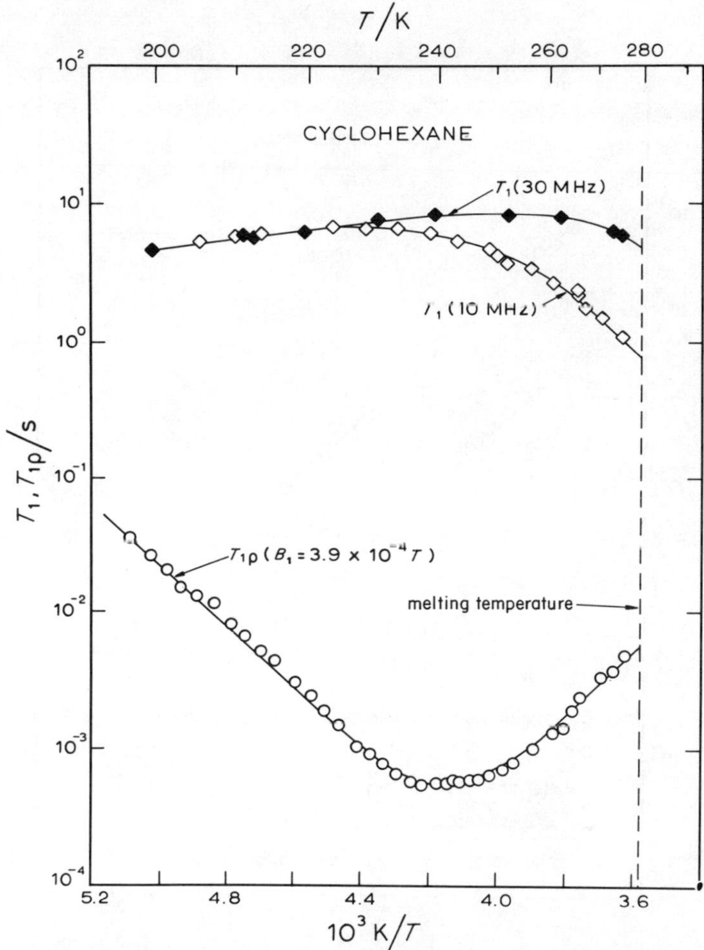

Figure 5.23 Semi-log plots of the proton relaxation times T_1 and $T_{1\rho}$ as a function of reciprocal temperature in polycrystalline plastic cyclohexane. The solid times drawn through the $T_{1\rho}$ measurements and high-temperature T_1 measurements were calculated from the τ in figure 24 using equations (5.3.36) and (5.3.35).[97] (Reproduced by permission of *Mol. Phys.*)

two sets of measurements for the low ΔS_f solids. Sherwood[103] has argued that the tracer measurements for both high and low ΔS_f solids are internally consistent with $E_a \cong 2L_s$ (L_s in the latent heat of sublimation) and indicative of a monovacancy diffusion mechanism as observed for the rare gas solids. Yet, for the NMR measurements, the ratio E_a/L_s gradually decreases from 2 to 1 in going from the upper to the lower end of the entropy of fusion range. In view of these discrepancies and the fact that the NMR measurements in the low ΔS_f solids are described by equations (5.3.35), (5.3.36) and (5.3.37) which hold for isotropic diffusion in the limit $\langle r^2 \rangle \gg d^2$, it would seem expedient to examine this model

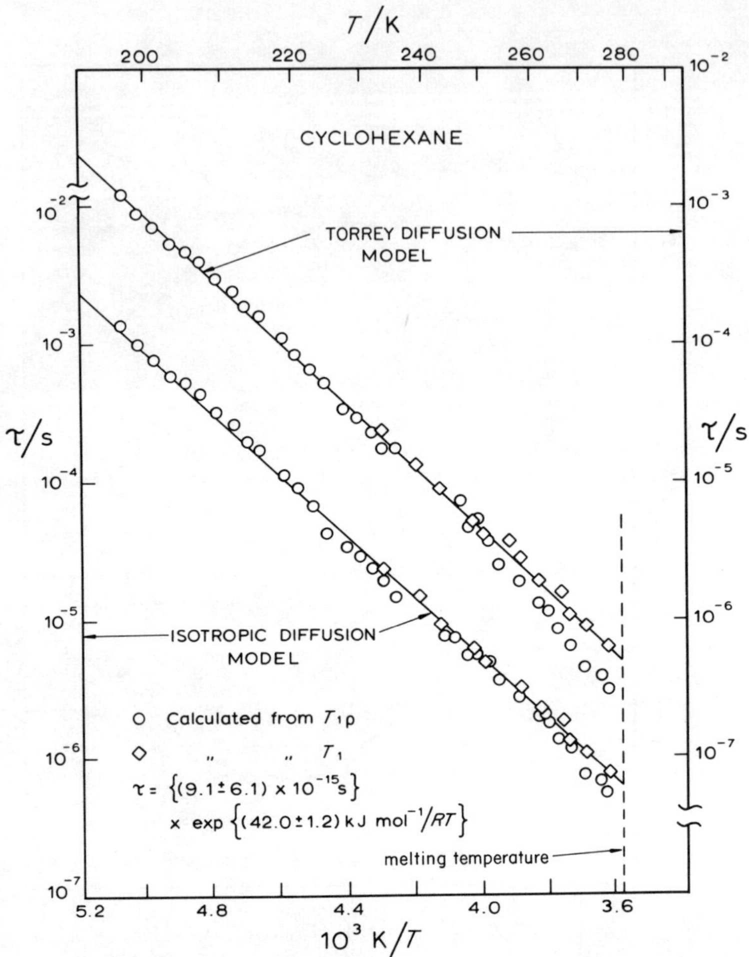

Figure 5.24 Semi-log plots of τ as a function of reciprocal temperature in polycrystalline plastic cyclohexane.[97] (Reproduced by permission of *Mol. Phys.*)

more closely to determine whether the disagreement is due to the identification of the NMR τ with that in equation (5.3.26).

Torrey's treatment[81] of spin relaxation by isotropic translational diffusion gives

$$J^{(1)}(\omega) = \frac{1}{6} J^{(0)}(\omega) = \frac{1}{4} J^{(2)}(\omega)$$

$$= \frac{8\pi n\tau}{15\, d^3} \int_0^{\infty} J_{3/2}^2 (\rho d) \frac{1 - A(\rho)}{\{1 - A(\rho)\}^2 + (\omega\tau/2)^2} \frac{d\rho}{\rho} \tag{5.3.39}$$

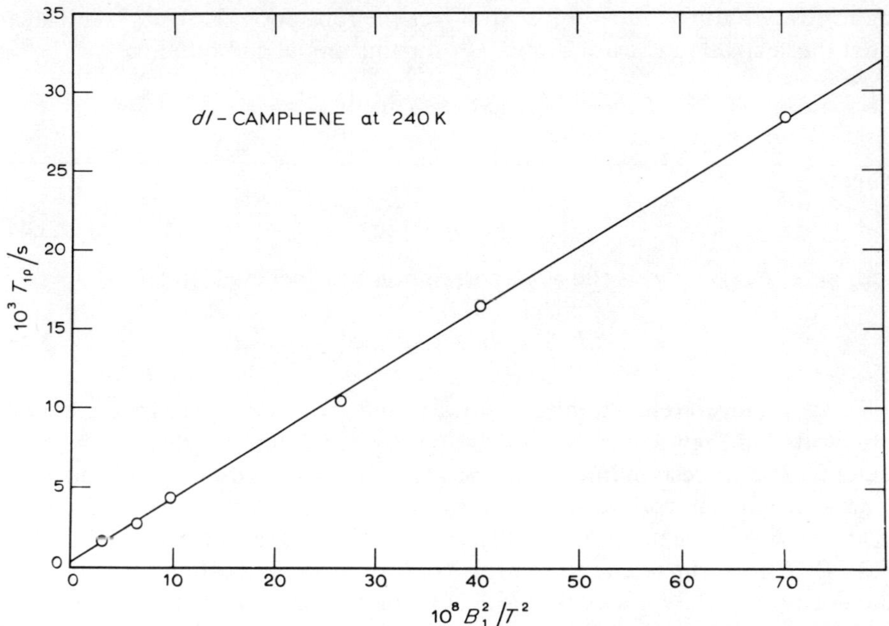

Figure 5.25 T_1 versus B_1^2 in polycrystalline *dl*-camphene at 240 K. The non-zero intercept on the vertical axis demonstrates the failure of the high-field theory in the limit $B_1^2 \to 0$. The value of the intercept on the horizontal axis is $B_1^2 = B_L^2$, here, $B_L^2 = M_{2r}/3$ in good agreement with the Ailion–Slichter theory.[49] (Reproduced by permission of *Mol. Phys.*)

where

$$A(\rho) = \int P_1(r) \exp (i\rho r) \mathrm{d}r \qquad (5.3.40)$$

Here $P_1(r)$ describes the distribution of the molecular jumps, d determines the lower limit of the integration and is interpreted as the distance of closest approach of two interacting spins, and the other quantities are as described previously. The choice of $P_1(r)$ must be both physically acceptable and computationally feasible. Torrey considered

$$P_1(r) = (4\pi D\tau r)^{-1} \exp\{-r/(D\tau)^{1/2}\} \qquad (5.3.41)$$

which gives

$$A(\rho) = (1 + D\tau\rho^2)^{-1} \qquad (5.3.42)$$

Equation (5.3.41) corresponds to a model which predicts that a spin spends most of its time in a bound state, but makes random transitions into a thermally activated state where it may undergo a large number of small diffusive steps, of an arbitrary nature, but too fast to affect the relaxation. Krüger[104] has calculated the relaxation rates T_1^{-1} and T_2^{-1} for this model and obtains

$$\frac{1}{T_{1,2}} = \frac{8\pi n}{15d^3}\gamma^4\hbar^2 I(I+1)\frac{1}{\omega}g_{1,2}(\alpha, y) \tag{5.3.43}$$

where

$$\alpha = \langle r^2\rangle/12d^2 \tag{5.3.44}$$

and, here, $y = \omega\tau_c$. τ_c is the NMR correlation time which is given by

$$\tau_c = (d^2/5 + \langle r^2\rangle/12)/D = \frac{2+5\alpha}{10\alpha}\tau. \tag{5.3.45}$$

The normalized relaxation rates $g_1(\alpha, y)$ and $g_2(\alpha, y)$ are given in reference[104] and plotted in Figure 5.26 as a function of $\omega\tau_c$ for different values of the parameter α. The corresponding values for $1/T_{1\rho}$ have not been calculated, but their general behaviour will be the same as that for $1/T_1$. When $\alpha \to \infty$, i.e. $\langle r^2\rangle \gg d^2$ and $\tau_c = \tau/2$, equations (5.3.43) become identical to equations (5.3.35) and (5.3.37). This is true whatever the form of $P_1(r)$ since, in this limit, $1 - A(\rho) \to 1$ and equation (5.3.39) reduces to (5.3.34). Figure 5.25 shows that when $\omega\tau_c \ll 1$, $1/T_{1,2} \propto \tau_c$ independent of the value of α, so that if τ_c is an Arrhenius function of temperature, E_a is obtained from the slope of the log $T_{1,2}$ versus T^{-1} plot. There should, therefore, be no uncertainties in the values of E_a determined from the high-temperature side of the $T_{1\rho}$ minimum or from T_2. For $\omega\tau_c \gg 1$, $1/T_2 \propto \tau_c$ while $1/T_1 \propto 1/\omega^2\tau_2$, but as $\alpha \to 0$ the latter is only true for very large values of $\omega\tau_c$; in the limiting case $\alpha = 0$, $1/T_1 \propto 1/\omega^{3/2}\tau_c^{1/2}$ so that the apparent activation energy is only one-half that for τ_c. This will not, however, explain why E_a (NMR) $\cong E_a$ (tracer)/2 since, first, the shape of the log $T_{1\rho}$ versus T^{-1} plot (Figure 5.21) and, secondly, $T_{1\rho} \propto \omega_1^2$ for $\omega_1\tau \gg 1$ (Figure 5.25) require $\alpha \to \infty$. It may, therefore, be concluded that the values obtained for τ are reliable and in particular their identification with the τ in equation (5.3.26) is correct. On the other hand, the only mechanistic information obtainable from the spin relaxation measurements is that the molecules undergo random displacements with an undeterminable distribution of jump lengths about a mean which must be greater than a molecular diameter.

The NMR measurements in *dl*-camphene and cyclohexane clearly require $\alpha \to \infty$, i.e., $\langle r^2\rangle \gg d^2$; equation (5.3.45) therefore reduces to equation (5.3.26). Using the tracer D and NMR τ in this latter equation, values have been calculated for $\langle r^2\rangle$ at both the melting point and the lower temperature limit of the tracer measurements in the low ΔS_f solids. In *dl*-camphene, at the melting point, 323 K, $\langle r^2\rangle^{1/2} = 2.13\ l$ while at 270 K, $\langle r^2\rangle^{1/2} = 0.50\ l$; in cyclohexane, at 280 K, $\langle r^2\rangle^{1/2} = 2.62\ l$, while at 240 K $\langle r^2\rangle^{1/2} = 0.44\ l$; in pivalic acid, at 310 K $\langle r^2\rangle^{1/2} = 2.00\ l$, while at 280 K, $\langle r^2\rangle^{1/2} = 1.11\ l$. Furthermore, assuming the tracer D is valid at the lower temperature limits of the NMR measurements,

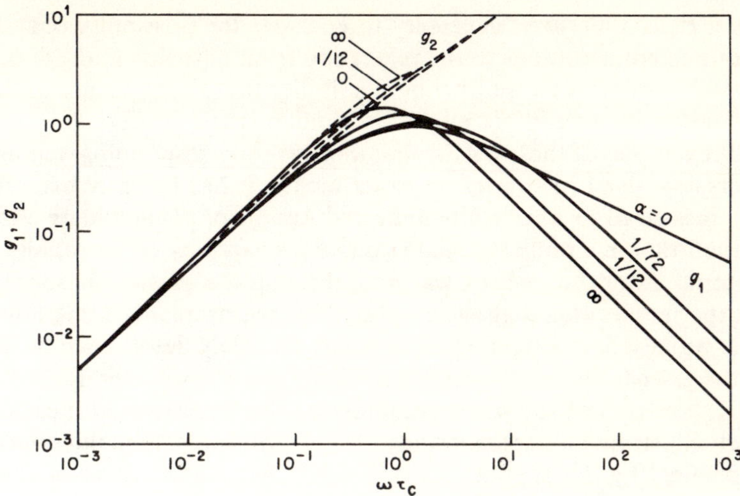

Figure 5.26 $1/T_1$ and $1/T_2$ (in units of $(\mu_0/4\pi)^2\, 8\pi n\gamma^4\hbar^2 I(I+1)/15d^3\omega$) as a function of $\omega\tau_c$.[104] (Reproduced by permission of *Z. Naturforsch.*)

for camphene at 220 K, $\langle r^2\rangle^{1/2} = 0.067\, l$, while for cyclohexane at 200 K, $\langle r^2\rangle^{1/2} = 0.035\, l$. The general pattern of behaviour is similar for all three materials; the values of $\langle r^2\rangle$ are compatible with the requirements of the n.m.r. measurements at the melting points, but not so at lower temperatures where $\langle r^2\rangle^{1/2} \ll l$. The range of compatibility of τ and D could be extended to lower temperatures by choosing a smaller value than l for d, but there is no *a priori* reason for doing this. It is, therefore, unlikely that the differences in the values of E_a for D and τ may be accounted for in terms of a temperature dependent value of $\langle r^2\rangle$.

There is no obvious way of reconciling the two sets of measurements in terms of relaxation theory. The fact that the measurements are compatible near to the melting point, but not at lower temperatures might be a pointer to the cause of the discrepancy. We also note that there is reasonable agreement between the tracer value of 7.0×10^{-13} m^2 s^{-1} for D at the melting point in cyclohexane and the corresponding value of 2.4×10^{-13} m^2 s^{-1} as measured by Tanner[105] using the pulsed-gradient spin-echo technique which gives a direct measure of D. The values of D at the melting points of plastic crystals, 10^{-12}–10^{-13} m^2 s^{-1}, are at the lower limit of those accessible by this technique; it is, therefore, of limited utility for studying self-diffusion in plastic solids and will not be described further here. It is clearly necessary to consider the possibility of systematic errors in both sets of data; we will only consider NMR measurements here. Errors in tracer experiments have been discussed in Chapter 2.

The NMR measurements have all been made in polycrystalline sample so as to avoid complications from the anisotropy in the relaxation rates for single crystals.[79] It is, therefore, necessary to consider the possibility of systematic errors due to contributions to the relaxation from diffusion along dislocations and grain boundaries as well as the possibility that the structure of the point defect and/or its migration might be fundamentally different in the polycrystalline samples of the low ΔS_f solids than in the corresponding well-annealed single crystals used in the tracer experiments. It has recently been shown[106] that $T_{1\rho}$ measured in a carefully annealed sample of *dl*-camphene was no different than that in a polycrystalline sample prepared by either rapidly cooling the material from the melt or warming through the plastic phase transition. These experiments also suggest that the NMR measurements are unaffected by dynamic processes at extended defects and are solely determined by diffusion within the grains. This is an important advantage of NMR over tracer measurements. The effects of sample crystallinity on the two types of measurements are nicely illustrated by a recent study[107,108] of diffusion in the solid solution 90% *dl*-camphene/10% *d*-camphor (T_m = 328 K). The NMR values of τ, given by

$$\tau = \{(8.0 \pm 2.0) \times 10^{-17}\ \mathrm{s}\} \exp\{(56.2 \pm 0.6)\ \mathrm{kJ\ mol^{-1}}/RT\},$$

are found to be identical to those obtained for the pure material when expressed in terms of the reduced temperature T/T_m. In contrast, the tracer values of D[107] are characteristic of the pure polycrystalline solid with E_D decreasing from 96 to 50 kJ mol^{-1}. The explanation is that although the camphor is uniformly distributed through the crystal on freezing, it tends to segregate from the lattice on cooling and thereby generates disorder, dislocations and grain boundaries which affect the tracer but not the NMR measurements. It could of course, be argued that in these experiments the NMR measurements are dominated by diffusion along dislocations and grain boundaries. If this were the case, then, in accordance with the observed exponential relaxation, assuming that either the rate of exchange of molecules between the grains and boundaries or the rate of spin diffusion between their respective spin subsystems is greater than their separate relaxation rates $1/T_{1\rho}^g$ and $1/T_{1\rho}^b$,

$$1/T_{1\rho} = g/T_{1\rho}^b + (1-g)/T_{1\rho}^g,$$

where g is the fraction of molecules in the boundaries. Assuming, that, similar intermolecular interactions occur in the grains and boundaries, we may calculate $1/T_{1\rho}^b$ and $1/T_{1\rho}^g$ as a function of f_b for *dl*-camphene using the NMR and tracer values for τ. To account for the NMR measurements, which are undoubtedly dominated by a single relaxation process over the temperature range 220 to 323 K (Figure 5.22), in terms of this hypothesis would require $g \geqq 0.9$, which is physically unrealistic.

There thus seems no reason whatsoever to doubt that the NMR experiment is monitoring intrinsic, thermally activated, molecular diffusion within the crystallites. Moreover, we have seen that the NMR measurements are not very sensitive to the microscopic details of molecular diffusion so that the values of

τ are unlikely to be significantly in error due to incorrect modelling. The only unexplained feature of the NMR measurements is that the values of M_2 determined from the $T_{1\rho}$ minima are typically 5–20 per cent smaller than the corresponding ones obtained from the frequency spectrum or calculated from the lattice parameter: hexamethylethane, 15%[22] perfluorocyclohexane, 9%[91]; triethylenediamine, 5%[96]; norbornane, 10%[96]; cyclohexane, 2%[97]; *dl*-camphene, 5%[97]; hexamethyldisilane, 10%[101]; pivalic acid, 11%[57]; norbornene 20%[96] and norbornadiene, 33%[96]. These are small though significant discrepancies which have not yet been explained, but they do not appear to be related in any way to ΔS_f. The most probable explanation is that the intermolecular dipole interactions are partially averaged by the lattice vibrations.

Finally, we compare NMR[108] and tracer[109] measurements which have been made on the same very pure sample of plastic white phosphorus ($\Delta S_f = 0.95\,R$). The results are interesting in that they illustrate the microscopic nature of NMR measurements. This solid is structurally unique amongst plastic crystals. X-ray diffraction measurements[110,111] show that it has a complicated body-centred cubic structure with 58 molecules per unit cell of edge $a_0 = 1.85$ nm. The molecular centres are distributed on four different symmetry sites, containing 2, 8, 24, and 24 molecules, respectively, in space group 143(T_d^3). The actual orientations of the individual molecules is not yet known, but with a molecular diameter of 0.50 nm (P–P = 0.221 nm, $\langle$(PPP) = 60° and a van der Waals radius of 0.19 nm for P the P_4 molecules cannot be arranged in the unit cell without appreciable mutual overlap implying a complex packing arrangement.[111] The molecules on the different sublattices should exhibit diffusional kinetics reflecting the differences in their environments. The ^{31}P $T_{1\rho}$ measurements of Boden and Cohen[108] in the range 200–317 K are characteristic of a solid with fast spin exchange between two spin sub-systems. There must, therefore, be two distinct 'molecular-environments' with characteristic jump frequences which are given in Figure 5.27. The radiotracer measurements by Hampton and Sherwood[109] cover the range 393–316 K (T_m = 317.4 K) and give

$$D = 3.6 \times 10^5\ \mathrm{m^2\,s^{-1}} \exp(-114.4\ \mathrm{kJ\,mol^{-1}}/RT)$$

with again $E_D \approx 2 \times L_S$ ($L_S = 58.6$ kJ mol^{-1}) in agreement with the tracer data for the simple face- and body-centred cubic solids. In contrast, the average E_a obtained from the NMR measurements is close to L_S as found for other low entropy of fusion solids. At the melting point, $1/\tau_1 = 9.5 \times 10^5\ \mathrm{s^{-1}}$ and $1/\tau_2 = 3.4 \times 10^5\ \mathrm{s^{-1}}$. These values are at the lower extreme of the values found for face-centred-cubic solids suggesting close packing of the molecules. Comparison of the average τ and D gives $\langle r^2\rangle^{1/2} \approx 1.4$ molecular diameter at T_m, but $\langle r^2\rangle^{1/2} \ll$ a molecular diameter at $T \ll T_m$. Thus, despite the structural complexity of this solid, the NMR and tracer data are typical of the corresponding data for other low entropy of fusion materials with the same apparent discrepancies between the two techniques. The fact that the NMR measurements

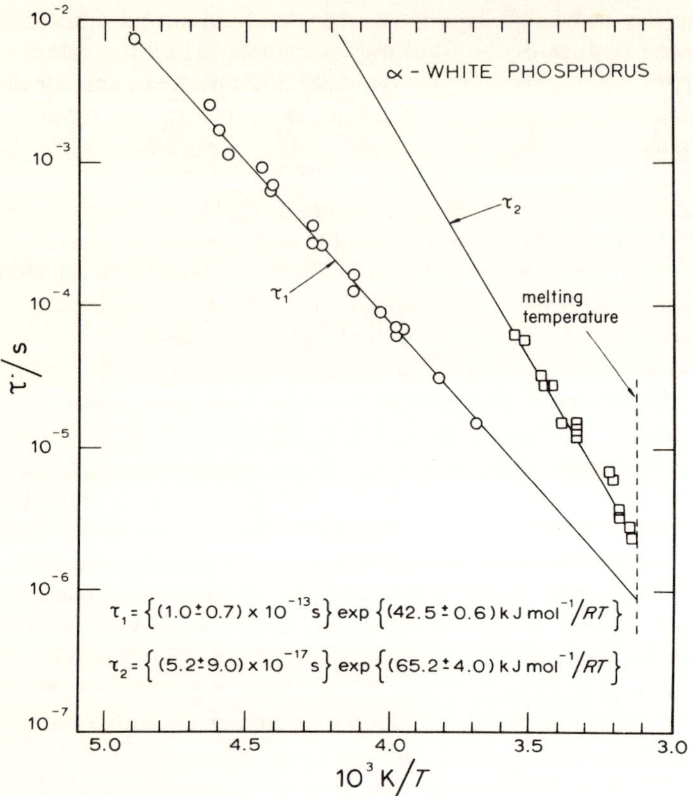

Figure 5.27 Semi-log plots of τ as a function of reciprocal temperature in polycrystalline plastic white phosphorus. (N. Boden and J. Cohen, unpublished results.)

distinguish molecular motion on different sublattices implies that translational diffusion must involve a discrete point defect: the results are not consistent with the diffusion of extensively relaxed vacancies or 'relaxions'.[123]

5.3.5(d) Pressure Dependence of τ

The difference in diffusion mechanisms between the high and low ΔS_f solids as indicated by the shapes of the log $T_{1\rho}$ versus T^{-1} plots, should also be reflected in the pressure dependence of $T_{1\rho}$. This is indeed the case and is nicely shown by the pressure dependence of τ in hexamethylethane and cyclohexane as measured by Strange and his coworkers.[112–114]

Adopting an activated state model for the diffusing molecule

$$1/\tau = z\nu \exp(-\Delta G/RT) \tag{5.3.47}$$

where z is the coordination number, ν is a notional lattice vibration frequency considered to be of the order of the Debye frequency,[115] and ΔG is the Gibbs free energy of activation. The volume of activation ΔV is then given by

$$\begin{aligned}\Delta V &= (\partial \Delta G/\partial P)_T \\ &= RT\{(\partial \ln \tau/\partial P)_T + (\partial \ln \nu/\partial P)_T\} \\ &\cong RT(\partial \ln \tau/\partial P)_T \end{aligned} \qquad (5.3.48)$$

Since the second term is only expected to be of the order 5% of the measured activation volume.[112] The activation volume may, therefore, be obtained from the slope of the log τ versus P plot. Furthermore, setting $\Delta G = \Delta H - T\Delta S$ and assuming ΔS to be temperature independent, we get

$$\Delta H = R\{\partial \ln \tau/\partial(1/T)\}_P \qquad (5.3.49$$

enabling the enthalpy of activation to be obtained as a function of pressure.

Figure 5.28 shows isotherms of log τ versus P for hexamethylethane.[114] Except at low pressures and high temperatures, the activation volume is found to be independent of temperature and pressure and of value 1.12 Ω_m (Ω_m is the molar volume in the solid). The activation enthalpy is found to increase linearly with pressure from 70 kJ mol^{-1} at atmospheric pressure to 106 kJ mol^{-1} at 2.07 kbar; the activation enthalpy at low pressure is different from the value of 82 kJ mol^{-1} obtained for E_a by Chezeau *et al.*,[22] possibly reflecting an error in the former value. The measurements at high temperature and low pressure are characterized by a higher activation volume 1.6 Ω_m and enthalpy (93 kJ mol^{-1}). The values of the activation volume are higher than usually encountered in metals,[115] but Burton and Jura[116] have calculated values of 1.2 Ω and 1.8 Ω, respectively, for the activation volumes for vacancy and divacancy diffusion in argon. It, therefore, appears that the pressure dependence of τ in hexamethylethane is consistent with a monovacancy diffusion mechanism with contributions from divacancies at high temperatures and low pressures. McKay and Sherwood[117] have obtained similar results from studies of the pressure dependence of the tracer self-diffusion coefficient in the range 293–354 K.

The pressure dependence of τ in cyclohexane[118] is more complex than in hexamethylethane. The isotherms, Figure 5.29, show that the activation volume decreases continuously with pressure ($0.3\,\Omega \lesssim \Delta V \lesssim 0.9\,\Omega$). At a given pressure, tangents to the isotherms are approximately parallel indicating ΔV increases linearly with temperature. The isobars in Figure 5.30 show that at low temperatures and pressures ΔH is independent of temperature and pressure and of value 40.7 kJ mol^{-1} in good agreement with the value of E_a[97], while at higher temperatures and pressures there is a transition to a higher value. Note how, at the melting point, τ is constant independent of temperature or pressure.

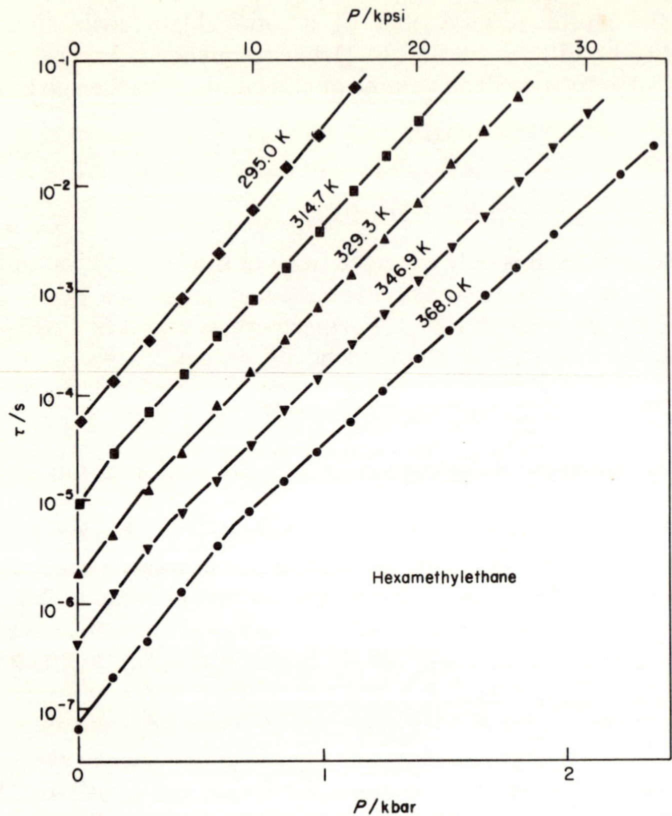

Figure 5.28 Isotherms for log τ as a function of pressure in polycrystalline hexamethylethane. The values of τ were obtained from $T_{1\rho}$ measurements using the Torrey model.[114] (Reproduced by permission of *Mol. Cryst. Liq. Cryst.*)

For an activated state model ΔH and ΔV are related by[119]

$$\left(\frac{\partial \Delta H}{\partial P}\right)_T = \Delta V(1 - \alpha T) \tag{5.3.50}$$

where

$$\alpha = \frac{1}{\Delta V}\left(\frac{\partial \Delta V}{\partial T}\right)_p.$$

For hexamethylethane, α is of the order of the expansivity of the crystal lattice consistent with a discrete monovacancy mechanism, while for cyclohexane it is at least in order of magnitude larger indicating a more complex mechanism.

Strange has also studied, though in less detail, the pressure dependence of τ

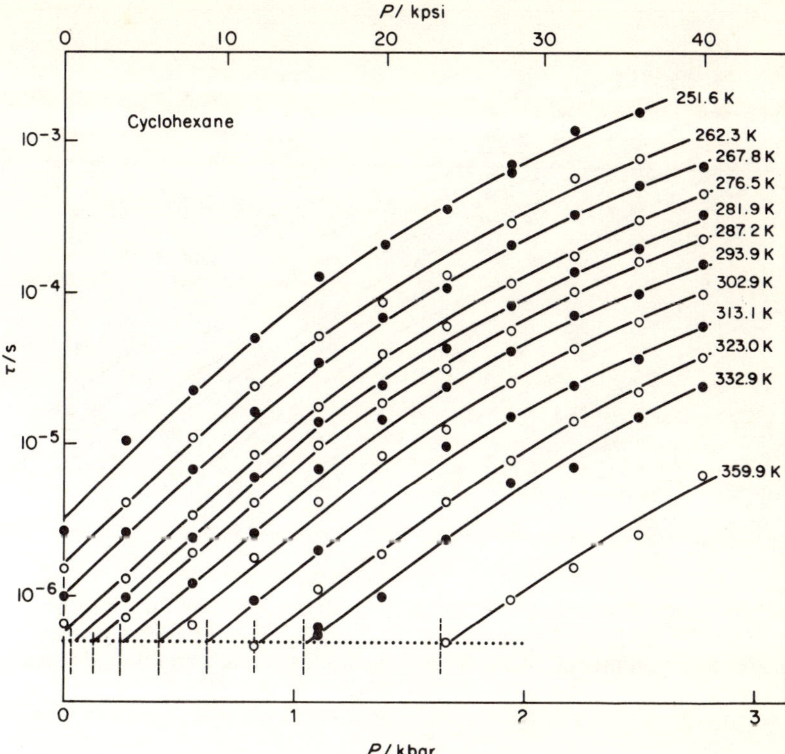

Figure 5.29 Isotherms for log τ as a function of pressure in polycrystalline cyclohexane. The values of τ were obtained from $T_{1\rho}$ measurements using the isotropic diffusion model. (J. H. Strange and S. M. Ross, unpublished results)

in several other plastic solids. In norbornane (f.c.c., $\Delta S_f = 1.53\ R$), $\Delta V \approx \Omega$ and independent of temperature and pressure as observed for hexamethylethane,[112] while for the lower entropy of fusion materials norbornene (h.c.p., $\Delta S_f = 1.21\ R$) and norbornadiene (h.c.p., 0.79 R) the behaviour is similar to that in cyclohexane with activation volumes of the order 0.8 Ω at low pressures.[112,113] Thus, the pressure dependence of τ appears to be characteristically different for high and low entropy of fusion crystals.

5.3.5(e) Self-diffusion in Relation to Melting

We see from Table 5.6 above that the values of τ at the melting points (τ_m) are approximately constant and characteristically shorter for body-centred cubic than for face-centred cubic solids; the average values are $(1.2 \pm 0.6) \times 10^{-7}$ s for the former and $(6.2. \pm 2.7) \times 10^{-7}$ for the latter. The values of τ_m

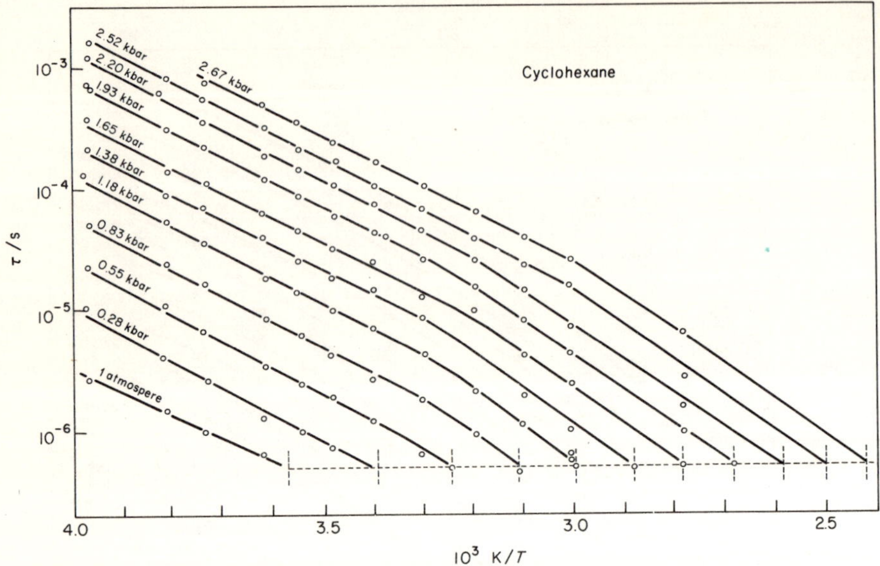

Figure 5.30 Log τ as a function of reciprocal temperature for given pressures in cyclohexane. The values of τ were obtained from the $T_{1\rho}$ measurements using the isotropic diffusion model. (J. H. Strange and S. M. Ross, unpublished results.)

are obviously determined by the crystal structure; they are apparently insensitive to the degree of orientational disorder in the plastic phase. The observations that τ_m is independent of pressure, at least up to several kbar, in the cubic phases of ethylene,[120] hexamethyldisilane[114] and cyclohexane,[114] and independent of composition in solid solutions of UF_6 in MoF_6[121] and *d*-camphor in *dl*-camphene[108] are also in keeping with the constancy of τ_m.

It is interesting to consider the implications of the above results for the applicability of the various semi-empirical models which have been used to related diffusion parameters to the thermodynamic properties of melting for metals[122–124]. The simplest and most widely employed relationships for metals are due to Van Liempt[122]

$$E_a/T_m = 142 \text{ J mol}^{-1}\text{K}^{-1} \tag{5.3.51}$$

and Nachtrieb and Handler[123]

$$E_a = 16.5\,\Delta H_f \tag{5.3.52}$$

where ΔH_f is the enthalpy of fusion. From the data in Table 5.6 above we find

$$E_a/T_m = 187 \pm 37 \text{ J mol}^{-1}\text{K}^{-1}$$

and

$$E_a = (13.4 \pm 5.1)\Delta H_f.$$

Heither relationship gives a good representation of the data. A more recently proposed relationship, due to Gibbs,[124] which seems to be far more successful than either equations (5.3.51) or (5.3.52) in correlating self-diffusion measurements in metals, is

$$\Delta G_m / T_m = \text{constant.} \tag{5.3.53}$$

Values for ΔG_m, the free energy of activation for self-diffusion at the melting point, were calculated from equation (5.3.47) with values for ν estimated using the modified Lindeman relation[125]

$$\nu_L = 2.55 \times 10^{12}\ \text{Hz}(T_m/MV^{3/2})^{1/2}, \tag{5.3.54}$$

where M and V are, respectively, the molar mass in grammes and volume in ccs. The values of ΔG_m obtained are plotted as a function of T_m in Figure 5.31 and are seen to exhibit distinct correlations for face-centred and body-centred cubic solids. Gibbs has published[124] a very similar set of plots for metals, the values for ΔG_m having been calculated from radiotracer measurements of D;

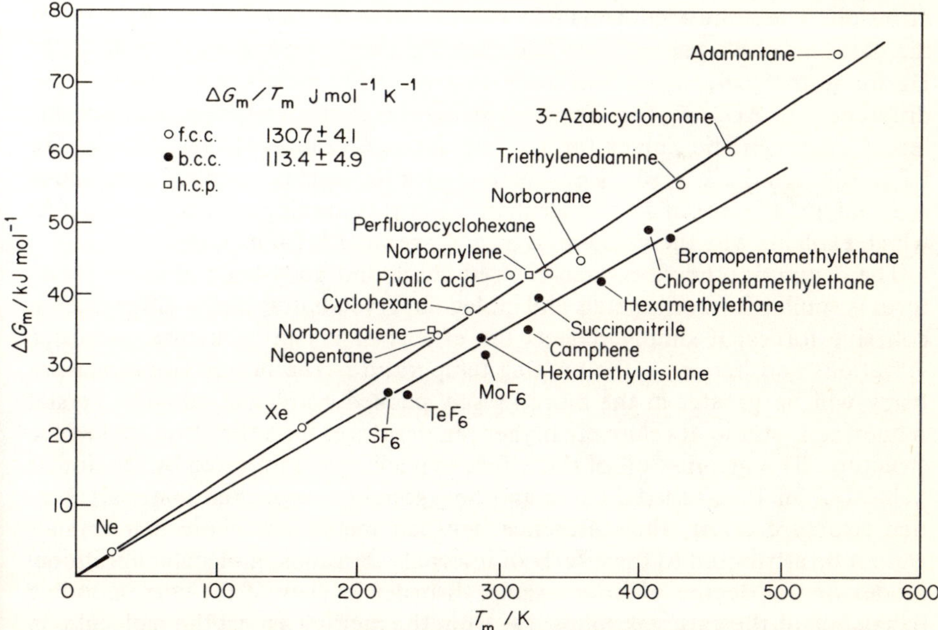

Figure 5.31 Free energy of self-diffusion at the melting point ΔG_m plotted as a function of the melting temperature T_m

for face-centred cubic metals he obtained

$$\Delta G_m / T_m = 116 \pm 9 \text{ J K}^{-1} \text{ mol}^{-1}$$

while for body-centred cubic metals

$$\Delta G_m / T_m = 96 \pm 8 \text{ J K}^{-1} \text{ mol}^{-1}.$$

Equations (5.3.51) and (5.3.52) are really special cases of Equation (5.3.53). This is seen on identifying E_a with the enthalpy of activation for diffusion which is related to T_m by

$$\Delta H = \{(\Delta G_m / T_m) + \Delta S\} \, T_m \tag{5.3.55}$$

Equation (5.3.51) follows from (5.3.53) if the entropy of diffusion ΔS is constant, while equation (5.3.52) is obtained from (5.3.51) by assuming the entropy of fusion ΔS_f is a constant. Equation (5.3.52) would not, therefore, be expected to hold for molecular crystals and the fact that equation (5.3.51) does not represent the data very well simply implies that ΔS is not a constant.

It is important to emphasize that the fact that $\Delta G_m / T_m$ in equation (5.3.53) is a constant for a particular crystal structure is simply a consequence of the observation that τ_m is a constant, since the value of ν_L is approximately constant. Thus, although equation (5.3.53) may be useful, as for example in the prediction of values for ΔS, it has no significance whatsoever for the mechanism of self-diffusion. The same arguments may be applied to the results for metals. From the data in Gibb's paper,[124] we find that the Debye frequency $\theta_D \cong 6 \times 10^{12}$ Hz for both face-centred and body-centred cubic metals, implying that the differences in $\Delta G_m / T_m$ for these two structures is again a consequence of different characteristic values for τ_m. The average values calculated for τ_m are 1.7×10^{-8} s and 2.2×10^{-9} s, respectively, for fcc and bcc metals. These values are roughly a factor of 40 shorter than the corresponding ones for non-metals which explains why the values of $\Delta G_m / T_m$ are smaller for metals.

The distinction between face-centred cubic and body-centred cubic structures is similar for both metals and molecular crystals despite the differences in cohesive forces; it simply reflects the effects of crystal structure on lattice vibrations and hence on the melting temperature. The lattice vibrational entropy will be greater in the more loosely packed body-centred cubic crystal rendering it stable at relatively higher temperatures than the close packed fcc structure. The assignation of this effect to packing is supported by the similar behaviour of face-centred-cubic and hexagonal-close-packed materials. Lattice structure apart, the difference between metals and molecular crystals cannot be attributed to the effects of internal vibrations, molecular librational modes or the degree of orientational disorder in view of the corresponding behaviour of the rare gas solids; i.e. from the melting aspect the molecules in orientally disordered cubic crystals behave as though they experience an approximately spherical intermolecular potential. On the other hand, the mechanism of self-diffusion is dependent on the orientational disorder of the solid and this is reflected by the dependence of the values of ΔS, estimated from equation (5.3.55), on ΔS_f.

The relationship between translational diffusion and melting is simply a consequence of the way the two processes are linked by the lattice dynamics. Similar relationships are also evident between molecular rotation frequences and the solid-plastic and melting transitions as outlined in Section 5.3.4.

It is interesting to note that there is much better agreement between the NMR and tracer values of $\Delta G_m/T_m$ than of E_a/L_s.

5.4 Conclusions

Studies of frequency spectra have been shown to be useful in that they enable the mechanisms of molecular motion occurring faster than the width of the 'rigid' lattice spectrum, typically 10^5 to 10^6 s^{-1}, to be identified. The rates of these motions may be determined by analysis of the shape of the spectrum in the line narrowing region, $\Delta\omega \cong 1/\tau$, but the activation parameters are not reliable as values of τ are only obtainable over one, or, at most, two orders of magnitude. In contrast, nuclear spin-relaxation measurements provide accurate values for τ in the interval 1 to 10^{-12} s.

In the ordered solid phase, translational diffusion is too slow to be detected. The reorientational motion is strongly influenced by the molecular symmetry. Molecules with octahedral or tetrahedral symmetry undergo quasi-isotropic reorientation, while those with an axis of symmetry undergo anisotropic reorientation about this axis; in both cases, $\tau_2 \simeq 5 \times 10^{-9}$ s at the solid-plastic transition temperature. For asymmetric molecules the rotational motion is quenched. In contrast, in the plastic phase there is an apparent uniformity of behaviour irrespective of molecular shape and both rotational and translational motions are easily studied. At the plastic-phase–liquid transition temperature $\tau_2 \cong 1$ to $2\,\tau_f$. This result implies that at least one-half the nearest neighbour molecules must be simultaneously rotating which, in view of the molecular packing in the plastic phase, requires the rotational motion to be correlated. There is obviously a need to study the collective rotational motion in relation to the 'coherence length' of the correlations in the orientation of the molecules, and, in particular, how this varies with temperature to see whether there is a gradual transition to single molecule reorientation as the temperature is reduced. Speculatively, it might be that the transition in the orientational ordering observed in crystals such as *dl*-camphene is driven by the requirements for correlated rotational motion as the melting temperature is approached. NMR experiments unfortunately monitor the behaviour of single molecules and it will be necessary to call on other techniques, such as coherent neutron scattering, to pursue this problem further.

The determination of the mean time interval τ between consecutive translational jumps of the molecules in plastic phases has been the major application of NMR in recent years. This has been a consequence of the development of $T_{1\rho}$ measurements which may be used to determine values of τ over some four decades down from the melting point; this corresponds to a temperature interval of typically 100 K and is to be compared with the much more restricted range, at best 50 K, available to radiotracer measurements. It is also important to emphasize that NMR measurements monitor molecular diffusion within the crystallites and there are no serious systematic errors due to sample crystallinity. This is a distinct advantage over the radiotracer method, but the NMR experiment monitors τ not D and it is important to measure both quantities in mechanistic studies. The values derived for τ and the activation parameters as obtained from measurements of $T_{1\rho}$ as a function of T and P are considered to be fairly reliable; this is because spin relaxation by translational diffusion is not very sensitive to the details of the motion and, therefore, to the model used. On the other hand, it is possible to distinguish between certain motional models from the shape of the $T_{1\rho}$ versus temperature plot; in this way it has been possible to demonstrate that the details of diffusion mechanism are dependent on the extent of the orientational disorder in the plastic phase as reflected by the value of ΔS_f.

For high ΔS_f solids ($\Delta S_f \geqq 2R$), diffusion is dominated by a monovacancy diffusion mechanism; in the case of hexamethylethane contributions from divacancies are detected at high temperatures. This conclusion is verified by the good agreement obtained between the NMR and tracer measurements when the nearest-neighbour lattice distance is used for $\langle r^2 \rangle^{1/2}$ in $D = \langle r^2 \rangle / 6\tau$. In the case of the low ΔS_f materials ($\Delta S_f \cong R$), the diffusion mechanism is far more complex as is indicated by both the pressure and temperature dependence of $T_{1\rho}$. There is an indeterminable distribution of jump lengths about a mean which must be greater than a molecular diameter; from the tracer D at the melting point it is estimated that, typically, $\langle r^2 \rangle^{1/2} \cong 2.5\, l$. This is not an implausible result for a low ΔS_f solid where the translational jumps of the molecule must be dependent on the relative orientation of a molecule with respect to its neighbours, i.e. there is a much greater degree of cooperation between the diffusing molecule and its neighbours than between an atom and its neighbours in a monatomic solid. The high ΔS_f solids are much more ordered and their compliance with a discrete monovacancy mechanism is quite acceptable. Between these two extremes, a gradual change in mechanism is indicated by the value of the ratio E_a/L_s which gradually decreases from 2 to 1 in going from the upper to the lower end of the ΔS_f range. For the low ΔS_f solids, the NMR and tracer measurements appear compatible at temperatures close to the melting points, but not at lower temperatures. There is no obvious explanation for these discrepancies. An attractive solution would be a temperature dependent value of $\langle r^2 \rangle$, but there is no evidence to support this hypothesis. Accepting that both the NMR and radiotracer measurements are free from any serious systematic errors, then it appears from the data available to date that the two techniques are monitoring distinct processes in different 'orientationally disordered forms' of the material. This is possible (see Chapter 2.6) though, to date, there is no direct experimental evidence that the different histories of the sample employed in the two experiments produce crystals in which the orientational packing is different.

An interesting result is that the translational jump frequencies converge to an approximately constant value at the melting point. This value is characteristically a factor of five greater for body-centred cubic than face-centred cubic solids and is independent of the degree of orientational disorder. This behaviour is due to the effect of crystal structure on lattice vibrations and hence on the melting temperature: the lattice vibrational entropy is greater in the more loosely packed body-centred cubic crystal rendering it stable at relatively higher temperatures than the close packed face-centred cubic or hexagonal close-packed structures. This relationship between translational diffusion and melting is simply a consequence of the way the two processes are linked by the lattice dynamics; similar, though less regular, relationships are also evident between molecular rotation frequencies and the ordered solid–plastic crystal and plastic crystal–liquid transitions.

REFERENCES

1. A. Abragam, *The Principles of Nuclear Magnetism*, Oxford U.P., London, 1961.
2. C. P. Slichter, *The Principles of Magnetic Resonance*, Harper and Row, New York, 1973.
3. M. Goldman, *Spin Temperature and Nuclear Magnetic Resonance in Solids*, Oxford U.P., London, 1970.
4. T. C. Farrar and E. D. Becker, *Pulse and Fourier Transform* NMR, Academic Press, New York, 1971.
5. D. C. Ailion, Adv. Magnetic Resonance, **5**, 177 (1971).
6. P. S. Allen, in *MTP International Review of Science–Physical Chemistry Series One*, ed. A. D. Buckingham, 1972 Vol. 4, p. 43.
7. D. I. Hoult and R. E. Richards, *Proc. Roy. Soc.* **A344**, 311 (1975).
8. A. J. Dianoux, S. Sykora, and H. S. Gutowsky, *J. Chem. Phys.*, **55**, 4768 (1971); G. W. Parker, *J. Chem. Phys.*, **58**, 3274 (1973).
9. J. G. Powles and J. H. Strange, *Proc. Phys. Soc.* **82**, 6 (1963).
10. J. Jeener and P. Broeckaert, *Phys. Rev.,* **157**, 232 (1967).
11. W. K. Rhim, A. Pines, and J. S. Waugh, *Phys. Rev.,* **B3**, 684 (1971).
12. I. J. Lowe, K. W. Vollmers, and M. Punkkien, *Pulsed Nuclear Magnetic Resonance and Spin Dynamics in Solids* Ed., J. W. Hennel, p. 70, Institute of Nuclear Physics, Krakow, Poland, 1973.
13. D. E. Barnaal and I. J. Lowe, *Rev. Sci. Instrum.,* **37**, 428 (1966); *Phys. Rev. Letters*, **11**, 258 (1963).

14. J. H. Van Vleck, *Phys. Rev.,* **74**, 1168 (1948).
15. N. Boden, M. Gibb, Y. K. Levine, and M. Mortimer, *J. Magn. Resonance*, **16**, 471 (1974).
16. H. S. Gutowsky and G. Pake, *J. Chem. Phys.,* **18**, 162 (1950).
17. E. R. Andrews and J. Lipofsky, *J. Magn. Resonance,* **8**, 217 (1972).
18. D. W. McCall and D. C. Douglas, *J. Chem. Phys.,* **33**, 777 (1960).
19. G. W. Smith, *J. Chem. Phys.,* **35**, 1134 (1961).
20. G. W. Smith, *J. Chem. Phys.,* **43**, 4325 (1963).
21. G. W. Smith, *J. Chem. Phys.,* **54**, 174 (1971).
22. J. M. Chezeau, J. Dufourcq, and J. H. Strange, *Mol. Phys.,* **20**, 305 (1971).
23. E. R. Andrew and R. G. Eades, *Proc. Roy. Soc.,* **A216**, 398 (1953).
24. A. Fratiello and D. C. Douglas, *J. Chem. Phys.,* 41, 974 (1964).
25. J. E. Anderson and W. P. Slichter, *J. Chem. Phys.,* **41**, 1922 (1964).
26. L. V. Dmitrieva and V. V. Moskalev, *Sov. Phys. Solid State,* **5**, 1623 (1964).
27. P. W. Anderson, *J. Phys. Soc. Japan,* **9**, 316 (1954).
28. R. Kubo and K. Tomita, *J. Phys. Soc. Japan,* **9**, 888 (1954).
29. M. Mehring, in '*NMR, Basic Principles and Progress*', *ed. P. Diehl*, Springer-Verlag, 1976, **11**, pp. 1–243.
30. D. E. O'Reilly, E. M. Peterson, C. E. Schele, and E. Seyfarth, *J. Chem. Phys.,* **59**, 3576 (1973).
31. U. Haeberlen, *High-Resolution NMR in Solids, Adv. Magn. Resonance*, Supplement 1, 1976.
32. H. W. Spiess, R. Grosescu, and U. Haeberlen, *Chem. Phys.,* **6**, 226 (1974).
33. A. Pines, M. G. Gibby, and J. S. Waugh, *J. Chem. Phys.*, **59**, 569 (1973).
34. H. W. Spiess, *Chem. Phys.,* **6**, 217 (1974).
35. S. Kaplan, private communication.
36. A. G. Redfield, *Adv. Magn. Resonance*, **1**, 1 (1965).
37. F. Noack, in '*NMR, Basic Principles and Progress*', ed. P. Diehl, Springer, New York, **3**, 83 (1971).
38. J. Jeener, *Adv. Magn. Resonance*, **1**, 205 (1965).
39. N. Boden, in '*Determination of Organic Structures by Physical Methods*', 1971, Vol. 4, p. 51.
40. N. Bloembergen, E. M. Purcell, and R. V. Pound, *Phys. Rev.*, **73**, 679 (1948).
41. R. Kubo and K. Tomita, *Proc. Phys. Soc. Japan,* **9**, 888 (1954).
42. D. C. Look and I. J. Lowe, *J. Chem. Phys.*, **44**, 2995 (1966).
43. S. Albert, H. S. Gutowsky, and J. A. Ripmeester, *J. Chem. Phys.,* **56**, 1332 (1972).
44. R. L. Hilt and P. S. Hubbard, *Phys. Rev.,* **A134**, 392 (1964); M. F. Band and P. S. Hubbard., *Phys. Rev.,* **A170**, 384 (1968).
45. J. Wendt and F. Noack, *Z. Naturforsch,* **29a**, 1660 (1974).
46. N. Boden, P. P. Davis, C. H. Stam, and G. A. Wesselink, *Mol. Phys.,* 25, 87 (1973).
47. D. C. Douglas and G. P. Jones, *J. Chem. Phys.*, **45**, 956 (1966).
48. D. W. McCall and D. C. Douglass, *Polymer,* **4**, 433 (1963).
49. C. P. Slichter and D. C. Ailion, *Phys. Rev.,* **A135**, 1099 (1964).
50. D. Wolf and P. Jung, *Phys. Rev.,* **B12**, 3596 (1975).
51. P. Rigny and J. Violet, *J. Chem. Phys.,* **51**, 3807 (1969).
52. N. Boden and R. Folland, *Mol. Phys.,* **21**, 1123 (1971).
53. S. Albert, H. S. Gutowsky, and J. H. Ripmeester, *J. Chem. Phys.,* **56**, 1332 (1972).
54. A. Zussman and S. Alexander, *J. Chem. Phys.,* **48**, 3534 (1968).
55. S. B. W. Roeder and D. C. Douglas, *J. Chem. Phys.,* **52**, 5525 (1970).

56. R. van Steenwinkel, *Z. Naturforsch,* **24a**, 1526 (1969).
57. R. L. Jackson and J. H. Strange, *Mol. Phys.,* **22**, 313 (1971).
58. H. A. Resing, *Mol. Cryst. Liq. Cryst.,* **9**, 101 (1969).
59. C. E. Nordman and D. L. Schmitkons, *Acta Cryst.,* **18**, 764 (1965).
60. W. J. Dunning, *J. Phys. Chem. Solids,* **18**, 21 (1961).
61. N. Boden, S. Hanlon, M. Mortimer, and S. Ross, unpublished results.
62. J. R. Green and D. R. Wheeler, *Mol. Cryst. Liq. Cryst.,* **6**, 1 (1969).
63. K. Adachi, H. Suga, and S. Seki, *Bull. Chem. Soc. Japan,* **43**, 1916 (1970).
64. M. Davis, *J. Chim. Phys.,* **63**, 67 (1966).
65. J. E. Anderson and W. P. Slichter, *J. Chem. Phys.,* **44**, 1797 and 3647 (1966).
66. R. Folland, S. M. Ross, and J. H. Strange, *Mol. Phys.,* **26**, 27 (1973).
67. N. I. Liu and J. Jonas, *Chem. Phys. Letters,* **14**, 555 (1972).
68. P. S. Hubbard, *Phys. Rev.,* **131**, 1155 (1963).
69. W. A. Steele, *Adv. Chem. Phys.*, **34**, 1 (1976).
70. N. Boden and R. Folland, *Chem. Phys. Letters,* **10**, 167 (1971).
71. N. Boden and R. Folland, *Chem. Phys. Letters,* **32**, 127 (1975).
72. J. G. Powles and M. Rhodes, *Phys. Letters*, **24A**, 523 (1967).
73. F. J. Bartoli and T. A. Litovitz, *J. Chem. Phys.*, **56**, 413 (1972).
74. S. Sunder and R. E. D. McClung, *Chem. Phys.*, **2**, 467 (1973).
75. M. Bloom and G. A. de Witt, *Can. J. Phys.,* **43**, 986 (1965).
76. D. E. O'Reilly, E. M. Peterson, and D. L. Hogenboom, *J. Chem. Phys.,* **57**, 3969 (1972).
77. C. Brot and B. Lassier-Govers, *Ber. Bunsen-Gesellschaft fur Phys. Chem.,* **80**, 31 (1976).
78. J. G. Powles, *Ber. Bunsen-Gesellschaft fur phys. Chem.,* **80**, 259 (1976).
79. D. Wolf. *Phys. Rev.,* **B10**, 2710 (1974).
80. D. Wolf, *Proc. 18th Ampere Congress,* p. 251, Nottingham (England).
81. H. C. Torrey, *Phys. Rev.,* **92**, 962 (1953).
82. H. C. Torrey, *Phys. Rev.,* **96**, 690 (1954).
83. H. A. Resing and H. C. Torrey, *Phys. Rev.,* **131**, 1102 (1963).
84. C. A. Sholl, *J. Phys. C,* **7**, 3378 (1974).
85. C. A. Sholl, *J. Phys. C,* **8**, 1737 (1975).
86. D. Wolf, *J. Magn. Resonance,* **17**, 1 (1975).
87. M. Eisenstadt and A. G. Redfield, *Phys. Rev.* **132**, 635 (1963).
88. T. G. Stoebe, T. O. Ogurtani, and R. A. Huggins, *Phys. Stat. Sol.,* **12**, 649 (1965).
89. R. N. Baughman and D. Turnbull, *J. Phys. Chem. Solids,* **33**, 121 (1972).
90. N. C. Lockhart and J. N. Sherwood, *Faraday Symp. Chem. Soc.,* **6**, 57 (1972).
91. N. Boden, J. Cohen, and P. P. Davis, *Mol. Phys.*, **23**, 819 (1972).
92. P. Bladon, N. C. Lockhart, and J. N. Sherwood, *Mol. Phys.*, **20**, 577 (1971).
93. P. Bladon, N. C. Lockhart, and J. N. Sherwood, *Mol. Cryst. Liq. Cryst.*, **19**, 315 (1973).
94. A. V. Chadwick, J. W. Forrest, A. A. Grimwood, and J. H. Strange, *Mol. Phys.,* **32**, 1773 (1976).
95. F. Volino, *Mol. Phys.,* **27**, 821 (1974).
96. R. Folland, R. L. Jackson, J. H. Strange, and A. V. Chadwick, *J. Phys. Chem. Solids,* **34**, 1713 (1973).
97. N. Boden, J. Cohen, and R. T. Squires, *Mol. Phys.,* **31**, 1813 (1976).
98. G. K. Krüger, *Z. Naturforsch,* **A24**, 560 (1969).
99. L. Pasemann and H. Schneider, *J. Magn. Resonance*, **9**, 255 (1973).
100. N. Boden, *Chem. Phys. Letters*, **46**, 141 (1977).

101. A. V. Chadwick, J. M. Chezeau, R. Folland, J. W. Forrest, and J. M. Strange, *J. C. S. Faraday 1*, **71**, 1610 (1975).
102. P. W. Salthouse and J. N. Sherwood, *J. C. S. Faraday I*, **73**, 1845 (1977).
103. J. N. Sherwood, *Specialist Periodical Report*, '*Surface and Defect Properties of Solids*', p. 250, Chemical Society, 1973.
104. G. K. Krüger, *Z. Naturforsch.,* **24A**, 560 (1969).
105. J. E. Tanner, *J. Chem. Phys.,* **56**, 3850 (1972).
106. N. Boden and S. Ross, unpublished results.,
107. N. T. Corke, N. C. Lockhart, R. S. Narang and J. N. Sherwood, *Mol. Cryst. Liq. Cryst.,* **44**, 45 (1978).
108. N. Boden and J. Cohen, *unpublished results.*
109. E. Hampton and J. N. Sherwood, *Phil. Mag.,* **30**, 853 (1974).
110. T. Sugawara, Y. Sakamoto, and E. Kanda, Sci. Rep., *Res. Insts. Tokyo Univ., 1949, Ser. A,* **1**, 29; T. Sugawara and E. Kanda, *Ibid*, 1949, A, **1**, 153.
111. D. E. C. Corbridge and E. J. Lowe, *Nature*, 1952, **170**, 629.
112. R. Folland and J. H. Strange, *J. Phys. C,* **5**, L50 (1972).
113. R. Folland S. M. Ross, and J. H. Strange, *Mol. Phys.,* **26**, 27 (1973).
114. J. H. Strange and S. M. Ross, *Mol. Cryst. Liq. Cryst.,* **32**, 67 (1976).
115. C. P. Flynn, *Point Defects and Diffusion*, Oxford, U. P., London 1972.
116. J. J. Burton and G. Jura, *J. Phys. Chem. Solids*, **28**, 705 (1976).
117. P. McKay and J. N. Sherwood, *J. C. S. Faraday 1*, **71**, 2331 (1975).
118. J. H. Strange and S. M. Ross, *private communication.*
119. L. C. Chhabildas and H. M. Gilder, *Phys. Rev.,***85**, 2135 (1972).
120. N. J. Trappeniers and F. A. S. Lightart, *Chem. Phys. Letters,* **19**, 465 (1973).
121. J. Virlet and P. Rigny, *J. Magn. Res.,* **19**, 188 (1975).
122. J. van Liempt, *Z. Phys.,* **96**, 534 (1935).
123. N. H. Nachtrieb and G. S. Handler, *Acta Met.,* **2** 797 (1954).
124. G. B. Gibbs, *Acta Met.,* **12**, 673 (1964).
125. J. J. Gilvarry, *Phys. Rev.,* **102** 308 and 317 (1956).
126. H. A. Resing, *J. Chem. Phys.,* **37**, 2575 (1962).
127. J. Virlet and P. Rigny, *Chem. Phys. Letters,* **4**, 501 (1970).
128. E. O. Stejskal, D. E. Woessner, T. C. Farrar, and H. S. Gutowsky *J. Chem. Phys.,* **31**, 55 (1959).
129. J. G. Powles, A. Begum, and M. O. Norris, *Mol. Phys.,* **17**, 489 (1969).
130. W. M. Yen and R. E. Norberg, *Phys. Rev.,* **131**, 269 (1963).
131. R. Henry and R. E. Norberg, *Phys. Rev.,* **B6**, 1645 (1972).
132. R. N. Bughman and D. Turnbull, *J. Phys. Chem. Solids,* **33**, 121 (1972).
133. N. C. Lockhard and J. N. Sherwood, *Faraday Symposia. Chem. Soc.*, **6**, 57 (1972).
134. E. M. Hampton, N. C. Lockhart, and J. N. Sherwood. *Chem. Phys. Letters,* **21**, 191 (1973).
135. E. M. Hampton and J. N. Sherwood *Private Communication.*
136. H. M. Hawthorne and J. N. Sherwood, *Trans. Faraday Soc.,* **66**, 1783 (1970).
137. H. M. Hawthorne and J. N. Sherwood, *Trans. Faraday Soc.,* **66**, 1792 (1970).

6

Rayleigh and Brillouin Scattering from Plastic and Related Crystals

A. J. Hyde

6.1 INTRODUCTION

Whilst experiments on the scattering of light by crystals have been carried out for some fifty years or more now, the number of experiments on plastic crystals is small. All but one have been carried out since the advent of the laser as a light source. The exceptional experiment was carried out in 1941 by Gross, who in collaboration with Raskin had been the first to experimentally confirm Brillouin's[11] predictions. In this experiment, broadening of the Rayleigh line was observed for solid camphene and hexachloroethane. For both materials the broadened line was depolarized and was almost as broad as in the liquid state.

6.2 EXPERIMENTAL METHODS

Brillouin scattering experiments have all been carried out with what is now regarded as conventional apparatus as used first by Chiao and Stoicheff[2] and described in at least one review article[3] and two books.[4,5] In the Brillouin scattering experiment, light from a laser source (usually Helium-Neon or Argon ion) is incident upon the crystal sample. The light scattered at some preselected angle (frequently 90°) passes into a Fabry–Perot interferometer. This can be either plane[2] or spherical[6] and scanned piezo-electrically[6] or by gas pressure changes.[2] The light from the interferometer passing through a small aperture falls onto a photomultiplier the output from which can be fed to a chart recorder or a C.R.O. display or some other mechanism for recording the spectrum. The Rayleigh components have been observed using either standard optical homodyne techniques[7,8,9] or, for very broad lines,[10] standard multiple monochromators. The appropriate polarized components can be obtained by insertion of a polarizer and/or an analyser in the apparatus.

The spectrum of light scattered from a plastic crystal is shown schematically in Figure 6.1. The central (Rayleigh) line is centred on the frequency of the incident radiation but may consist of three or four components:

(A) A component of the same polarization and same width as the incident light which arises from scattering from static impurities and imperfections in the crystal.

(B) A broadened component of the same polarization as the incident light arising from non-propagating entropy waves and related to the thermal conductance of the system.

(C) A considerably broadened depolarized component arising from molecular rotation or relaxation processes.

(D) A depolarized component arising if the polarizability is not isotropic.

Symmetrically placed on either side of the Rayleigh line lie the Brillouin lines.[11,12] These consist of a peak arising from interaction of the incident light with a longitudinal phonon (thermal acoustic wave) in the crystal and two peaks caused by transverse phonons. The longitudinal peak is more intense than the transverse ones. Depending on the crystal symmetry, orientation with respect to the incident and scattered beams, and the values of the elastic constants of the crystal, some of these three peaks will be absent or very weak. The Brillouin peaks will be broader than the incident radiation due to attenuation of the thermal acoustic waves in the crystal but the attenuation has to be very large to be accurately measurable by the change of peak width.

Peak A is usually totally unwanted and can yield almost no information.

Peak B has a width at half height related to the thermal diffusivity of the material.

Peak C has a width related to the relaxation time of the molecular motion (rotation or otherwise) taking place.

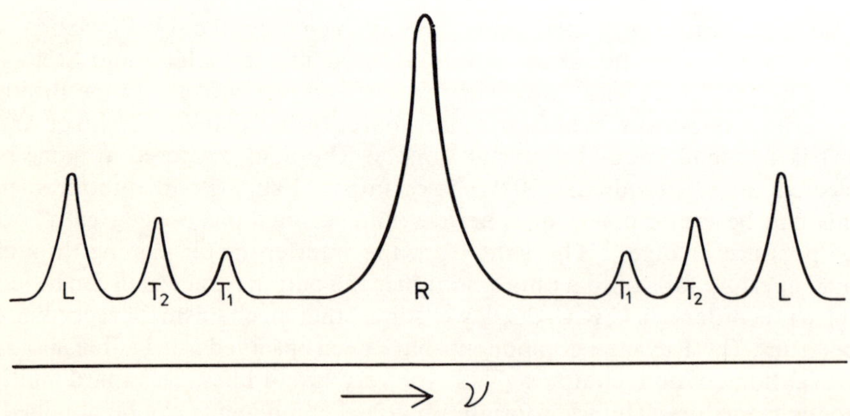

Figure 6.1 Schematic representation of a typical Rayleigh/Brillouin spectrum of a plastic crystal

Peak D gives information on the anisotropy of polarizability — if any exists.

The *full*-width at half-height (Γ) of peak B is given by[8,9]

$$\Gamma = \frac{2\Lambda}{\rho C_p} \cdot \mathbf{K}^2$$

where

$$\mathbf{K} = \frac{4\pi}{\lambda} \sin \frac{\theta}{2}$$

Λ is the thermal conductance, ρ the density, C_p the heat capacity per unit mass, θ the angle of observation and λ the wavelength of light in the medium.

The frequency shift of any Brillouin peak from the exciting line is related to the velocity of the relevant phonon as follows

$$\Delta\nu = \pm \frac{2V_s}{C} n\nu_0 \sin \frac{\theta}{2}$$

where $\Delta\nu$ is the shift, V_s the velocity of the phonon, ν_0 the incident light frequency, C the velocity of light in vacuum and n the refractive index of the material under experiment.

By suitable choice of the incident and scattered light directions and the orientation of the crystal axes with respect to these, the velocity of phonons in specific crystallographic directions may be obtained. The relationship of the acoustic velocities in specific directions to the elastic constants (C_{ij}) of the crystal are well known for all crystal symmetries.[13] e.g. for cubic symmetry†

$$\rho\, V^2_{\text{long}}\,(110) = \tfrac{1}{2}(C_{11} + C_{12} + 2C_{44})$$
$$\rho\, V^2_{\text{long}}\,(100) = C_{11}$$
$$\rho\, V^2_{\text{trans}}\,(110) = C_{44}$$

where V_{long} and V_{trans} are the longitudinal and transverse phonon velocities in the specified directions. C_{11}, C_{12} and C_{44} are the three independent elastic constants for the cubic system.

It is therefore possible in some cases to evaluate the elastic constants for the material.

If the intensities (relative to some standard—say toluene) of the appropriate longitudinal and transverse phonon Brillouin components and the polarized components of certain other lines are measured, the magnitudes of some of the Pockels (strain optical) coefficients for the material may be evaluated.[14]

6.3 RESULTS

The earliest experiments were those of the author and his colleagues[15] who measured the total scattered intensity and its horizontally and vertically

† The terminology for crystal properties will be that used by Nye.[13]

polarized components and low resolution Brillouin spectra for single crystals of succinonitrile and pivalic acid. Hypersonic sound velocities and Landau—Placzek Ratios[12] (L.P.R. = Central line (Rayleigh) Intensity/Total Brillouin Intensity) were measured from the Rayleigh-Brillouin spectra. In the total intensity measurements a dip in scattered intensity occupying a 10° range below the melting point was found for succinonitrile.

Experiments appear to have been made on only six molecular plastic crystals and the series of rare gases and so it becomes convenient to review the results compound by compound.

The rare gases are of necessity rotator phase molecular crystals but since the molecules are monatomic one might expect their behaviour to be quantitatively different from the globular molecular crystals.

Pivalic acid and succinonitrile are atypical rotator phase molecules from a chemical structural point of view. The former contains hydrogen bonds, i.e. much stronger intermolecular forces than the usual dispersion and dipolar forces present in plastic crystals. The latter is almost certainly dependent on the existence of rotational isomers for its plastic behaviour and is complicated by the intramolecular rotational isomerization process concerned. For these reasons the compounds will be discussed in the following (non chronological) order: firstly the globular hydrocarbon molecules adamantane and norbornylene (bicyclo [2,2,1]hept-2-ene) followed by the atypical pivalic acid and succinonitrile, then the most recently measured carbon tetrabromide and carbon tetrachloride and finally the group of rare gases.

6.3.1 Adamantane

This material exists in a face-centred cubic rotator phase from 208.6 K up to the melting point at 543 K (7 atmos. pressure). An investigation of adamantane has been made over a wide temperature range by Krasser, Koglin, Schneider and Stockmeyer.[16] Elastic constants have been evaluated from the acoustic velocities derived from Brillouin splittings as follows:

$$C_{11} = 0.99 \times 10^9 \,\mathrm{N\,m^{-2}}$$
$$C_{12} = 0.33 \times 10^9 \,\mathrm{N\,m^{-2}}$$
$$C_{44} = 0.27 \times 10^9 \,\mathrm{N\,m^{-2}}$$

The velocities obtained (*ca.* 900 $\mathrm{ms^{-1}}$) are clearly in error for solids, being about two to three times too low.

At the end of a recent paper by Damien[17] on acoustic (4 MHz) measurements of the elastic constants of adamantane, mention is made of a measurement of the hypersonic velocities using Brillouin scattering. This is reported to show agreement with the acoustic measurements which give elastic constants 6 to 10 times (and hence velocities 2.5 to 3 times) higher than those quoted by Krasser *et al.*

Dr. Krasser and his coauthors have reinterpreted their results[18] to give hypersonic velocities 1680 ms^{-1} and elastic constants

$$C_{11} = 3.1 \times 10^9 \text{ N m}^{-2}$$
$$C_{44} = 0.51 \times 10^9 \text{ N m}^{-2}$$
$$C_{12} = 1.9 \times 10^9 \text{ N m}^{-2}$$

Dr. Damien has also informed the author that his Brillouin scattering experiments give

$$C_{11} = 7.6 \times 10^9 \text{ N m}^{-2}$$
$$C_{12} = 4.5 \times 10^9 \text{ N m}^{-2}$$
$$C_{44} = 3.4 \times 10^9 \text{ N m}^{-2}$$

These last results when compared with 4 MHz acoustic values show strong dispersion for C_{11} and C_{12} but not C_{44}.

From the Brillouin spectrum published by the German authors and the quoted free spectral range of their interferometer the author estimates a sound velocity of *ca.* 2540 m^{-1} which is in good agreement with the French acoustic and Brillouin values.

6.3.2 Norbornylene (Bicyclo[2,2,1]hept-2-ene)

This material exists in a hexagonal close-packed rotator phase from 129 K up to the melting point at 320 K. It is the most complicated crystal (as opposed to molecule) yet studied and has been examined by Jackson and his colleagues at the University of Kent.[10,19] It is unusual for a plastic crystal in that it is hexagonal instead of cubic. This means that it is uniaxial birefringent and its elastic properties are described by five independent elastic constants. The birefringence may be determined by observation of the change in polarisation of a laser light beam as it passes through the crystal. This manifests itself as visual changes in the amount of light scattered at right angles to the beam, being a maximum when the direction of polarization is perpendicular to the direction of observation. Measurement of the distance (d) between maxima in the beam gives the birefringence using[20]

$$d = \frac{\lambda}{n_E - n_o}$$

For norbornylene this was found to be $n_E - n_o = 8.6 \pm 0.7 \times 10^{-4}$ for $\lambda =$ 514.5 nm at 296 K.

In order to determine all five elastic constants measurements have to be made with three different crystal orientations instead of the two for the cubic case. Both V_v and V_h spectra (the capital letter indicates the state of polarization of the incident beam and the subscript that for the scattered beam) were recorded and the intensities of the Brillouin peaks relative to those in toluene were measured.

The analysis for obtaining the elastic constants C_{12} and C_{13} is complicated and is given in considerable detail in the paper; C_{11}, C_{33}, and C_{44} are obtained directly from phonon velocities in the x $[2\bar{1}\bar{1}0]$ and z $[0001]$ directions. The constants obtained are:

$$C_{11} = 4.10 \pm 0.08 \times 10^9 \text{ N m}^{-2}$$
$$C_{33} = 4.01 \pm 0.08 \times 10^9 \text{ N m}^{-2}$$
$$C_{12} = 3.59 \pm 0.18 \times 10^9 \text{ N m}^{-2}$$
$$C_{13} = 3.00 \pm 0.32 \times 10^9 \text{ N m}^{-2}$$
$$C_{44} = 0.91 \pm 0.03 \times 10^9 \text{ N m}^{-2}$$

The equivalence, within experimental error, of C_{11} and C_{33} means that the two phonons are pure modes in this material and not mixed modes as they would be in general.

From the scattered intensities, relative to toluene, assuming an effective Pockels coefficient for toluene and neglecting the weak birefringence, values of the magnitudes of three of the five independent Pockels coefficients (p_{ij}) are obtained:

$$p_{12} = 0.31 \pm 0.02$$
$$p_{13} = 0.28 \pm 0.02$$
$$p_{44} = 0.03 \pm 0.005$$

Molecular reorientation in the crystal was studied by measuring the broad depolarized Rayleigh line out to $\sim 100\ \text{cm}^{-1}$. Up to $\sim 5\ \text{cm}^{-1}$ the line shape can be approximated quite well by a Lorentzian having a width corresponding to reorientational correlation times obtained from Raman line shapes and NMR measurements—once some allowance has been made for leakage of V_v light arising from the low scattering power. Between $5\ \text{cm}^{-1}$ and $100\ \text{cm}^{-1}$ however the scattered intensity is very much greater than the theoretical Lorentzian—by some four or five times in some cases–and the line shape seems exponential rather than Lorentzian.

The authors consider two possible explanations—these being that (a) the observations are of the free rotor part of the relevant correlation function or (b) intermolecular effects are present. The former seems unlikely since the depolarized Raman lines are Lorentzian in shape up to $15\ \text{cm}^{-1}$. On the other hand the theoretical second moment for a line broadened by rotation alone can be calculated to be $\sim 280\ \text{cm}^{-2}$ whereas the experimental value is $400\ \text{cm}^{-2}$ indicating intermolecular interaction. Bucaro and Litovitz[21] have pointed out the importance of collision-induced anisotropy of polarizability for liquids in giving rise to depolarized scattering and have predicted a frequency dependence of the intensity arising from this as:

$$I(\omega) \propto \omega^{12/7} \exp(-\omega/\omega_0)$$

for $\omega > \omega_0$, ω_0 being a constant. For $\omega > 40$ cm^{-1} the liquid norbornylene follows this behaviour with $\omega_0 \sim 13$ cm^{-1} a value similar to those found by Bucaro and Litovitz for molecular liquids. The line shape for the solid is almost identical with that for the liquid which suggests that not only is the reorientation motion little affected by solidification but the process responsible for the higher frequency spectrum is also unaffected.

6.3.3 Pivalic Acid (Trimethylacetic acid)

Pivalic acid exists from 280.1 K to the melting point at 309.7 K in a rotator phase face-centred cubic form. The acid molecule can hydrogen bond (both bonds being formed with one neighbour) with any of its twelve nearest neighbours.

The earliest measurements[15] on this substance gave phonon velocities varying from 2033 to 1870 ms^{-1} as the temperature increase from 296 K up to the melting point.

Some years after the original measurements pivalic acid was reinvestigated by the French groups at Montpellier and Lille.[22] Working at higher resolution they observed Brillouin components from both longitudinal and transverse phonons. The velocity of the longitudinal phonons, obtained from the Brillouin shift, varied from 1644 to 1757 ms^{-1} and of the transverse from 868 to 635 ms^{-1} depending on crystal orientation. No Rayleigh line broadening was found, unlike succinonitrile, and only scatter from impurities and entropy fluctuations was observed. The width of the Brillouin lines was around 310 MHz i.e. less than half that previously observed in succinotrile.

A more extensive investigation was made by the group at the University of Kent about the same time.[14] For the polarized spectra (Brillouin lines) two crystal orientations were used (a) incident light along $[1\bar{1}0]$; 90° scattered light along $[\bar{1}\bar{1}0]$ has components associated with phonons propagating along $[100]$ (b) incident light along $[100]$, scattered light along $[0\bar{1}0]$ contains components associated with phonons along $[110]$. Brillouin line intensities for various states of polarization of incident and scattered light were measured as well as the splittings, and the Brillouin spectrum of toluene was measured under identical conditions in order to evaluate the Pockels coefficients (p_{ij}). Measurements were made from 282 K up to 302 K. The $[100]$ phonon is longitudinal and its velocity ranges from 1891 to 1828 ms^{-1} as the temperature rises but the Brillouin line half-width is constant at 140 MHz. Two phonons propagate along $[110]$, the transverse one being observed in the V_h spectrum and the longitudinal in the V_v. The longitudinal phonon velocity decreases from 1818 to 1747 ms^{-1} as the temperature rises from 282 K to 302 K and the half-width falls from 138 to 110 MHz. The transverse phonon velocity falls from 819 to 805 ms^{-1} in the same range but the peaks are too small for widths to be measured accurately. The range of values of the elastic constants over the temperature range are:

$$C_{11} = 3.54 \text{ to } 3.30 \times 10^9 \text{ N m}^{-2}$$
$$C_{12} = 1.68 \text{ to } 1.46 \times 10^9 \text{ N m}^{-2}$$
$$C_{44} = 0.664 \text{ to } 0.642 \times 10^9 \text{ N m}^{-2}$$

From the Brillouin line intensities relative to the toluene Brillouin line intensities and assuming an effective Pockels coefficient (p_{ij}) for toluene, the values of the magnitude of p_{12}, p_{44}, $p_{11} - p_{12}$ are obtained. Only the magnitudes are available since the scattered intensity is related to p^2. The values obtained in the same temperature range are:

$$p_{12} = 1.405 \text{ to } 1.392$$
$$p_{44} = 0.188 \text{ to } 0.226$$
$$p_{11} - p_{12} = 0.420 \text{ to } 0.343$$

For amorphous solids $p_{44} = \frac{1}{2}(p_{11} - p_{12})$; At about 18°C this is so for pivalic acid but presumably accidentally, since, as expected, the relation does not hold at other temperatures. The width of the depolarized part of the Rayleigh line was found to increase with increasing temperature. The line shape could be fitted with a Lorentzian, from the half-width of which reorientational correlation times (τ_R) were found. The value of τ_R at 306.4 K was 0.62×10^{-8} s and exhibited an Arrhenius type temperature dependence from which an activation energy for the reorientational process of 60 ± 4 kJ mol^{-1} was found. The reorientational correlation time is much greater than that for liquids composed of anisotropically polarizable molecules and for the plastic crystal studied previous to this one, succinonitrile; $\tau_R \sim 10^{-1}$ s. The value of the activation energy is quite close to the known energy of dimerisation of pivalic acid, by formation of two hydrogen bonds, namely 56 k*J* mol^{-1}.[23] In contrast to succinonitrile (see below) no trace of a very broad depolarized spectral component was found.

More recently, the French group at Saclay have examined more closely the Rayleigh line for pivalic acid.[24] They have resolved the line into three parts all centred at the incident frequency: (a) a very narrow line due to elastic scattering from static impurities, (b) a somewhat broadened polarized line arising from non-propagating entropy fluctuations and (c) a considerably broadened depolarized line arising from anisotropy fluctuations. Component (b) is Lorentzian and from the width at half height a value of 6.7×10^{-8} m^2s^{-1} is found for the thermal diffusivity of pivalic acid. This is of the expected magnitude for soft organic materials as evaluated from the very few thermal conductance measurements available for these materials.

From the depolarized part of the spectrum, (c), a value of the reorientational correlation times of the molecules is found: $\tau_R = (\pi\Gamma)^{-1}$ (Γ being the full-width of the line at half height). From the Arrhenius type variation of τ_R with temperature in both the plastic crystal and liquid range, activation energies of 31 kJ mol^{-1} and 15.4 kJ mol^{-1} respectively are found. The molecular reorientational correlation time of the molecules at 286 K is found as 1.3×10^{-8}s (cf. 3.2×10^{-8}s found by the Kent group).

The authors make some comparisons with activation energies and reorientational correlation times obtained by NMR[25] and acid proton radiotracer diffusion[26] measurements and find rough agreement with their results. There is however great disagreement between the two light scattering measurements, the reorientational correlation times (τ_R) differing by a factor of about two and a half and the temperature variation of τ giving rise to activation energies differing by a factor of two. Whilst it does not follow that the different experiments are all measuring the same process the discrepancy between the light scattering experiments (which do measure the same thing) is a serious one.

6.3.4 Succinonitrile (1.2 Dicyanoethane)

This material exists in a body-centred cubic rotator phase from 233 K to the melting point at 331.3 K. It is the most complicated molecule of those considered here by virtue of its existence in three rotational isomeric forms (two *gauche* and one *trans*). There is about 20% of the *trans* form and 40% of each *gauche* form in the plastic phase. This means that in some ways the crystal can be regarded as being a three component mixed crystal since the two *gauche* forms are not superimposable and the *trans* form has quite a different shape. It seems possible that the existence of the plastic phase of succinonitrile is due to this isomerism since the similar rigid molecules, maleonitrile and fumaronitrile do not appear to have a plastic phase. The isomerism gives rise to an extra effect, in that, as well as rotation of the whole molecule occurring, there is rotation of the nitrile groups around the central C–C bond.

The earliest Brillouin scattering[15] measurements gave a hypersonic velocity of 2353 ms^{-1}. A dip (of about 10%) in the total scattered intensity measurements followed by a similar rise was observed over the eight or nine degrees below the melting point.

Subsequent work has all been done by the groups from Montpellier, Saclay, and Lille and the University of Kent. In their first experiments[27] the French observed Brillouin lines, a Rayleigh component due to static impurities and entropy fluctuations and a broad depolarized Rayleigh component. The last had a constant half-width at half-height of 2.7 GHz independent of angle of observation. This corresponds to a rotational correlation time of 5.9×10^{-11}s.

The first short paper from the English group studied the polarized and depolarized spectra.[28] A pair of polarized Brillouin lines had a splitting which corresponded to a hypersonic sound velocity of 2180 ± 25 m s^{-1}. No shear phonons were observed and from the Lorentzian profile of the depolarized Rayleigh component a relaxation time of 4×10^{-11} s was found [cf. NMR $\tau_R = 4.4 \times 10^{-11}$ s].

The French published two further papers in the same year. In one[29] they reported polarized and depolarized Rayleigh components, a Brillouin doublet and parasitic light at ν_0. They also observed a continuous background of unknown origin. A depolarization factor, $\rho = 0.55 \pm 0.15$ independent of angle, temperature and sample orientation was found. The depolarized Rayleigh line had a Lorentzian profile the full width at half height of which ($\Gamma = 2.5$ GHz at 20 °C) was independent of scattering angle. The temperature variation could be expressed in Arrhenius form:

$$\Gamma = \Gamma_0 \, e^{-E/RT}$$

with

$$E = 20.8 \pm 1.2 \text{ kJ mol}^{-1}$$

$$\Gamma_0 = 2.50 \pm 0.6 \times 10^{13} \text{ Hz}$$

in the plastic phase and

$$E = 12.5 \pm 3 \text{ kJ mol}^{-1}$$

$$\Gamma_0 = 2.8 \pm 1.8 \times 10^{12} \text{ Hz}$$

in the liquid. The peak increases in intensity about 2.3 times on melting. The above results agree with values found from dielectric relaxation measurements.[30]

A strongly depolarized Rayleigh component with width independent of scattering angle is characteristic of a molecular rotation process—in this case of relaxation time 6.4×10^{-11} s at 20 °C. With succinonitrile, two types of motion are possible: (a) rotation of the CN group around the three-fold axis (central C–C bond direction) or (b) rotation of the molecule in the *trans* conformation about the fourfold axis (roughly the nitrile group axis direction). The polarizability ellipsoid of the molecule in the *trans* conformation is almost an ellipsoid of revolution with its principal axis almost along the fourfold axis. If rotation of type (b) occurred there would be almost no change of induced dipole moment. Since there is a change, it would appear that process (a) is more likely.

In the second paper[31] the Brillouin scattering was studied and hypersonic velocities and attenuations found from the shifts ($\Delta\nu$) and line widths using the usual equations:

$$V = \frac{C}{2n}\frac{\Delta\nu}{\nu}\left(\sin\frac{\theta}{2}\right)^{-1}$$

and

$$\alpha = \pi\Gamma/V$$

No transverse Brillouin components were observed, the elastic anisotropy ($C_{11} - C_{12}/2C_{44}$) was found to be 1.18. The maximal acoustic velocities along [100] and [111] were 2218 m s^{-1} and 2183 m s^{-1} respectively. No variation of acoustic velocity with angle of observation (i.e. frequency or wave vector **K**) was found, but the velocity at 10 GHz shows marked dispersion at lower temperature (273 K to 233 K) when compared with direct acoustic measurements at 6 MHz—the velocity difference being ~10% at 233 K. The velocity varies rapidly with temperature and the thermoelastic coefficient

$$\frac{1}{C_{11}} \cdot \frac{dC_{11}}{dT} \frac{1}{\rho V^2} \frac{d(\rho V^2)}{dT} = 4.5 \times 10^{-3} K^{-1}$$

On melting there is a discontinuity in the hypersonic velocity as expected. The Brillouin linewidth varies linearly with the square of the wave vector ($\mathbf{K}^2$) and around 10 GHz varies very little with temperature over the range 233 to 293K having a value $\Gamma = 0.35 \pm 0.03$ GHz. The associated relaxation time $\tau = (\pi\Gamma) = 4.5 \pm 0.4 \times 10^{-10}$ s is suggested to arise from relaxation between rotational isomeric forms of the molecule.

The group from the University of Kent also produced a longer paper in 1971.[32] The $V_v + V_h$ and V_h spectra which they obtained exhibit the following characteristics, some of which are normally associated with solids and some with liquids: Solid like: Low Landau–Placzek Ratios (0.7 to 0.35), a high longitudinal hypersonic velocity (lying between glasses and soft plastics), and a shear phonon velocity with a value about half that of the longitudinal velocity. Liquid like: A broad depolarized Rayleigh component whose width is strongly temperature dependent and Brillouin linewidths about five times broader than those for normal solids. The longitudinal phonon velocity increases linearly with decreasing temperature down to about 15 °C where there is a discontinuity in slope, the new rate of increase being constant but greater. It is suggested that the relaxation effects below 15 °C cause the extra increase in velocity. Assuming that only one relaxation process is taking place and extrapolating to find the 'zero frequency' velocity at low temperatures, the authors find a value of $\tau = 8.6 \times 10^{-12}$ s for the relaxation process.

At low temperatures no shear phonons are evident from the spectra but above 283 K two small peaks appear. From the splitting an acoustic velocity of 713 ms^{-1} at 2.8 GHz is found.

It was found that the depolarized Rayleigh component could not be fitted to a single Lorentzian but by a Lorentzian superimposed on a broad line of unknown profile. Reorientational correlation times obtained from the Lorentzian part of the profile were around 10^{-10} s and showed an Arrhenius type temperature behaviour from which an activation energy of 16 kJ mol^{-1} was found.

The rotational correlation times as a function of T^{-1} are compared graphically for light scattering, NMR and dielectric measurements and are found to be in the same region of the time scale, but since the NMR measurements mirror the effects of reorientational motion on the inter and intramolecular proton–proton nuclear dipole interactions, the dielectric measurements follow the electric dipole reorientations and the light scattering depends on the reorientation of the molecular polarizability ellipsoid, the reorientational correlation times from the different experiments need not necessarily be equal.

The activation energy for reorientation (16 kJ mol^{-1}) is compared with those from the dielectric and NMR studies on the plastic and liquid phases:

Dielectric	11.3 kJ mol^{-1}
NMR	25 kJ mol^{-1} in solid, 16.4 kJ mol^{-1} in liquid.

In 1972 the French workers published a note on the polarised Rayleigh component.[7] The spectra measured at different angles (i.e. wave vector **K**) were fitted with Lorentzians and the line width found linear in $\mathbf{K}^2$. The thermal diffusivity was evaluated as $1.5 \times 10^{-7}\,m^2s^{-1}$. (cf. pivalic acid $6.8 \times 10^{-8}\,m^2s^{-1}$).

In a further paper[33] that year, the Brillouin spectra are again discussed. It is noted that the dispersion of the hypersonic acoustic velocity at −40 °C is three times that between the [100] and [111] directions at 6 MHz and the attenuation is about twenty times that expected from the theory of Bommel and Dransfeld.[34] The linewidth *vs*. $\mathbf{K}^2$ can be fitted either with a straight line or a curve of the form $\Gamma = (\omega^2\tau)/(1 + \omega^2\tau^2)$ where ω is the acoustic frequency and τ the relaxation time.

Components due to transverse phonons were not observed and the longitudinal phonon velocity shows, as might be expected, some variation with crystal orientation; 50 m s^{-1} at most. At 238 K there is considerable velocity dispersion (from 2350 m s^{-1} at low frequencies up to 2550 m s^{-1} at *ca*. 15 GHz) but at 293 K very little dispersion is observed. From the linewidths, relaxation times are evaluated—these are shorter than the reorientation times found previously e.g. τ_R (at 331 K) $= 7 \times 10^{-12}$ s i.e. six times less than τ at room temperature. They show Arrhenius type behaviour from which an activation energy of 12.5 kJ mol^{-1} is found.

The calculated attenuation is similar to the experimental value giving a maximum at 273 K (The hypersonic velocity at 9 GHz and 331 K $\approx V_0$, i.e. there is no dispersion).

The most recent paper[35] from the French group at Montpellier is mainly concerned with problems of crystal orientation. From the spectra obtained with a crystal in 38 different orientations a procedure is devised for locating the axes of the crystal. The absolute orientation having been established, experiments were carried out with the light incident along several crystal directions in several states of polarization to check the optical behaviour. The Rayleigh component profiles were fitted with the sum of two Lorentzians, a small amount of parasitic elastic scatter being neglected. For two different crystal orientations the relaxation times found from the Rayleigh linewidths are 6.2×10^{-11} s and 4.4×10^{-11} s (c.f. earlier value of 6.4×10^{-11} s). It is suggested that the former is associated with the residence time of the nitrile group in the *gauche* and *trans* positions and the latter with the relaxation of the nitrile group and a relaxation connected with orientational disorder created by hypersonic waves.

There seems to be satisfactory agreement between the rotational relaxation times obtained, these lying largely between 4×10^{-11} s and 10×10^{-11} s. Two directly obtained values stand apart: the 7×10^{-12} s and the 4.5×10^{-10} s obtained by the French both from Brillouin linewidths.

Whether these two very different values arise due to the greater difficulties of measuring the Brillouin linewidths is not clear, but they should be in much closer agreement.

The value of 8.6×10^{-12} s obtained indirectly by the group at Kent may not refer to an orientational relaxation at all.

The two activation energies obtained from the Rayleigh broadening are in considerable disagreement and the very short relaxation time found from the Brillouin linewidths would seem to cast doubt on the energy value obtained.

6.3.5 Carbon Tetrachloride

Carbon tetrachloride exhibits two 'plastic' phases, the first (A) face-centred cubic in structure and the other (B) rhombohedral.

All available information[36,37] fits the following scheme

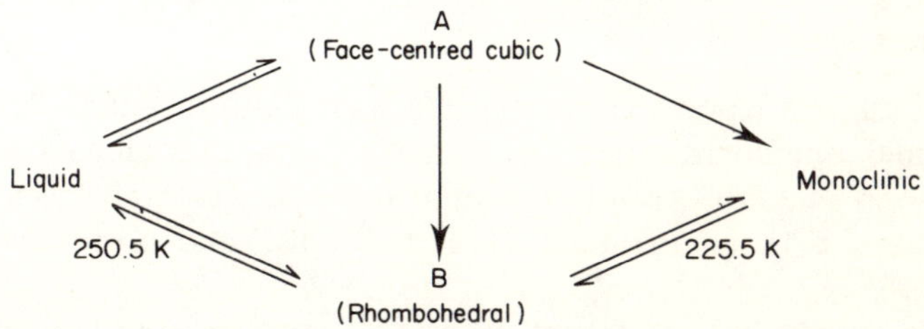

A is metastable with respect to B at all temperatures between 225.5 K and 250.5 K.

Rayleigh–Brillouin spectra of polycrystalline samples of the two 'plastic' phases have been obtained;[38] for A from the solid which first appears on cooling the liquid and for B by allowing the solid state conversion A → B to occur, melting all but a seed of B and then allowing this to grow (apparently not in such a way as to form a single crystal nor even a polycrystal of large crystallite size).

Phase A exhibits the usual pair of longitudinal Brillouin lines and a pair of transverse lines with a frequency shift about one third that for the longitudinal lines. Phase B shows a pair of longitudinal Brillouin lines. One measurement only of the Brillouin shift in phase A is shown and this seems to be several per cent smaller than for phase B. In phase B the hypersonic velocity falls roughly linearly from about 1720 m s^{-1} at 228 K to *ca.* 1590 m s^{-1} at the melting point where it falls abruptly to about 1240 m s^{-1} and then slowly to *ca.* 1050 m s^{-1} at 293 K. The linewidth of the Brillouin lines in phase B increases as the temperature is raised: from 0.2 GHz at −55 °C to 0.3 GHz at the melting point where there is a sudden jump to *ca.* 0.55 GHz in the liquid.

There is considerable scatter on the points in the solid phases and some hysteresis is observed near the melting (freezing) point which arises with supercooling and the formation of phase A at 245.2 K which transforms to phase B which then melts at 250.5 K on heating.

6.3.6 Carbon Tetrabromide

Carbon tetrabromide has only one plastic phase which exists between 319 K and 365 K.

The Brillouin spectra obtained[39] in this phase exhibit a longitudinal and two transverse pairs of phonon peaks (c.f. CCl_4). The transverse velocity is *ca.* 650 m s^{-1} and the longitudinal velocity decreases from 1239 m s^{-1} to 1170 m s^{-1} over the temperature range 330 K to 362 K. As with other materials there is a sharp fall (to 791 m s^{-1}) on melting and then a slow decrease comparable with that below the melting point.

No softening of the longitudinal mode was observed as the melting point was approached and no change in linewidth observed at the transition (c.f. CCl_4).

6.3.7 Rare Gases

The rare gases are perforce rotator phase molecules. With the exception of helium (which one might expect to be an exception) they all have entropies of fusion close to 14 J mol^{-1} K^{-1} (c.f. pivalic acid 6.52 J mol^{-1}K^{-1}, succinonitrile 11.20 J mol^{-1} K^{-1}, adamantane 21 J mol^{-1} K^{-1} and norbornylene 10.1 J mol^{-1}K^{-1}).

All these substances crystallize with face-centred cubic lattices and have melting points and entropy of fusion as follows:

	T_m/K	$\Delta S_f/J$ mol^{-1}K^{-1}
Ne	24.56	13.84
Ar	83.75	14.09
Kr	115.9	14.15
Xe	161.3	14.26

They form the only group of similar materials on which light scattering measurements have been made. These measurements have all been made by Stoicheffs group at the University of Toronto.

The first materials examined were xenon and neon.[40,41] Xenon was examined in eight orientations at 186 K, in six of these orientations a pair of longitudinal and a pair of transverse phonon Brillouin lines were observed. In the other two, a second pair of transverse lines appeared. Elastic constants were evaluated by two independent forms of analysis to yield:

$$C_{11} = 2.98 \pm 0.04 \times 10^9\,\mathrm{N\,m^{-2}}$$
$$C_{12} = 1.9 \pm 0.03 \times 10^9\,\mathrm{N\,m^{-2}}$$
$$C_{44} = 1.48 \pm 0.04 \times 10^9\,\mathrm{N\,m^{-2}}$$
$$A = \frac{2C_{44}}{C_{11} - C_{12}} = 2.74 \pm 0.3$$

For neon, only one longitudinal and one transverse phonon were observed and from measurements on these at two suitable crystal orientations the following elastic constants were found:

$$C_{11} = 1.18 \pm 0.06 \times 10^9\,\mathrm{N\,m^{-2}}$$
$$C_{12} = 0.77 \pm 0.06 \times 10^9\,\mathrm{N\,m^{-2}}$$
$$C_{44} = 0.56 \pm 0.06 \times 10^9\,\mathrm{N\,m^{-2}}$$

In the following year, some more work was published on neon and argon.[42] An argon ion laser was used for experiments on neon because of its low scattering power. For neon at 24.3 K the following elastic constants were found:

$$C_{11} = 1.175 \pm 0.02 \times 10^9\,\mathrm{N\,m^{-2}}$$
$$C_{12} = 0.74 \pm 0.02 \times 10^9\,\mathrm{N\,m^{-2}}$$
$$C_{44} = 0.595 \pm 0.015 \times 10^9\,\mathrm{N\,m^{-2}}$$

For argon at 82.0 K they found:

$$C_{11} = 2.33 \pm 0.08 \times 10^9\,\mathrm{N\,m^{-2}}$$
$$C_{12} = 1.49 \pm 0.06 \times 10^9\,\mathrm{N\,m^{-2}}$$
$$C_{44} = 1.17 \pm 0.07 \times 10^9\,\mathrm{N\,m^{-2}}$$

A short publication in 1973[43] points out the difference between the 'first sound' elastic constants at 114 K found from Brillouin scattering for krypton and the 'zero sound' constants from neutron scattering:

'First sound'–Brillouin scattering
(sound frequency x phonon lifetime $\ll 1$)

$$C_{11} = 2.657 \pm 0.03 \times 10^9\,\mathrm{N\,m^{-2}}$$

$$C_{12} = 1.725 \pm 0.02 \quad \times 10^9\,\mathrm{N\,m^{-2}}$$
$$C_{44} = 1.261 \pm 0.015 \times 10^9\,\mathrm{N\,m^{-2}}$$

'Zero sound'—Neutron scattering
(sound frequency x phonon lifetime $\gg$ 1)

$$C_{11} = 2{,}89 \pm 0.04 \times 10^9\,\mathrm{N\,m^{-2}}$$
$$C_{12} = 1.85 \pm 0.04 \times 10^9\,\mathrm{N\,m^{-2}}$$
$$C_{44} = 1.44 \pm 0.01 \times 10^9\,\mathrm{N\,m^{-2}}$$

More recently, further measurements have been made to find the Pockels coefficient for krypton.[44] The intensities of the lines from the solid at 115.5 K was measured by comparison with those from the liquid at 116.5 K which in turn were compared with a well characterized silica suspension (Ludox). The following values are quoted:

$$(p_{11})_{\text{liquid}} = 0.32 \pm 0.02$$
$$V_{\text{liquid}} = 694 \pm 5\ \mathrm{m\,s^{-1}}$$

$$\text{Solid}\quad \begin{aligned} p_{11} &= 0.34 \pm 0.04 \\ p_{12} &= 0.34 \pm 0.05 \\ p_{44} &= 0.037 \pm 0.005 \end{aligned}$$

Comparison of p_{11} with a value found using $p = (\epsilon - 1)(\epsilon + 2)/3\epsilon^2$ assuming $\epsilon = n^2$ and using known refractive index values gives agreement within experimental error.

The intensities of the polarized and depolarized Brillouin components for suitable crystal orientations are proportional to the squares of p_{12}, p_{44} and $(p_{11} - p_{12})$ and hence one can in general only find magnitudes of these quantities. Presumably the zero intensity of H_h component of the (100) longitudinal phonon Brillouin line implies $p_{11} = p_{12}$ and therefore in this case the individual values can be obtained.

6.4 CONCLUSIONS

6.4.1 Brillouin Scattering

The measurement of the shifts of the Brillouin components arising from longitudinal and transverse phonons would seem to constitute a completely valid and practical method of determining elastic constants of transparent materials. In the case of fragile materials at low temperatures it may be a preferred method. The case of adamantane clearly requires further study.

Additional measurement of intensities relative to a standard yield values for some strain optical Pockels coefficients and once again at low and high temperatures this might be difficult by other methods.

The broadening of the Brillouin lines is more difficult to measure and the two experiments done on the same material yield relaxation times which are in very poor agreement with one another.

6.4.2 Rayleigh Scattering

The measurements of the polarized Rayleigh scattering have yielded values of the thermal diffusivity of two materials which are similar to those calculated from thermal conductance, density and heat capacity for soft organic solids.

Measurements of the depolarized component have not yielded consistent values of relaxation times or molecular rotation times and in fact the general fitting to single relaxation mechanisms does not seem to be good (*cf.* The unexplained background in Reference 29 and the fitting of a Lorentzian to only part of the line shape in reference 10). The case of succinonitrile seems particularly complex—almost certainly due to the internal rotation possibilities. The relaxation times grouped around 7×10^{-11} s may well be associated with the internal reorientation of the $-$CN group but the very short (10^{-12} s) and very long (4×10^{-10} s) values are not easily explained—especially the former. On the whole it would seem that the separation of internal motion and whole molecular reorientation may not be possible in such a complex case.

REFERENCES

1. E. Gross and A. Raskin, *Acta Physicochem.*, U.R.S.S., **17**, 127 (1942).
2. R. Y. Chiao and B. P. Stoicheff, *J. Opt. Soc. Am.*, **54**, 1286 (1964).
3. H. Z. Cummins, *Light Scattering in Solids*, Ed. M. Balkanski, Flammarion, Paris, (1971), p. 1.
4. B. Chu, *Laser Light Scattering,* Academic Press, New York, (1974).
5. B. J. Berne and R. Pecora, *Dynamic Light Scattering*, Wiley New York, (1975).
6. A. Durand and A. S. Pine, *IEEE J. Quant. Electronics,* **4**, 523 (1968).
7. M. Adam, G. Searby, and P. Berge, *Phys. Rev. Lett.* **28**, 228 (1972).
8. G. B. Benedek in *Polarisation, Matiere et Rayonnement*, Presses Universitaires de France, Paris (1969).
9. H. Z. Cummins and H. L. Swinney, in *Progress in Optics*, Ed. E. Wolf, North Holland, Vol. VIII (1970).
10. R. Folland. D. A. Jackson, and S. Rajagopal, *Mol. Phys.,* **30**, 1063 (1975).
11. L. Brillouin, *Ann. Phys.,* **17**, 88 (1922).
12. L. Landau and G. Placzek, *Z. Phys. Sowjetunion*, **5**, 172 (1934).
13. J. R. Nye, *Physical Properties of Crystals*, Oxford U.P. London (1964).
14. M. J. Bird, D. A. Jackson, and J. G. Powles, *Mol. Phys.,* **25**, 1051 (1973).
15. A. J. Hyde, J. Kevorkian, and J. N. Sherwood, *Disc. Faraday Soc.*, **48**, 19 (1969).
16. W. Krasser, E. Koglin, O. Schneider, and R. Stockmeyer, *Ber. Bunsenges Phys. Chem.*, **77**, 386 (1973).
17. J. C. Damien, *Solid State Comm.*, **16**, 1271 (1975).

18. W. Krasser, E. Koglin, O. Schneider, and R. Stockmeyer, *Ber. Bunsenges Phys. Chem.*, **80**, 574 (1976).
19. R. Folland, D. A. Jackson, and S. Rajagopal, *Mol. Phys.*, **30**, 1053 (1971).
20. J. N. Gayles, A. W. Lohmann, and W. Peticolas, *App. Phys. Lett.*, **11**, 310 (1967).
21. J. A. Bucaro and T. A. Litovitz, *J. Chem. Phys.*, **54**, 3846 (1970).
22. R. Vacher, L. Boyer, and W. Longueville, *Compt. Rend.*, **B274**, 1383 (1972).
23. E. W. Johnson and L. K. Nash, *J. Amer. Chem. Soc.*, **72**, 547 (1950).
24. D. Beysens, R. Vacher, G. M. Searby, L. Boyer, and M. Adam, *Rev. Phys. Appliquées*, **9**, 465 (1974).
25. R. L. Jackson and J. H. Strange, *Mol. Phys.*, **22**, 313 (1971).
26. G. M. Hood, N. C. Lockhart, and J. N. Sherwood, *J.C.S. Faraday I*, **68**, 736 (1972).
27. L. Boyer, R. Vacher, L. Cecchi, M. Adam, and P. Berge, *Phys. Rev. Lett.*, **26**, 435 (1971).
28. D. A. Jackson, M. J. Bird, H. T. A. Pentecost, and J. G. Powles, *Phys. Letters*, **35A**, 1 (1971).
29. M. Adam, L. Boyer, R. Vacher, P. Berge, and L. Cecchi, *J. de Physique*, **32**, **C5**, a233 (1971).
30. W. Longueville, H. Fontaine, and A. Chapoton, *J. Chem. Phys.*, **68**, 436 (1971).
31. L. Boyer, R. Vacher, M. Adam, and L. Cecchi, *Light Scattering in Solids*, ed. M. Balkanski, Flammarion, Paris (1971) p. 498.
32. M. J. Bird, D. A. Jackson, and H. T. A. Pentecost, *Light Scattering in Solids*, ed. M. Balkanski, Flammerion, Paris, p. 493 (1971).
33. H. Fontaine, R. Fouret, L. Boyer, and R. Vacher, *J. Physique,* **33**, 1115 (1972).
34. H. E. Bommel and K. Dransfeld, *Phys. Rev.,* **117**, 1245 (1960).
35. L. Boyer, R. Vacher, and L. Cecchi, *J. Physique,* **36**, 1347 (1975).
36. R. Rudman and B. Post, *Science*, **154**, 1009 (1966).
37. J. P. Badialy, J. Bruneauz-Pollee, and A. Defrain, *J. Chim. Phys.,* **73**, 113 (1976).
38. M. Djabourov, C. Levy-Mannheim, J. Leblond, and P Papon, *J. Chem. Phys.,* **66**, 5748 (1977).
39. V. J. Tekippe and L. L. Abels, *Phys. Lett.,* **60**, 129 (1977).
40. B. P. Stoicheff, W. S. Gornall, H. Kiefte, D. Landheer, and R. A. McLaren, *Light Scattering in Solids*, Ed. M. Balkanski, Flammarion, Paris, p. 450 (1971).
41. W. S. Gornall and B. P. Stoicheff, *Phys Rev.*, **B4**, 4518 (1971).
42. S. Gewurtz, H. Kiefte, D. Landheer, R. A. McLarne, and B. P. Stoicheff, *Phys. Rev. Letters,* **29**, 1454 (1972).
43. H. E. Jackson, D. Landheer, and B. P. Stoicheff, *Phys. Rev. Letters,* **31**, 296 (1973).
44. Y. Kato and B. P. Stoicheff, *Phys. Rev.,* **B11**, 3985 (1975).

7

Infra-Red and Raman Studies of Molecular Motion in Plastic Crystals

R. T. Bailey

7.1 Introduction

Rotator phases are characterized by a dynamic disorder of the orientations of the molecules in the crystal lattice, although the barrier to reorientation can vary greatly in different systems. Indeed, reorientations can exist without rotational disorder when only rare jumps between indistinguishable lattice orientations occur. In this case, the molecule must posses at least an axis of symmetry. An example of this type is crystalline benzene where the molecules perform 60° jumps at a temperature dependent rate of $10^5\,s^{-1}$ at 100 K.[1] Also, in some cases, the molecules have cubic symmetry and the reorientations between indistinguishable positions are endospherical. Hexamethylene tetramine belonging to the point group T_d is an example of this type.[2] Where only indistinguishable orientations are involved, the crystal space group need not have the same symmetry elements as the molecule, e.g. crystalline benzene is orthorhombic.

Less common, is the situation where orientational disorder exists without apparent reorientational jumps, i.e. the disorder is frozen. In this case, the potential wells of the allowed orientations are of about equal depth and separated by very high potential barriers. This type of disorder either sets in at crystallization or is the result of 'freezing in' the dynamical disorder present at higher temperatures. In both cases, the zero point entropy is finite, e.g. CO where the disorder is two-fold[3] or with a residual entropy of nearly $R \ln 2$[4] or pentachlorotoluene at low temperatures where it is six-fold.[5]

In this chapter, emphasis will be placed on rotator phases where there is orientational disorder which is dynamic in nature, and a comparison made, where appropriate, with the disorder in the liquid phase.

7.2 SPECTROSCOPIC THEORY

Spectroscopic studies of molecular motion in condensed phases can be developed along two lines;[6] the conventional Schrödinger approach dealing with transitions between quantum states in the molecule, and the more recently developed Heisenberg picture where attention is focussed instead on the time development of the system. In the former, the vibrational frequency, the half-band width and the integrated band intensity can be interpretated in terms of molecular motion and intermolecular interactions. The latter approach which has developed considerable popularity recently considers the detailed band contour associated with a molecular transition. The spectrum is considered as the Fourier transform of an appropriate time correlation function. Experimentally, the correlation function is usually obtained by computing the Fourier integral over a complete experimental frequency spectrum thus allowing the short and long term behaviour of the correlation function to be treated separately. The short time behaviour can be discussed fairly rigorously in terms of the dynamics of the many-bodied system, whereas statistical arguments are usually employed to establish the form of the correlation function at long times. Using these techniques, information on the intermolecular forces and torques which govern the reorientational motion as well as the motion itself can be extracted from infrared and Raman bandshapes.

7.2.1 Schrödinger Approach

It is well known that phase transitions in molecular crystals are accompanied by changes in the vibrational spectra of the crystal. Changes in symmetry and structure in the crystal are invariably accompanied by parallel changes in multiplicity and splitting, integrated intensity, or line widths. Quantitative studies of these phenomena can lead to information regarding the structures and intermolecular potentials. First order 'rotational' transitions in rotator phase crystals generally lead to marked changes in the vibrational spectra. The characteristic times for the process of absorption (infrared) or scattering (Raman) of a photon are short compared with free molecular rotation times (10^{-10} s) so that actual rotational transitions may be observed in appropriate cases. Also, if the molecules are undergoing librational lattice motions with only occasional jumps to a new orientation, direct observation is possible. A clear case of rotation in the crystal is provided by hydrogen[7] where discrete rotational transitions are observed.

7.2.1(a) Internal Vibrations

One of the most fruitful ways of studying rotator phase transitions is to relate the shifting and splitting of vibrational modes to changes in the intermolecular interactions.

The ground state vibrational wavefunction for the crystal is given by the product of the ground state vibrational functions

$$\Phi^0 = \prod_{qp} \psi^{\circ}_{qp} \tag{7.2.1}$$

where the site in the unit cell is identified by q and the unit cell by p. An excited state wavefunction corresponding to the excitation of a vibration r in molecule s can be written

$$\theta^{i}_{r,s} = \psi^{i}_{r,s} \prod_{q,p} \psi^{0}_{q,p} \tag{7.2.2}$$

Since this function is degenerate in the crystal, linear combinations of these functions must be treated by degenerate perturbation theory.

First the combinations

$$\Phi^{i}_{r} = \frac{1}{\sqrt{N}} \sum \theta^{i}_{r,s} \exp(i\mathbf{k} \cdot \mathbf{R}_s) \tag{7.2.3}$$

are formed. These represent motion of an excitation packet over sites $\mathbf{r}$ in the $\mathbf{N}$ units cells and are thus called 'one site excitons'. In this expression R_s is a vector representing the position of the sth unit cell. Since only $\mathbf{q} = 0$ states can interact with single photons, a further classification of the excited state wavefunctions can be made. In this case, all the unit cells move in phase and the factor group symmetry applies. The potential energy of the crystal V is written.

$$V = V_0 + \sum_{u,v} V_{u,v}\,, \tag{7.2.4}$$

where V_0 represents the potential of the isolated molecules and $V_{u,v}$ is the intermolecular potential between molecules u and v. If the interaction energies are small compared with molecular energies, that is $V_{u,v} \ll V_0$, the two molecule interaction term can be treated as a perturbation. Application of first order perturbation theory gives the secular determinant,

$$\| \mathbf{H}' + \mathbf{E}(\lambda - \lambda_0) \| = 0 \tag{7.2.5}$$

where λ is the matrix of the perturbed roots

$$\lambda_i = 4\pi^2 c^2 v_i^2 \tag{7.2.6}$$

and

$$H'_{i\alpha} = \int \Phi^{i\alpha} \sum_{u,v} V_{u,v}\, \psi^{i\sigma}\, \mathrm{d}\tau.$$

$\mathbf{E}$ is a diagonal unit matrix. The determinant is factored into blocks involving only one symmetry species α at a time. The first order solution for the frequency of the αth component of the multiplet associated with ith non-degenerate vibration of the free molecule is

$$hcv_{i,\alpha} = hcv_i^0 + \mathbf{D}^i + \mathbf{M}^{i,\alpha} \tag{7.2.8}$$

with

$$\mathbf{D}_i = \sum_{v} \int \{ |\psi^{i}_{u} \psi^{0}_{v}|^2 - |\psi^{0}_{u} \psi^{0}_{v}|^2 \} V_{u,v}\, \mathrm{d}\tau \tag{7.2.9}$$

and

$$\mathbf{M}^{i,\alpha} = \sum B^{*}_{\alpha r} B_{\alpha q} \int (\psi^{i}_{u} \psi^{0}_{v})^{*} V_{u,v} (\psi^{0}_{a} \psi^{i}_{v}) \mathrm{d}\tau$$

in which u identifies molecules on sites r, and v these on sites q. The term $\mathbf{D}^{i}$ is an energy shift term and gives the frequency shift of the vibrational band in the crystal compared with the free molecules. The term $\mathbf{M}^{i,a}$ represents a splitting term and has a different value for each $\mathbf{q} = 0$ component in the crystal exciton band. If there are Z molecules in the unit cell, each non-degenerate vibrational mode will split into Z components. For degenerate internal modes, depending on the local symmetry, further splitting may occur. The crystal selection rules are based on the symmetry of the excited state functions. These states are classified according to the factor group of the crystal. Kopelman[8] has recently summarized the general methods of determining selection rules in crystals and the formulation of crystal functions. The potential energy $V_{u,v}$ is the primary quantity which determines the shift and splitting that occurs when a rotator phase transition is traversed. For instance, in the 1036 cm^{-1} band of benzene, increasing temperature increases intermolecular coupling because of rapid reorientation about the symmetry axis. Two principal approaches have been used to express the intermolecular potential function. In the first, the potential is expanded in terms of a multipole series retaining the first term, the dipole-dipole term. The transition dipole moment which could be obtained from infrared intensity measurements was thus used for the calculations. Hexter[10] attempted to correlate intensity measurements with crystal splitting with varying success. Dipole–dipole interactions appear to be of primary importance only when the absorption bands are very intense.

The second type of potential expresses the interaction energy in terms of a sum of non-bonded pairs of atoms on neighbouring molecules in the crystal. Thus the pair potential is expressed as a sum of atom–atom interaction terms. Dows[11,12] has applied this potential for interactions between hydrogen atoms in methyl chloride and ethylene. All of the atom–atom potentials proposed are of the form

$$-A/R^{6} + B \exp(-CR) \qquad (7.2.11)$$

Sets of constants A, B and C have been derived by Kitaigorodskii for C–C, C–H and H–H interaction terms.[13] Williams[14] has also determined potentials of this type from the crystal structures and properties of hydrocarbon crystals.

7.2.1(b) Lattice Vibrations

Since the forces which cause vibrational coupling are closely related to those which control the structure of the crystal, the potential energy function $V_{u,v}$ can be used to predict the equilibrium structures, lattice frequencies and other crystal properties such as compressibilities. This has been demonstrated for ethylene[11] where the lattice frequencies and compressibility of the solid were calculated on the assumption of H---H forces derived from intermolecular crystal splittings.

Since many of the molecules in solids which form the rotator phase have cubic symmetry and also because of rotational disorder, plastic crystals are usually isotropic and consequently liquid phase selection rules apply. Vibrational bands are usually broader and more liquid-like than conventional ordered crystals, reflecting the increased rotational freedom. The lattice frequencies are influenced to a much greater extent by rotational disorder, especially the torsional modes. Lattice modes which are sharp in the ordered crystalline phases degenerate to broad features, or even vanish in some cases, above the rotational transition temperature. Thus, in the case of crystalline acetylene,[15] the two infrared active translational lattice modes found below the transition temperature of 133 K, are replaced by a single, very weak broad band above this temperature. This change was explained by a statistical averaging of the molecular quadrupole moment by the large amplitude torsional motions in the rotator phase. Intermolecular splitting of certain vibrational modes below the transition temperature often vanishes in the plastic phase.

7.2.2 *Heisenberg Approach*

An alternative to the Schrödinger approach, is to study the time dependence of an absorption or scattering process by means of an appropriate time correlation function. Time correlation functions have been known since the early work of Bloembergen *et al.*[16] but only recently has the approach been developed for infrared and Raman bandshapes.[6,17] Correlation functions provide a concise method of expressing the degree to which two dynamical properties are correlated as a function of time. In other works, they describe quantitatively how long some property of the system persists until it is averaged out by the thermal molecular motion.

An auto-correlation function is defined as

$$C(t) = \langle \alpha(0)\, \alpha(t) \rangle_0 \tag{7.2.12}$$

where the dynamical quantities $\alpha(t)$ are defined in such a way as to make the ensemble average of α equal to zero. Then if the system is ergodic (i.e. if time averages are the same as averages over the initial ensemble), $C(t)$ will approach zero as $t \to \infty$. The average $\langle\ \rangle_0$ is over an ensemble of systems at reference time 0. Normally this ensemble is a canonical Boltzman distribution characteristic of systems in thermal equilibrium.

Because the response of the system to a weak specific probe is related directly to a particular correlation function, many experiments have been devised to determine specific correlation functions. For example, infrared and Raman bandshapes for linear molecules can be used to compute the auto-correlation functions $\langle \mathbf{u}(0) \cdot \mathbf{u}(t) \rangle$ and $\langle P_2 [\mathbf{u}(0) \cdot \mathbf{u}(t)] \rangle$ where $\mathbf{u}$ is a unit vector pointing along the molecular axis[17] and $P_2(x)$ is the Legendre polynomial of index 2. These correlation functions measure the rate of reorientational motion of the molecules in a particular molecular environment and have the advantage over other techniques (such as NMR) that the form of $C(t)$ is obtained and not just the correlation time.

Orientational motion in a molecular system occurs in response to interaction with the molecular environment and so can be used to probe the overall structural dynamics occuring on the same characteristic time scale (10^{-10} -10^{-13} s). Some of the experimental techniques which can be used to probe particular aspects of orientational dynamics in plastic crystals are listed in Table 7.1. These techniques are divided into two groups, those which probe essentially one molecule correlation functions, and the second group which are $\mathbf{k} = 0$ coherent methods and probe the total orientational dynamics of a region which is usually much larger than the range of the intercorrelations of orientations. In the latter case, the nature of the motion, the amplitudes and relative orientations of the allowed angular steps are harder to determine. If time averaged, the basic problem amounts to the elucidation of the nature of the disorder which can be accomplished by single crystal x-ray or neutron diffraction. Only quasi-elastic neutron scattering yields simultaneously the nature of the orientational order and its rate of rearrangement.

The relationship between spectral bandshape and orientational correlation functions was established originally by Gordon.[17] No assumptions regarding the motion in the molecular system were made.

7.2.2(a) *Infrared Absorption*

The intensity of infrared absorption as a function of frequency ω in terms of transitions between quantum states (Bohr-Schrödinger) is given by

$$I(\omega) = \frac{3\hbar''(\omega)}{4\pi^2[1 - \exp(-\hbar\omega/kt)]} = 3 \sum_i \sum_f |\langle f | \hat{\epsilon} \cdot \mathbf{u} | i \rangle|^2 \delta(\omega_{fi} - \omega) \tag{7.2.13}$$

where $\hat{\epsilon}$ is a unit vector along the electric field of the radiation and the sums i and f run over all the quantum states of the molecules.

The transformation of this expression into the Heisenberg picture involves representing the δ function by its Fourier integral

$$\delta(\omega) = \frac{1}{2\pi} \int_{-\infty}^{+\infty} e^{i\omega t} \, dt \tag{7.2.14}$$

Table 7.1 Specific probes of reorientational dynamics

	Angular momentum	First spherical harmonic	Second spherical harmonic	All spherical harmonics
Self	NMR T_1 (spin rotation coupling)	IR vibrational–rotational line-shapes	Raman vibrational–rotational lineshapes, NMR T_1 (quadrupolar or dipolar coupling)	Incoherent Rayleigh scattering
Collective		Permanent dipole absorption (far IR, microwave)	Depolarized Rayleigh scattering	

giving

$$I(\omega) = \frac{3}{2\pi} \sum_{if} \rho_i \langle \mathbf{i} | \hat{\epsilon} \cdot \mathbf{u} | f \rangle \langle f | \hat{\epsilon} \cdot \mathbf{u} | \mathrm{i} \rangle$$

$$\times \int_{-\infty}^{+\infty} \mathrm{d}t \exp \mathbf{i} \{ (E_\mathrm{f} - E_\mathrm{i}) / (\hbar - \omega) t \tag{7.2.15}$$

The energy eigenvalues E_f and E_i may be expressed in terms of the Hamiltonian operator H for its rotational translation motion giving

$$I(\omega) = \frac{3}{2\pi} \int_{-\infty}^{+\infty} \mathrm{d}t \, \mathrm{e}^{-\mathrm{i}\omega t} \sum \rho_i \langle \mathbf{i} | \hat{\epsilon} \cdot \mathbf{u}(0) | f \rangle \langle f | \hat{\epsilon} \cdot \mathbf{u}(t) \mathbf{i} \rangle \tag{7.2.16}$$

Performing the sum over the complete set of final states $| f \rangle$ we have

$$I(\omega) = \frac{3}{2\pi} \int_{-\infty}^{+\infty} \mathrm{d}t \, \mathrm{e}^{-\mathrm{i}\omega t} \sum \rho_i \langle \mathbf{i} | \hat{\epsilon} \cdot \mathbf{u}(0) \hat{\epsilon} \cdot \mathbf{u}(t) | \mathbf{i} \rangle \tag{7.2.17}$$

The sum over the initial states i, weighted by ρ_i is simply an equilibrium average $\langle \; \rangle_0$:

$$I(\omega) = \frac{3}{2\pi} \int_{-\infty}^{+\infty} \mathrm{d}t \, \mathrm{e}^{-\mathrm{i}\omega t} \langle \hat{\epsilon} \cdot \mathbf{u}(0) \, \hat{\epsilon} \cdot \mathbf{u}(t) \rangle_0 \tag{7.2.18}$$

If the system is isotropic, which is usually the case for rotator phase solids, the same result is obtained for any direction of polarization $\hat{\epsilon}$ so that

$$I(\omega) = \frac{1}{2\pi} \int_{-\infty}^{+\infty} \mathrm{d}t \, \mathrm{e}^{-\mathrm{i}\omega t} \langle \mathbf{u}(0) \cdot \mathbf{u}(t) \rangle_0 \tag{7.2.19}$$

Thus, in the Heisenberg picture, the spectral distribution is given by the Fourier transform of the correlation function of the dipole moment operator of the absorbing molecules. Fourier inversion gives

$$C(t) = \langle \mathbf{u}(0) \cdot \mathbf{u}(t) \rangle_0 = \int_{\text{band}} I(\omega) \mathrm{e}^{\mathrm{i}\omega t} \, \mathrm{d}\omega \tag{7.2.20}$$

were the integration is over the complete vibration rotation bandshape. Usually, it is convenient to normalize the observed spectrum,

$$\hat{I}(\omega) = I(\omega) / \int_{\text{band}} I(\omega) \mathrm{d}\omega \tag{7.2.21}$$

so that

$$\langle \mathbf{u}(0) \cdot \mathbf{u}(t) \rangle_0 = \int \hat{I}(\omega) \, \mathrm{e}^{\mathrm{i}\omega t} \, \mathrm{d}\omega \tag{7.2.22}$$

where $\mathbf{u}(t)$ is a unit vector along the direction of the transition dipole. The use of a normalized spectrum largely eliminates dielectric effects of the local field due to the radiation. Only fluctuations about the average local field will effect the normalized spectrum. Vibrational perturbations (frequency shifts) are neglected in this treatment but this difficulty can be reduced by removing the mean frequency shift from the vibrational band origin. Then only fluctuations of the frequency shift distort the bandshape. Vibrational relaxation and other broadening mechanisms were also neglected in Gordons treatment.

Vibrational relaxation can contribute significantly to or even dominate a vibrational lineshape. This problem has been treated by several workers[18] all of whom break down the intensity of an absorption band into orientational and vibrational contributions. For a randomly oriented isotropic system, the intensity distribution is given by

$$I(\omega) = \frac{1}{2\pi}\int_{-\infty}^{+\infty} \langle \mathbf{u}(0) \cdot \mathbf{u}(t) \rangle_{\mathrm{tr}} \times \langle \hat{Q}^{v}(0) \cdot \hat{Q}^{v}(t) \rangle_{vib} \exp(\mathrm{i}\omega t)\mathrm{d}t \qquad (7.2.23)$$

In this expression u is a unit vector along the molecular axis which is parallel to the infrared transition moment for the normal coordinate of vibration Q. Thus, the intensity $I(\omega)$ is make up from contributions from both the orientational and vibrational correlation functions. The vibrational contribution has the same form in both infrared and Raman spectra but the orientational contributions are different. An exact separation of the vibrational and reorientational contributions can only be achieved for the infrared active totally symmetric modes $C_n(A)$ and $C_{nv}(A)$ where $\langle \hat{Q}^{v}(0) \cdot \hat{Q}(t) \rangle$ has been obtained from the corresponding Raman lineshape. If one assumes, however, that all the vibrational modes of a given molecule have the same vibrational correlation function the separation can be achieved for other symmetries.

7.2.2(b) Raman Scattering

Recently Gordon[19,20] obtained the expression,

$$\hat{C}(t) = \langle P_2 \left[\mathbf{u}^{(z)}(0) \cdot \mathbf{u}^{(z)}(t)\right] \rangle \qquad (7.2.24)$$

where $P_2(x) = \frac{1}{2}(3x^2 - 1)$ for the correlation function of the Raman active, totally symmetric vibration in a linear molecule or symmetric top. In the case of a degenerate vibration of a tetrahedral molecule

$$\hat{C}(t) = \langle P_2 \left[\mathbf{u}^{(2)}(0) \cdot \mathbf{u}^{(2)}(t)\right] \rangle \qquad (7.2.25)$$

where $\mathbf{u}^{(2)}$ is the unit vector along any of the twofold axes.[21] In the absence of coriolis coupling, the triply degenerate vibrations of a tetrahedral molecule also have the same correlation function.

In Gordon's treatment, vibrational relaxation was neglected, but subsequently several authors have derived expressions to include this mechanism. Both Bartoli and Litovitz[85] and Nafie and Peticolas[18] have used somewhat similar approaches in their treatments of vibrational Raman lineshapes. The expressions they obtained for the relevent correlation functions are

$$\langle \hat{Q}^{v}(0) \cdot \hat{Q}^{v}(t) \rangle_{\text{vib}} = \int_{-\infty}^{+\infty} \hat{I}_{\text{vib}}(\omega) \mathrm{e}^{-i\omega t}\, \mathrm{d}\omega \qquad (7.2.26)$$

$$\langle \mathrm{Tr}\hat{\beta}^{v}(0) \cdot \hat{\beta}^{v}(t) \rangle_{\text{tr}} \langle \hat{Q}^{v}(0) \cdot \hat{Q}^{v}(t) \rangle_{\text{vib}} = \int_{-\infty}^{+\infty} \hat{I}_{\perp}(\omega)\, \mathrm{e}^{-1\omega t}\, \mathrm{d}\omega \qquad (7.2.27)$$

and

$$\langle \mathrm{Tr}\hat{\beta}^{v}(0) \cdot \hat{\beta}^{v}(t) \rangle_{\text{tr}} = \frac{\int_{-\infty}^{+\infty} \hat{I}_{\perp}(\omega) \mathrm{e}^{-i\omega t}\, \mathrm{d}\omega}{\int_{-\infty}^{+\infty} \hat{I}_{\text{vib}}(\omega) \mathrm{e}^{-i\omega\tau}\, \mathrm{d}\omega} \qquad (7.2.28)$$

where

$$\hat{I}_{\text{vib}}(\omega) = \frac{\hat{I}_{\parallel}(\omega) - \frac{4}{3}\hat{I}_{\perp}(\omega)}{\int_{-\infty}^{+\infty} [I_{\parallel}(\omega) - \frac{4}{3} I_{\perp}(\omega)]\, \mathrm{d}\omega} \qquad (7.2.29)$$

and $\hat{I}_{\perp}(\omega)$ and $\hat{I}_{\parallel}(\omega)$ are the scattered intensities perpendicular and parallel to the electric vector of the exciting laser radiation. Equations (7.2.26) and (7.2.28) represent the correlation functions associated solely with vibration relaxation and reorientation of the anisotropy respectively. This separation depends on the trace of the tensor α^{v} being non-zero and is therefore only strictly applicable to vibrational modes belonging to the totally symmetric representations. Assuming, however, that the vibrational contribution is the same for all modes of a given molecule, separation may be accomplished using equation (7.2.28) where $\hat{I}_{\perp}(\omega)$ is obtained from any Raman band and $\hat{I}_{\text{vib}}(\omega)$ is obtained from a totally symmetric Raman band.

7.2.3 Pure Rotational Absorption

Absorption of polar liquids and plastic crystals in the far infrared–microwave region is due primarily to librational and relaxational motions. The primary absorption process in non-polar molecules is due to transient dipoles induced in a molecule by the neighbouring molecules and modulated by their motion. The intensity distribution resulting from these processes $\alpha(\bar{v})$ is the result of many bodied interactions and is best interpreted in terms of a Fourier transform of an appropriate time correlation function whose short and long time behaviour can be separated.

Neglecting the internal field and the effect of induced dipoles, $\alpha(\bar{v})$ is given by,[22,23]

$$\alpha(\bar{v}) = \frac{8\pi^{3} N \bar{v}^{2} c \mu^{2}}{3kTn(\bar{v})} \int_{-\infty}^{+\infty} \langle \mathbf{u}_{l}(0) \cdot \sum_{i} \mathbf{u}_{i}(t) \rangle \mathrm{e}^{-2i\pi\bar{v}ct}\, \mathrm{d}t \qquad (7.2.30)$$

where $\mathbf{u}_i$ is the unit vector along the permanent dipole μ_i of the ith molecule, $n(\bar{\nu})$ the refractive index, and N the molecular number density. Neglecting cross-correlations the correlation function becomes $\langle \mathbf{u}_1(0) \cdot \mathbf{u}_1(t) \rangle$.

For some purposes, the rotational velocity correlation function is a better probe of the molecular dynamics in these systems. This is defined as $\langle \dot{\mathbf{u}}_1(0) \cdot \Sigma_i \dot{\mathbf{u}}_i(t) \rangle$ and is related to the corresponding vectorial correlation function by.

$$\frac{d^2}{dt^2}(\langle \mathbf{u}_1(0) \cdot \sum \mathbf{u}_i(t) \rangle) = -\langle \dot{\mathbf{u}}_1(0) \cdot \sum \dot{\mathbf{u}}_i(t) \rangle \qquad (7.2.31)$$

This function is damped out to zero in about $2ps$ and is thus the 'time domain equivalent' of the far infrared (2–200 cm^{-1}) region of the frequency domain.

The absorption intensity then becomes,

$$\alpha(\bar{\nu}) = \frac{2\pi N\mu^2}{3kTcn(\bar{\nu})} \int_{-\infty}^{+\infty} \langle \dot{\mathbf{u}}_1(0) \cdot \sum_i \dot{\mathbf{u}}_i(t) \rangle \exp(-2\pi i\bar{\nu}ct)\mathrm{d}t \qquad (7.2.32)$$

In principle both correlation functions $f_0(t) = \langle \mathbf{u}_1(0) \cdot \Sigma_i \mathbf{u}_i(t) \rangle$ and $f_2(t) = \langle \dot{\mathbf{u}}_1(0) \cdot \Sigma_i \dot{\mathbf{u}}_i(t) \rangle$ contain the same information, but in practice, frequency data above about 10 cm^{-1} have little effect on $f_0(t)$. The converse is true for $f_2(t)$. The short-time motion (libration) is thus is best studied *via* $f_2(t)$.

When a correction is applied for the internal field the following expression is obtained for the correlation function $f_2(t)$.[23]

$$f_2(t) = \frac{2}{\pi} \times \frac{3kT}{2\pi N\mu^2} \left[\frac{9n_0}{(n_0^2 + 2)^2} \right] c^2 \times \int_0^{\infty} \alpha(\bar{\nu}) \cos(2\pi\bar{\nu}ct)\mathrm{d}\bar{\nu} \qquad (7.2.33)$$

Generally a normalized value of $f_2(t)$ is computed by dividing by $f_2(0)$ at each t, so that $f_2(t)$ is independent of the non-dispersive correction.

7.2.4 The Relationship between Correlation Functions

The dipole and polarizability correlation functions are only two of many correlation functions which describe molecular motion. Different correlation functions, such as those associated with depolarization of fluorescence and certain NMR experiments explore different aspects of the molecular motion. It is important therefore to relate the various correlation functions in order to correlate the information derived from the various experimental techniques.

For the specific case of an ensemble of linear free-rotor molecules, Gordon[24] has given an exact relationship between the dipole $\langle \mathbf{u}(0) \cdot \mathbf{u}(t) \rangle$ and polarizability (Raman) $\langle P_2(\mathbf{u}(0) \cdot \mathbf{u}(t)) \rangle$ correlation functions. Also, in some cases, a quantitative relationship exists between Raman bandshapes and nuclear spin relaxation measurements.[20] For fluids consisting of linear molecules, and for nuclear spin relaxation controlled by nuclear quadrupole interaction, or by intramolecular nuclear magnetic-dipole spin–spin interactions, the nuclear spin relaxation times (T_1 and T_2) may be calculated from the spectral distribution of the depolarized component of a totally symmetric vibration-rotation Raman band. Thus,

$$T^{-1} = \frac{3(2I+3)}{80I^2(2I-1)}\left[\frac{Q}{\hbar}\right]^2 \int_{-\infty}^{+\infty} \langle P_2[\mathbf{u}(0)\cdot\mathbf{u}(t)]\rangle \mathrm{d}t \qquad (7.2.34)$$

where I is the nuclear spin and $[Q/\hbar]^2$ is the quadrupole coupling constant. An isotropic relaxation process ($T = T_1 = T_2$) was assumed. The same relationship applied to symmetric-top molecules provided the relaxing nuclei lie on the molecular symmetry axis, i.e. the intramolecular electric field gradient has cylindrical symmetry. Thus, the consistency of results derived from several experiments may be tested. T_1 and T_2 may be obtained from steady state or pulsed NMR measurements and $\langle P_2[\mathrm{U}(0)\cdot\mathrm{U}(t)]\rangle$ from Raman scattering or by electric field induced infrared absorption. The Raman experiments have the advantage of determining the complete time dependence of the correlation function whereas the spin experiments determine only the area under the function. For spherical top molecules, the correspondence between light scattering and spin relaxation is not exact but is close enough for practical comparison since the two correlation functions become identical in the opposing limits of rotational diffusion and free rotation. These relationships do not depend on dynamical models and hold for both classical and quantum systems. A relationship between dielectric dispersion and the dipole correlation function has been derived.[25] The connection between infrared and Raman bandshapes has also been discussed by Berne, Pechukas, and Harp.[26] They outlined an approximate method whereby a knowledge of the dipole correlation function would enable a reasonable prediction of $\langle P_2[\mathbf{u}(0)\cdot\mathbf{u}(t)]\rangle$ to be made and vice versa. Two procedures were used for estimating the distribution of $\langle\mathbf{u}(0)\cdot\mathbf{u}(t)\rangle$ and thereby predicting Raman bandshapes from observed infrared bandshapes. In the first of these, the information entropy of the distribution, $P(\theta, \phi, t)$, under the constraint imposed by the known $\langle\mathbf{u}(0)\cdot\mathbf{u}(t)\rangle$ was maximized; and the second minimized the mean square difference between the distribution and the equilibrium ($t \to \infty$) distribution under the same constraint. Using the first of these procedures, the following approximate relationships were obtained:

$$\langle\mathbf{u}(0)\cdot\mathbf{u}(t)\rangle \sim \left[\coth\beta(t) - \frac{1}{\beta(t)}\right] \qquad (7.2.35)$$

and

$$\langle P_2[\mathbf{u}(0)\cdot\mathbf{u}(t)]\rangle \sim 1 - \frac{3}{\beta(t)}\left[\coth\beta(t) - \frac{1}{\beta(t)}\right] \qquad (7.2.36)$$

where $\beta(t)$ is a Lagrange undetermined multiplier. Thus if $\langle\mathbf{u}(0)\cdot\mathbf{u}(t)\rangle$ is known at time t_1 $\beta(t)$ and hence $\langle P_2[\mathbf{u}(0)\cdot\mathbf{u}(t)]\rangle$ can be obtained.

7.3 MODELS OF REORIENTATIONAL MOTION

Many of the reorientational models proposed for the liquid state can also be envisaged for rotator phase solids since the degree of local order near the melting point for the two phases are comparable.[27] In contrast to liquids however, reorientational freedom in the rotator phase seems to be an intrinsic property of each molecular site and a large increase in local volume is not a prerequisite. Free volume models developed for liquids may thus be discarded for most plastic crystals.[28,29] If the presence of a neighbouring vacancy were sufficient for a molecule to reorientate, the activation energies for orientation and translation should be comparable. In practice, however, this is not found and in norborylene, for example, the relevant energies are 6 and 84.7 kJ mol^{-1} while for adamantane they are 13 and 150 kJ mol^{-1}.[30] Near the melting point of adamantane, the translational diffusion jump rate which is governed by vacancy jumps is about 10^6 s^{-1}, whereas the rotational jump rate is about 5×10^{11} s^{-1}.[2] The motion of a single vacancy is unlikely to relax 5×10^5 molecules. Further evidence against a vacancy triggered mechanism for reorientation is provided by variable pressure NMR measurements which indicate a small activation volume for reorientation.[30,31]

Two main groups of models of molecular motion in plastic crystals may be distinguished. The first of these constitute what may loosely be termed diffusion models where the molecule may undergo free rotation, classical rotational diffusion or collision interrupted rotation. The second group of models are stochastic in nature. One type of stochastic theory has been developed to relate reorientational motions to vibrational lineshapes, while a second type was developed primarily to explain the far infrared-microwave absorption in plastic crystals. This model was centred on the classical correlation function of an ensemble of harmonic oscillators subject to random adiabatic collisions.[32,33] The results were later extended to include torsional oscillations[34] and the angular motions of molecules submitted to thermal collisions in a multipotential well.[35]

These models of rotational motion are best discussed in terms of increasing barrier to molecular reorientation.

7.3.1 Rotational Diffusion

7.3.1(a) Free Rotation

The classical mechanical correlation function for the first two spherical harmonics have been calculated for linear[24] and symmetric top[36] molecules. In all cases, the corresponding correlation times should be of the order of $(I/kT)^{1/2}$

Classical free rotation is very unlikely in pure rotator phase crystals, except for hydrogen, although nearly free rotation can occur in rare gas matrices or clathrates. Even near melting in a very globular molecule such as adamantane,[2] the correlation time for the second spherical harmonic is 2×10^{-12} s whereas $(I/kT)^{1/2}$ is 0.8×10^{-12} s. In most cases, moreover, the volume available to a molecule at a lattice site is less than the van der Waals molecular diameter. For example, in tertiary butyl chloride the intermolecular distance in the fcc crystal is 0.61 nm whereas the molecular diameter is 0.72 nm.

7.3.1(b) Classical Rotational Diffusion

This model was originally proposed by Debye[37] to describe dielectric relaxation phenomena in liquids. The orientational motion of the molecules is assumed to follow a rotational diffusion equation with the rotational diffusion coefficient given by a Stokes–Einstein relationship. More elaborate treatments of the rotational diffusion approach have been proposed[38,39] but generally an adequate description of the reorientational process is only obtained if each diffusive step is limited to small angles. When classical rotational diffusion occurs the time correlation functions for the spherical harmonics are exponential and the Hubbard relationship is valid. This is

$$l(l + 1)\tau_l\tau_j = I/kT$$

where l is the order of the spherical harmonic. In crystals, the necessary fast fluctuating random torques, implicit in this model, would be produced by phonon interactions arising from the large arharmonicity of the motions. The Debye model is inadequate in dealing with rotational diffusion in liquids composed of spherical molecules.[40–42] It has not been shown to apply to plastic crystals. Even at the low temperature end of the rotator phase of neopentane[43] the τ_1/τ_2 ratio is closer to 2 than to 3 predicted by the Hubbard relation. In nonadecane,[44] in its rotator phase, neutron quasi-elastic scattering indicates that the number of steps in each 360° orientation is at least 8, but even this cannot be considered as rotational diffusion in view of the rather large diffusive step.

7.3.1(c) Collision Interrupted Reorientation–M- and J-Diffusion

The Debye model of rotational diffusion was generalized by Gordon[24] to allow reorientation through angular steps of arbitrarily large size. In this extended diffusion model, the molecules are assumed to rotate freely in the intervals between collisions. Two limiting cases of the extended diffusion model were considered by Gordon—the J-diffusion and M-diffusion models. Both of these models allow angular diffusion steps of arbitary size. In J-diffusion, both the magnitude and direction of the angular momentum vector of the molecule are randomized into a Boltzman distribution by a collisional process. The M-diffusion model terminates a free rotational step by randomizing the orientation of the angular momentum vector, the magnitude of this vector remaining unchanged. The time between collisions is a random variable characterized by its mean, the correlation time for the rotational angular momentum τ_J, which appears as an adjustable parameter in the extended diffusion model. In Gordon's[24] original treatment only linear molecules were considered. The general diffusion equation was found to obey exact classical mechanics at short times. Thus the power series expression at short times begins,

$$\langle \mathbf{u}(0) \cdot \mathbf{u}(t) \rangle = 1 - t^2kT/I + O(t^3) \tag{7.3.1}$$

which is the correct classical result.[45] This shows that the initial curvature of a classical dipole correlation function is independent of the intermolecular interactions. The usual Debye exponential formulae for small angle diffusion do not have the correct initial time dependence. At longer times, the form of the dipole correlation function is determined in Gordon's[24] model by giving the relation between the rotation frequencies, ω_R, during successive diffusion steps. The M-diffusion model specifies the ω_R to be unchanged in successive steps while in the J-diffusion model the rotational frequencies ω_R are completely uncorrelated in successive steps so that the ω_R are distributed independently over a Boltzman distribution. The dipole correlation function in the M-diffusion limit is given by,[24]

$$\langle \mathbf{u}(0) \cdot \mathbf{u}(t) \rangle = \exp(-t/\tau) \sum_{n=0}^{\infty} (n+1)\,(t/2\tau)^n \times \sum_{k=0}^{(n+1)/2} F_k(t)/2^k\mathrm{k}!\,(n+1-2k)! \tag{7.3.2}$$

where $F_0(t)$ the correlation function for a free linear molecule is given by,

$$F_0(t) = \langle \cos \omega t \rangle = \int_0^\infty \cos(\omega t)\, \omega \exp(-\tfrac{1}{2}\omega^2) d\omega \tag{7.3.3}$$

The results of the calculations for some typical parameters are shown in Figure 7.1. These curves sometimes become negative suggesting that after a certain time it is more probable to find a dipole in the opposite half of the sphere to which it began at a time t earlier. Molecules with the most probable thermal angular momentum will have rotated through about 180° at the negative minima. Large angle reorientations cannot occur in the Debye treatment. This model was also generalized, by allowing the rate $(1/\tau)$ at which diffusion steps end to depend on the rotation frequency ω. A frequency dependence of the angular momentum relaxation time is suggested by several simple physical models. In this modification, the rate at which the *direction* of the angular moment diffuses decreases with increasing rotation frequency. Thus,

$$1/\tau = (1/\tau_0)\, f(\omega) \tag{7.3.4}$$

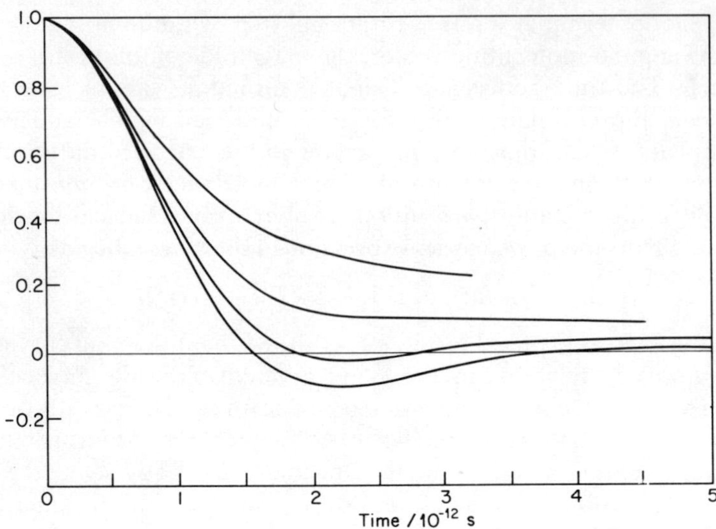

Figure 7.1 Dipole correlation functions for the M-diffusion model, with $\tau = 3.0, 1.8, 1.0,$ and 0.5 (from bottom to top). The time scale is in reduced units of $(I/kT)^{1/2}$

where $f(\omega)$ is a function which decreases at large ω. The dipole correlation function was integrated numerically for a number of dispersion functions $f(\omega)$. Qualititatively, these dispersion functions produce a secondary maximum in the correlation function and show a damped oscillatory approach to rotational equilibrium, rather than the monotonic decay usually assumed in statistical mechanics.

The corresponding expression for the J-diffusion model is more complicated. Dipole correlation functions for some typical values of τ are shown in Figure 7.2. Qualitatively, they resemble the curves for the M-diffusion case but they differ in that the minimum shifts to longer time as τ decreases and in general the decay is more rapid at longer times.

The general diffusion models may also be used to calculate Raman correlation functions. The Raman correlation function for a thermal distribution of free linear molecules is given by,

$$\langle P_2[\cos \omega t\} \rangle_{\text{free}} = \langle \tfrac{3}{4} \cos 2\omega t \rangle + \tfrac{1}{4} \qquad (7.3.5)$$

This approaches $\frac{1}{4}$ at long times rather than zero. This constant term in the P_2 correlation function for free rotation corresponds to the depolarized component of the Q-branch in the Raman spectrum. Any interactions, however small, will however, eventually randomize the direction of the angular momentum vector and thus the P_2 correlation function will also approach zero at long times.

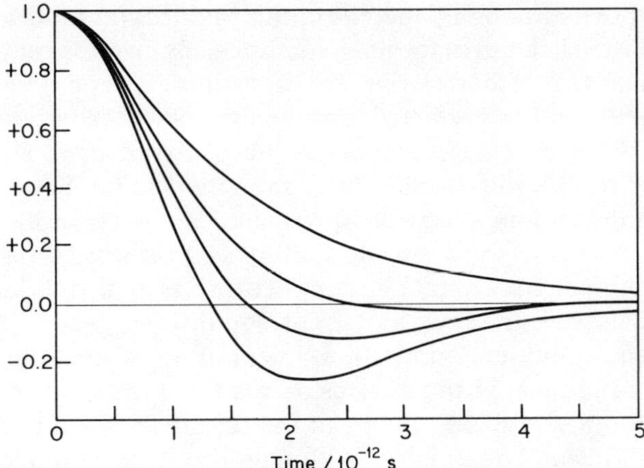

Figure 7.2 Dipole correlation functions for the *J*-diffusion model, with τ = 30, 1.75, 0.8, and 0.4

The applicability of the extended diffusion model to rotational diffusion was tested by comparing the predicted infrared and Raman correlation functions with those obtained by Fourier transformation of experimental infrared and Raman bandshapes. Good agreement was obtained for CO and N_2.[24]

The extended-diffusion approach has been used by McClung[46] to calculate correlation functions for classical spherical top molecules. It was shown that the Debye model can be recovered from the J-diffusion limit of the extended diffusion model when the angular momentum correlation time, τ_J is short compared to the average rotation period of the molecule. At the other extreme, where τ_J is much longer than the average rotational period, both M and J-diffusion models yield results closely resembling those obtained with a perturbed free rotor model.[42]

In subsequent publications McClung[47] applied the expressions derived for a spherical top to calculate correlation functions for CH_4 and CD_4 in both the pure liquid form and in dilute solutions in liquid noble gases. These molecules represent an intermediate case ($\tau_J \sim 1$) between the two limiting cases $\tau_J \ll 1$ and $\tau_J \gg 1$. The M-diffusion limit of the extended diffusion model predicts spectral line shapes with a very sharp central peak when $\tau_J \leq 1.0$, whilst the line shapes predicted in the J-diffusion limit were broader in the central region. This behaviour resulted from the persistence of orientational correlation over a large number of collisions. The M-diffusion model however, requires that the molecules do not have the magnitude of the angular momenta changed by a collision. Clearly there must be collisional modulation of the magnitude of the angular momentum if the populations of the rotational states remain

Boltzman. In real situations, therefore, the M-diffusion model should be applied with caution in band shape analysis, especially near the centre of the band, and in the analysis of correlation functions at very long times. Satisfactory agreement between the predicted and the experimental correlation functions obtained by Fourier transformation of the observed band shapes were obtained. These results will be considered in more detail in Section 7.4.

From a similar semi-classical approach, (using classical mechanics to describe the motion of the molecule and Boltzman statistics to obtain the correlation functions) McClung[48] has applied the extended M and J-diffusion models to the reorientational motion of liquids composed of symmetric top molecules. The rotational correlation functions and their Fourier transforms were shown to reduce, in the J-diffusion model in the limit of very-short angular momentum correlation times, to the results obtained by solution of the anisotropic diffusion equation.[39,49] Although this treatment allows some physical insight into the behaviour of the molecular system undergoing diffusion, its utility is limited essentially to the extended J or M-diffusion models. If a more sophisticated description of the collision process were adopted, an alternative approach which employs explicit functions or operators to describe the collisions would be preferred. Nevertheless, in spite of its deficiencies, the extended model although based on a greatly oversimplified picture of the rotational motion in fluids, does seem to be a valid approximation for many systems of linear, spherical top and symmetric top molecules. It seems unlikely that this model will be developed for asymmetric top molecules due to the complexity of the mathematical treatment of the motion. For some molecules however, the expressions for the symmetric rotor will provide a reasonable approximation.

Bliot and Constant[50] recently pointed out that expressions for the dipole correlation function equivalent to those of the M and J-diffusion models, could be obtained from the memory functions introduced by Berne and Harp.[15] The results were valid for linear, spherical top, and symmetric top molecules. The theory was illustrated by calculations for CH_3F.

An alternative, more general, approach to the problem of rotational motions in fluids has been developed by Fixman and Rider[52] and by St. Pierre and Steele.[53,54] These workers derived expressions for the k = 0 reorientational correlation functions and spectral densities. The approach of Fixman and Rider[52] was based on the assumption that the rotational motion could be treated as a Markoffian process (as are the extended diffusion models) and were able to derive a general expression for the time dependent orientational and angular momentum distribution function. The free motion of the molecule was described by a streaming operator $\mathscr{S}$ and the effects of the intermolecular interactions by a collisional operator $\mathscr{L}$. Construction of matrix representations of these operators in a suitable Hilbert space enabled these authors to obtain expressions for the $k = 0$ reorientational correlation functions and spectral densities for a general Markoff process. Changes in both orientation and angular momentum could be included in the specification of the collision. In the application of their general theory to specific collisional models, no changes in the orientation of the molecules during collisions were considered. Their relaxation model was thus essentially a modification of the *M* and *J*-diffusion models. As well as relaxation models, a Langevin model was also considered in which the angular momentum distribution function followed a Fokker–Planck equation. In the limit of small τ_J this model predicted rotational correlation functions in agreement with those from the rotational diffusion equation theory.

St. Pierre and Steele[53,54] approached the problem from the collision-interrupted free rotation model and calculated the conditional phase space distribution functions for symmetric top molecules undergoing *J* and *M*-diffusion. Starting from explicit distribution functions obtained for classical symmetric tops and linear rotors derived in a previous paper,[57] they showed how collisional events that change the angular momentum can be accounted for. The collisional distribution function was represented as a sum of contributions from those molecules undergoing a particular number of collisions. As in Gordons treatment of the linear molecule, the *n*th of these terms turns out to be an $(n - 1)$-fold integral. These integrals were computed in general form and for the specific case of a symmetric top. Using Fourier transform techniques the integrals were expressed as convolutions which were analogous to the expressions derived by Fixman and Rider.[52] The results obtained indicated that the M and J-diffusion models provided plausible predictions for experimental spectra and correlation functions over a wide range of collision parameter.

A modification of the M-diffusion model in which fewer implicit assumptions and restrictions were made has been described by Frenkel, Wegdam, and van der Elsken.[55] A matrix description of the rotational motion of a dipolar molecule undergoing frequent collisions was given. The probability that the molecule has undergone an *n*-collisional process is contained explicitly in the matrix. By taking the ensemble average, an analytical expression for the dipole moment and angular momentum correlation functions was obtained. Other collisional distributions apart from the usual Poisson distribution can be employed. Thus the model can also be applied to describe systems where strong correlation exists between successive collisions. Librational motions are naturally included in this model.

The Debye diffusion model has been generalized to include arbitary reorientational angles by Cukier and Lakatos–Lindenburg.[56] Any reorientational process including large angle reorientations, reorientations between distinct sites appropriate to continuous random walks, and the usual diffusion equation approximation, can be described by this model. Inertial effects were neglected The form of the correlation functions for infrared and Raman spectroscopy were found to depend only on the medium isotropy and molecular geometry. Rotational motion in non-isotropic systems such as liquid crystals was also considered.

A somewhat different model of orientational motion in liquids, has been developed by Kivelson and Keyes.[57] This theory, which was developed in terms of Mori's[58] formalism for generalized hydrodynamics, was developed for three orientational variables interrelated by three coupled linear transport equations. In appropriate limits, the theory reduces to that for rotational diffusion, gas-like extended rotational diffusion, and solid-like oscillatory rotations or cell model motions. This theory therefore, gives a unified picture, encompassing all limits of molecular motion in liquids and plastic crystals. Most of the interesting features of the rotational diffusion theory, Gordon's extended rotational diffusion theory, Steeles inertial effects and Ivanov's jump theory[59] are shown. Isotropic rotational motion was assumed and all coupling between rotational motions and collective hydrodynamic transport modes neglected. This three variable theory is however only accurate at relatively long times so that the results may not be reliable at very short times corresponding to the wings of the spectral band. Kometani and Shimizu[60] have also presented a three variable theory and applied it to a discussion of correlation times primarily in the rotational diffusion limit. They were able to express τ_θ in terms of torques and angular momenta.

Using a somewhat different approach Shimzu[61] obtained essentially the same relationship as Gordon between the correlation function and spectral density. The correlation function for rotational Brownian motion of a molecular system was considered theoretically. Two limits of Brownian motion, the inertial limit and the Debye limit were assumed. A theory of infrared band-shapes in terms of the correlation function was subsequently developed.

7.3.2 STOCHASTIC MODELS

7.3.2(a) Vibrational Lineshapes

A stochastic-type theory for diatomic molecules in inert solutions was developed by Bratos *et al.*[62–64] This semi-classical theory considered one active molecule isolated in a bath of non-absorbing inert solvent molecules. The active molecule was assumed to execute anharmonic vibrations modified by a stochastic potential $V_s(r, t)$ due to the solvent. The active molecule also performed stochastic reorientations which were not correlated with the vibrations. Quantum theory was used to describe the vibrations while the reorientations were treated classically. The relaxation function $G(t)$ was split into two contributions, the vibrational relaxation time $G_v(t)$ and the reorientation relaxation time $G_R(t)$. The vibrational relaxation process was determined basically by the same statistical theories used in lineshape studies.[65,66] The bandshapes predicted by this theory were basically asymmetric with limiting symmetric cases. A continuous sequence of band forms were predicted including, Lorentzian, Gaussian, and Voight profiles as well as several forms of asymmetric profile. The form of a particular band was shown to depend on the nature of the dominant relaxation process. Recently, Bratos extended his theory of infrared bandshapes to include Raman band profiles. Band profiles were predicted for various contributions from vibrational and rotational relaxation. Bands associated purely with vibrational relaxation were all asymmetric although the asymmetry was small in certain cases. A continuous sequence of band contours were predicted which depending on the relative contributions from the two mechanisms ranged from an O–Q–S type band, a Lorentzian, a Gaussian, a Voight profile and several asymmetric profiles.

7.3.2(b) Low-frequency Absorption

The observed far infrared–microwave absorption in simple polar molecules in the liquid or rotator phases is due to three main processes:

(a) a non-resonant part arising from stochastic diffusion of the permanent dipole (well represented by the Debye equations) extending typically over the range 10–100 cm^{-1}: the so-called Debye plateau.[67,68]

(b) Additional absorption above the plateau generally attributed to the short time libration or torsional oscillation of the dipole. The restoring torque arises from the derivative of the potential barrier due to the neighbouring molecular fields—Poley absorption.

(c) Additional absorption due to transient dipoles induced by neighbouring molecular fields and modulated by their motion. This causes absorption in non-dipolar molecules in the 2–250 cm^{-1} region.

Processes (a) and (b) are essentially the long and short time parts of the same overall process, models for which have been developed by Brot.[34,35] and Wyllie.[69]

These have been extended and applied to the plastic and liquid phases of progressively anisotropic molecules by Larkin *et al.*[32,33, 70–71] In these models a molecular rotor undergoes librational motions in a potential well determined by neighbouring molecules, and which under the influence of strong collisions might reorient from one well to another. The librational angular frequency ω_0 of a polar molecule within a potential well of depth V(kJ mol^{-1}), of semi-angular aperture ξ (radians) and of the form

$$u(\theta) = V \sin^2\left(\frac{\pi}{2\xi}\,\theta\right) \tag{7.3.6}$$

has the small angle (harmonic) solution,

$$\omega_0 = \frac{\pi}{\xi}\left(\frac{V}{2I}\right)^{1/2}$$

where I is the relevant moment of inertia of the librator. This frequency is similar to that obtained from a molecular dynamics calculation using a Lennard–Jones–Devonshire potential.[73] Statistical calculations of barrier heights in *t*-butyl chloride have shown that because of the quasi-isotropy of the rotator phase, Lennard–Jones potentials give a very good representation of the potential well depths.[74] In Brots classical autocorrelation function treatment of a librator in a crystal lattice, the width of the librational absorption ($1/\tau$) is a function of both weak thermal collisions of frequency ($1/\tau_i$) perturbing the motion and of a natural width which includes contributions from the anharmonicity of the well and its departure from cylindrical symmetry. Relaxational absorption occurs when the librator receives sufficient energy in a strong collision to enable it to jump from one well to another. The mean times of residence within a potential well and of the duration of jump from one well to another are both functions of V, ξ and the temperature T. This arises from the Boltzman population of the potential wells

$$\frac{\tau_r}{\tau_j} = \frac{1 - \exp(-V/kT)}{\exp(-V/kT)} \tag{7.3.7}$$

and the fact that $\tau_r \approx \tau_D$, the Debye relaxation time. In some liquids, $\tau_j \approx 0.4 \times 10^{-12}$ s and angle of jump of 20–60° have been found.[75] In practice, an experimentally observed spectrum is fitted by deducing a value for the aperture angle ξ from the lattice structure of the rotator phase solid using trial values of V and τ_i. The trial values are adjusted until the best fit to ω_0, $\tau_r \approx \tau_D$ and the relative amplitudes of librational and relaxational absorption are obtained.

A second model of a librator with cooperative relaxation has been proposed by Hill[76] and developed by Wyllie.[69] In this itinerant oscillator model, a dipole with moment of inertia I performs damped librational motion within a cage of Z nearest neighbours. The damping arises from lack of rigidity of the cage. The cooperative diffusional reorientation of the cage molecules about their centres of gravity is a function of the moment of inertia of the cage and its damping coefficient. Since two damping coefficients are used in this model good agreement with experimental spectra can generally be obtained.

7.4 EXPERIMENTAL TECHNIQUES

To obtain meaningful information on molecular dynamics and intermolecular interactions from spectral bandshapes precise measurements of the spectral distribution $I(\omega)$ must be performed. Particular care must be taken to remove instrumental distortion and to obtain accurate values of $I(\omega)$ in the wings of the band where appropriate.

7.4.1 Infrared Absorption

Suitable bands for study are usually difficult to find in liquids and plastic crystals. This problem is due to the complexity of condensed phase infrared spectra arising from overtones and combination bands, hot bands and overlapping fundamentals necessitating the use of artificial curve resolving techniques. Intense bands should also be avoided if possible due to distortion and interference effects in the thin films necessary. These difficulties are compounded by the very large changes in refractive index that can occur in the traversal of the absorption band. Conventional double beam spectrometry is difficult where accurate transmission intensities on pure liquids and solids are required. The spectra must be obtained with an open reference beam and the radiation losses in the cell computed by the application of classical optical theory. These problems have been treated in a recent review and a series of papers by Jones *et al.*[77–79] Methods based on interferometry, simple reflection or total attenuated reflection (ATR) can also be used. With modern high resolution infrared spectrometers, instrumental factors should not influence the absorption profile significantly except perhaps for very sharp bands. Various techniques are however available for reducing the effect of slit distortion on the observed band profile. These techniques were reviewed by Seshadri and Jones in 1963[80] but recently the availability of computors has simplified all aspects of data processing from the initial smoothing of new data to sophisticated methods of bandshape analysis.[81–83]

7.4.2 Computation of Dipole Correlation Functions

The experimental dipole correlation function is computed from the observed spectral density $I(\omega)$ according to

$$\hat{C}(t) = \int_{\text{band}} \hat{I}(\omega) \cos[(\omega - \omega_s)t]\,d\omega \tag{7.4.1}$$

Normally only the real part of $C(t)$ is computed, the imaginary part contributing only a small quantum correction. The shape of the correlation function will depend on the choice of the reference frequency ω_s the band centre. If ω_s is chosen at the maximum intensity of the gas phase absorption (centre of Q-branch), then $C(t)$ will contain all the information about the induced frequency shift as well as the other broadening influences. If, however, ω_s is computed from

$$\omega_s = \int_{\text{band}} \omega\, \hat{I}(\omega)\, d\omega - M(1) \tag{7.4.2}$$

where $M(1)$ is the first moment of the band, (i.e. equal to the band centre frequency), $C(t)$ contains mainly the information about the broadening mechanisms, since the frequency shift has already been allowed for in selecting the reference frequency. The spectral density $I(\omega)$ should be computed from the absorption coefficient $\alpha(\omega)$.

$$I(\omega) = \frac{\alpha(\omega)}{1 - \exp(-\hbar\omega/kT)} \tag{7.4.3}$$

which corrects for induced emission. Normally, however, the absorbance, $I(\omega) = \ln(I_0/I)$ is a sufficiently good approximation.

7.4.3 Raman Scattering

Most of the requirements for infrared bandshapes also apply to Raman scattering but overtones and combination bands are generally less of a problem. Modern Raman instruments and high intensity stable laser sources enable the measurement of accurate band contours to be made with relative ease. High gain, low noise phototubes in combination with photon counting systems have made possible the recording of even weak Raman lines with the additional advantage of digital data output, greatly facilitating data processing. Furthermore, strong Raman fundamentals do not pose the problems that are present in the infrared.

Unlike infrared absorption, it is sometimes possible to separate the orientational relaxation processes from other line broadening mechanisms with the aid of accurate isotropic and anisotropic scattering measurements. The study of strongly polarized lines requires an optical system capable of separating the strong isotropic compound from the depolarised spectrum. A Glan–Thompson polarizing prism is useful to ensure that only light of the correct polarization enters the sample and a polarization scrambler placed before the entrance slit eliminates grating effects. Generally, the instrument functions are Gaussian and both the intrinsic lineshape and the orientational spectrum are Lorentzian, so that closed expressions can be used to remove instrument effects. Instrumental broadening effects can also be estimated by recording the Raman bandshape at smaller and smaller spectral slit widths until no change is observed in the band profile.

Another technique for obtaining orientational linewidths, and associated correlation times, for weakly polarized Raman bands, was described by Rakov[84] and recently revived.[85] The depolarized Raman linewidths $\omega(T)$ are measured over a wide temperature range. At the high temperatures, $\omega(T)$ has a large orientational contribution, whereas at low temperatures $\omega(T)$ approximates to the vibrational width. Assuming the orientational linewidth (ω_{OR}) and vibrational linewidth (ω_{VIB}), are both Lorentzian, the observed spectrum will also be Lorentzian whose width will be the sum of ω_{OR} and ω_{VIB}. Thus,

$$\omega(T) = \omega_{VIB} + \omega' \exp(-A/T) \tag{7.4.4}$$

where it is assumed that ω_{OR} has an Arrhenius dependence and ω_{VIB} is independent of temperature. This technique is particularly useful for plastic crystals where bandshapes can be measured over a wide range of temperature.

7.5 APPLICATIONS

7.5.1 Hydrogen

Heat capacity measurements show the presence of a transition in solid H_2 at 1.5 K increasing to higher temperatures with increasing concentration of ortho-H_2 and pressure. On cooling H_2 first crystallizes in the hexagonal close-packed structure but in normal H_2 this changes to face centred cubic structure at about 1.5 K. The problem of the onset of molecular rotation in solid hydrogen has been treated theoretically using an effective one-molecule potential.[86] Qualitative agreement was found with the dependence of the phase transition on temperature, pressure and ortho-H_2 concentration.

In solid H_2 there is virtually free rotation and the rotational constants remain virtually unchanged on solidification. Rotation in the solid is present at very low temperatures because the spacing of rotational levels is large ($\sim 350\ cm^{-1}$) compared with intermolecular binding energies. The presence of intermolecular interaction however ensures the delocalization of rotational excitation giving rise to a rotational exciton band 20 cm^{-1} wide. As a result, structure is observed in the spectrum.

The pure rotational Raman spectrum of solid para-H_2 contains three sharp lines at 351.8, 353.9 and 355.8 cm^{-1}.[87] Solid ortho-H_2 also contains three Raman lines at 176.8, 179.4 and 182.0 cm^{-1}. The spacing in para-H_2 calculated on the basis of quadrupolar forces is about 10% larger than the observed spacing of 2 cm^{-1}. All these features are associated with the transition $S_0(0)$ ($J = 0$ to $J = v$ in the ground vibrational state). Infrared absorption in solid para-H_2 results from quadrupole-induced transitions and was observed at 355.6 cm^{-1}. Calculation of the frequency and intensity of this absorption based on a hexagonal close-packed structure were in general agreement with experiment.[88]

Vibration–rotation Raman spectra of solid para-H_2 consists of two sharp lines at 4149.8 and 4485.8 cm^{-1}, assigned to the $Q_1(0)$ and $S_1(0)$ transitions respectively.[88] The theory for the infrared induced fundamental in solid H_2 has been developed by van Kranendonk.[89] Three sharp features are observed in the infrared,[90] two of which are assigned to $Q_1(1)$ and $Q_1(0)$ transitions. The third Q branch feature of 4155.0 cm^{-1} is attributed to the $v = 1$ vibrational exciton band of para-H_2 accompanied by an orientational transition in the pair of ortho-molecules on neighbouring sites.

The $S(0)$ branch of solid normal hydrogen consists of three sharp features assigned to $S_1(0)$, $Q_1(1) + S_0(0)$ and $Q_1(0) + S_1(0)$. When normal hydrogen is cooled through the transition at 1.5 K the infrared lines $S_1(0)$ and $S_1(1)$ vanish completely.[91] These lines appear in the I.R. due to quadrupolar intermolecular interactions which are only effective as long as the molecular sites are not centrosymmetric. The disappearance of these lines implies a centrosymmetric structure such as face-centred cubic. The far infrared absorption (50–100 cm^{-1}) shows in addition that orientational ordering takes place below the phase transition.[92]

7.5.2 Hydrogen Chloride

Solid HCl exists in two solid phases separated by a first order transition at 98.4 K. Above this temperature (T_λ), the crystal structure is face-centred cubic and the molecules undergo rapid tumbling. Below T_λ the structure is face-centred orthorhombic and the molecules form a zig-zag chain. The molecular motion of HCl in its condensed phases has been studied using the Raman scattering technique.[93] In the liquid and cubic phases, the depolarization ratio, the spectral lineshape, and the width of the H–Cl stretching vibration were measured as a function of temperature. Since HCl has a dipole moment, in addition to the short range intermolecular forces, the molecular dynamics are influenced by long-range, dipole–dipole interactions. Moreover, the presence of hydrogen bonding in liquid and solid HCl can modulate the polarizability and provide an additional light scattering mechanism. The polarized and depolarized spectra of HCl in the condensed phases are shown in Figures. 7.3 and 7.4. Despite an appreciable narrowing, the depolarization ratio remains essentially unchanged on both sides of the liquid–solid transition temperature (158.9 K). In the temperature range 170–120 K, the depolarization ratio remained constant at about 0.06 implying that the average value α and the anisotropy β of the polarizability tensor are approximately equal. Thus, in the solid there must be either considerable rotational freedom or else a disordered structure similar to that found in the ammonium halides. The lineshapes were not changed in any significant way in traversing the liquid-solid transition so that the molecular motion must be rather similar in both phases. When the data was analysed, assuming only rotational relaxation, the barrier hindering reorientation was found to be very small in both liquid and cubic solid phases. Below 120 K the shape of the stretching mode became asymmetric and satellite peaks developed indicating the evolution of a less symmetric symmetry structure. In the study of the temperature dependence of the low frequency spectrum in the 0–300 cm^{-1} region the very broad band in the liquid and cubic phases was interpreted as arising from charge transfer interactions which rapidly modulate the polarizability of the hydrogen bonded system. As the temperature is lowered below T_λ the broad lattice band breaks and splits up into several sharp lines characteristic of the ordered phase.

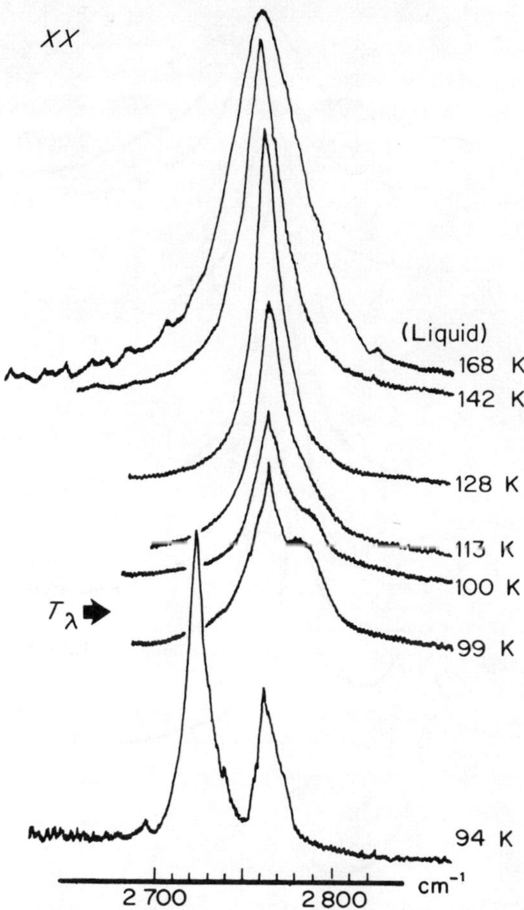

Figure 7.3 Temperature variation of the polarized Raman spectrum of the HCl internal stretching mode

In a subsequent study[94] Raman correlation functions and times were computed for the liquid and cubic solid phases of HCl as a function of temperature. The correlation times obtained from polarized and depolarized Raman spectra are compared with NMR correlation times as a function of temperature in Figure. 7.5. The values of τ_c obtained from depolarized scattering show a continuous temperature dependence above and below the melting point. Whereas the values of τ_c obtained from the polarized spectra are quasti-discontinuous at the liquid–solid transition temperature. The lineshapes of the polarized spectra reflect mainly the contributions from non-reorientational processes, such as vibrational relaxation and translational diffusion. Vibrational relaxation due to anharmonicity in the vibrational potential is expected to be important in hydrogen bonded HCl but it was not possible to distinguish the

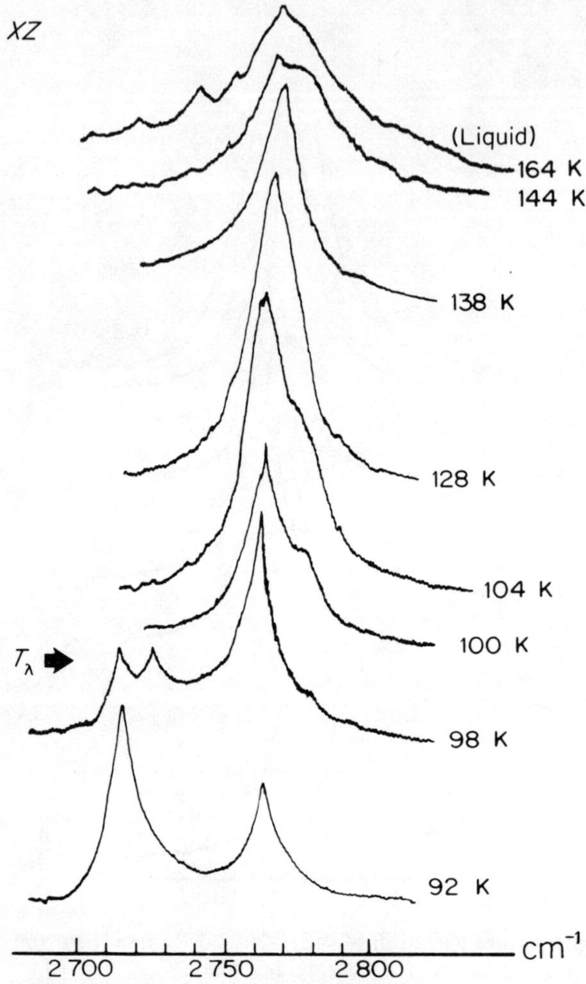

Figure 7.4 Temperature variation of the depolarized Raman spectrum of the HCl internal stretching mode

relative contributions of the two processes to the observed band profile. The change in the shape of τ_c against $1/T$ at the melting point was interpreted in terms of a rapid rupturing of the Cl----H hydrogen bond, leading to a large decrease of τ_c on melting.

From the depolarized scattering an Arrhenius activation energy of 1.7 kJ mol^{-1} was obtained.[94] This is a factor of 2 different from the value obtained from NMR measurements indicating that different physical processes are measured by the two techniques. In a strongly hydrogen bonded system such as

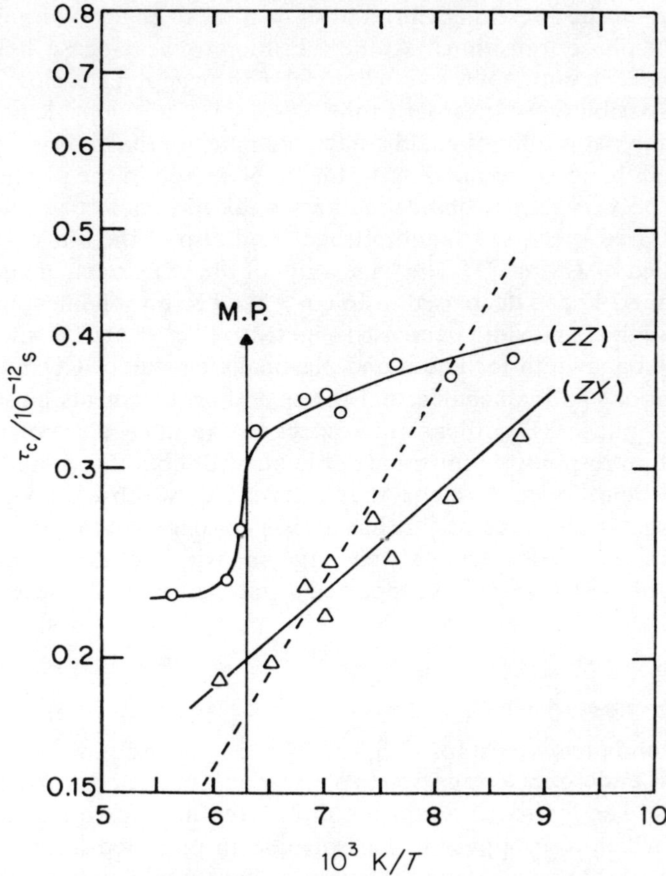

Figure 7.5 Correlation times, τ_c calculated from the polarized and depolarized Raman spectra as a function of of temperature. ----- NMR correlation time

condensed HCl, the commonly made assumption that the vibrational and reorientational correlation functions are uncorrelated is not valid, and the relationship between the depolarized Raman bandshape and the NMR T_1 time is not applicable. Nevertheless, all the data were consistant with large reorientational steps in both liquid and plastic HCl interrupted by rapid hydrogen bond formation–rupture processes. The far IR absorption[72] of the plastic phase at 100 K has been fitted satisfactorily by a librational model.

7.5.3 Carbon Monoxide and Nitrogen

The molecules CO and N_2 are isoelectronic and both have two solid phases α and β. The transitions occur at 35.6 K for N_2[95] and 61.6 K for CO.[96] The infrared spectra of the stretching vibration show marked changes on going through the rotator phase transition.[97] At the transition to the α-phase at 35 K the absorption at 2336 cm^{-1} splits into a doublet with an appreciable loss in intensity. This was attributed to disorder in the β-phase but it is possible however, that $k \neq 0$ combination bonds could contribute much of the observed absorption. The Raman band corresponding to the N–N stretch in the plastic phase was found to be very narrow indicating very weak intermolecular interactions.[98]

The infrared spectra of liquid nitrogen and also of the α and β solid phase were studied by Ewing.[99,100] The bandwidth of the CO stretch in liquid CO was 22 cm^{-1} at 80 K and decreased to 16 cm^{-1} at 69 K, on solidification to the rotator phase the bandwidth decreased slightly to 14 cm^{-1} at 67 K. Ewing[100] interpreted the bandwidth for liquid and plastically crystalline CO in terms of rotatory motions and concluded that the spatial arrangements must be similar in the two phases. Disorder with respect to the molecular polarity in solid CO could be responsible for some of the linewidth but this should not contribute more than 1.5 cm^{-1} which is the observed bandwidth in the α-phase where this disorder is also present.[99] It has also been suggested that this type of disorder is responsible for the relatively large line widths of the Raman spectrum of solid α-phase CO.[101] This is supported by measurements of the lattice spectra of the solid.

7.5.4 Hydrogen Sulphide

Three solid phases exist for H_2S,[102,103] phase 1 stable below the melting point at 186.6 K, undergoes a transition to phase II at 126.2 K and a further transition at 103.6 K. The hydrogen atoms are disordered in the two high temperature phases[104] which also appear to be isotropic to polarised light.[105] NMR work has indicated that reorientation is occurring in phases I and II which are almost certainly rotator phases.[106]

Infrared studies[107] in the region of the internal modes found that the bands in the high temperature phases were broad compared with those of the tetragonal phase III, confirming the presence of molecular reorientation.

7.5.5 Phosphine PH_3 and PD_3

Heat capacity measurements have established that PH_3 undergoes three solid–solid transitions at 88.1, 49.4 and 30.3 K.[108] These phase transitions have been confirmed by infrared measurements which have also been used to identify the corresponding transitions in PD_3.[109]

The infrared spectra of solid PH_3 has been studied by a number of workers,[110,111] but it seems probable that only two of the phases were studied. A more recent study of the infrared and Raman spectra of phosphine[112] included the four phases of pure PH_3 and PD_3 and some mixed isotopic crystals.

These measurements confirmed that the two higher temperature phases were disordered but it was not possible to distinguish between static or dynamic disorder. The vibrational bands were very sharp in the two ordered phases.

Very recently, Raman[113] and nuclear spin relaxation studies[114] of phosphine have appeared in the literature. Comparison of the reorientational correlation times obtained from the two techniques provides insight into the anisotropy of molecular reorientation in the liquid and plastic crystalline phases. Furthermore, the dependence of molecular rotation on moment of inertia may be investigated.

The polarized and depolarized Raman scattering in the temperature range 142 to 297 K was obtained for both the ν_1 (A_1) band at 2306 cm^{-1} and the ν_4 (E) band at 1110 cm^{-1} in liquid phosphine.[113] The depolarized linewidths were found to be at least an order of magnitude greater than the polarized widths at all the temperatures studied for the 2306 cm^{-1} band. The isotropic linewidths were found to remain constant at 2.3 cm^{-1} over the whole temperature range 142 to 297 K, but the depolarized linewidths varied from 24 cm^{-1} at 142 K to 58 cm^{-1} at room temperature.

When the reorientational linewidth was separated from the isotropic component following the procedure outlined in Section 2, Raman reorientational correlation times of 0.09 ps at room temperature up to 0.22 ps at 142 K were obtained. A fit of the linewidths to an Arrhenius type expression yielded an activation energy of 2.1 J mol^{-1} for the molecular reorientation process.

Reorientational and angular momentum correlation times were determined from ^2H and ^{31}P spin lattice relaxation times for both the plastic crystalline and liquid phases.[114] These results are shown as a function of reciprocal temperature, together with the Raman results in Figure 7.6. The corresponding values of $\tau_\theta^{(2)}$ (eff) and $\tau_\theta^{(2,0)}$ are very similar implying that the reorientational motion of PH_3 is essentially isotropic in the liquid. There is a discontinuity in $\tau_\theta^{(2)}$ at the melting point representing a subtle change in the rotational motion in passing from the liquid to the rotator phase while τ_j remains essentially unchanged. This behaviour is unlikely to arise from an increase in intermolecular interactions in the rotator phase since there is a decrease in apparent activation energy for $\tau_\theta^{(2)}$ from 10.5 kJ mol^{-1} in the liquid to 5.5 kJ mol^{-1} in the solid. Behaviour of this type is in marked contrast to that observed for other spherical or near spherical molecules where $\tau_\theta^{(2)}$ has a single Arrhenius temperature dependence over the entire liquid and rotator phases with no discontinuities. This implies that the collisional interactions and molecular field are unchanged in traversing the phase boundary. In the case of phosphine, the continuity in τ_j suggests that the collision process in unchanged whereas the discontinuity in $\tau_\theta^{(2)}$ implies a change in the molecular field. It is also somewhat surprising that the data for liquid PH_3 and PD_3 is more characteristic of M-diffusion in contrast to other small spherical and symmetric top molecules such as CH_4, SF_6, and CCl_4 which undergo J-diffusion in the liquid phase.

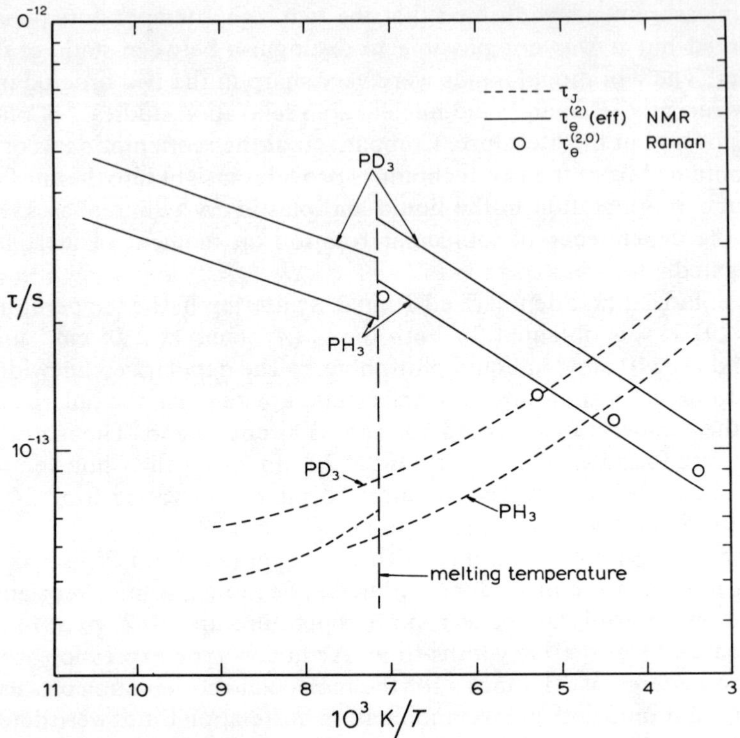

Figure 7.6 Reorientational and angular momentum correlation times as a function of reciprocal temperature in liquid and plastic crystalline phases of PH_3 and PD_3

7.5.6 Methanes CH_nD_{4-n}(n = 0–4)

Heat capacity measurements[115] show that while CD_4 undergoes two λ-type transitions at 27.0 and 22.1 K, CH_4 has only one at 20.4 K. Each of the partially deuterated species also have two transitions between 2 and 26 K.[116] In phase 1, the molecular orientations in CD_4 are completely disordered while in phase II, six out of the eight molecules in the unit cell are ordered. Phase 1, in CH_4 is also disordered.

The infrared absorption[118] of the v_3 band of CH_4 has been studied as a function of temperature from 105 to 28 K. It was concluded that the molecules are nearly freely rotating in the liquid as well as the rotator phase near the melting point. At lower temperatures the motion became librational. Raman studies[119] also indicated that there was almost free rotation in both liquid and plastic phases. The far infrared spectrum[120] of phase 1, in CH_4 and CD_4 also showed a single broad absorption characteristic of the plastic phase.

Extensive infrared and Raman studies of all isotopic modifications of methane as pure and mixed isotopic crystals have been carried out by Cabana *et al.*[121, 122] Site splitting and orientational effects have been used to obtain information on the site symmetries of molecules in phase II crystals. It was concluded that a portion of the molecules only undergo quantized rotation in the pure solids. Reorientational correlation functions for the ν_3 and ν_4 bands of liquid CH_4 and CD_4 have also been computed.[123] This suggests that free rotational effects dominate the short time dependence of $C(t)$ whereas intermolecular effects become important at long times. In a later publication, McClung[47] used the M-and J-diffusion models to compute reorientational correlation functions and spectral densities for CH_4 and CD_4 and compared them with the experimental values of Carbana *et al.*[123] The agreement was satisfactory for liquid CH_4 and CD_4 in the J-diffusion limit. These results were confirmed in a later study of liquid CH_3D, where the dipole correlation function for the ν_1 band at 2194 cm^{-1} was in good agreement with the J-diffusion model,[124] i.e. the collisions were sufficiently energetic to change both the magnitude and direction of the angular momentum. This model also successfully reproduced the behaviour of the plastic solid at 41 K by changing the parameter τ_J from 0.9 to 0.6. This reflects the increased intermolecular interactions in the plastic phase.

Gordon[22] used the infrared data of Ewing and the Raman data of Harrold and Plint to determine the reorientational correlation functions $\langle \mathrm{u}(0) \cdot \mathrm{u}(t) \rangle$ and $\langle P_2[\mathrm{u}(0) \cdot \mathrm{u}(t)] \rangle$. From the shape of these correlation functions he concluded that CH_4 reorientated by large angle free diffusion. From Gordon's data one can show that this corresponds to an angular step size of $\sim 50°$.

7.5.7 Substituted Methanes

7.5.7(a) Symmetrical Tetrahalides

The three tetrahalides, CF_4, CCl_4, and CBr_4 all exist in rotator phase modifications which have been studied in detail spectroscopically.

Solid CF_4 melts at 89.5 K and has an order–disorder phase transition at 76.2 K, above which limited molecular rotation occurs.[125,126] An extensive infrared and Raman study of CF_4 has been reported by Fournier *et al.*[127] who observed very little change in the spectra when the liquid crystallizes. The broad bands which show no splitting and the appearance of inactive modes are all indicative of rotational disorder in the high temperature solid phase. Furthermore, drastic changes in the spectra occur when the solid is cooled below 76.2 K. Vibrational linewidths are considerably reduced, most of the fundamentals become doublets, and ν_1 becomes inactive. The low temperature solid phase is thus ordered. More recently, the Raman spectra of the liquid and plastic phases of CF_4 have been investigated between 77 and 185 K.[128] Again, little difference was found between the Raman spectrum of the liquid at the melting point and that of the plastic crystal at the same temperature. Due to the large dipole derivative (4.8 DÅ^{-1}) of the C–F bond,[129] there is a strong transition dipole-transition dipole coupling between the degenerate ν_3 modes in the plastic phase. In cubic crystals, this coupling results in the longitudinal optical mode (LO), associated with a triply degenerate vibration, to appear at a higher frequency than the transverse (TO) modes[130] i.e.

$$\nu_{LO}^2 - \nu_{TO}^2 = \left(\frac{n_\infty^2 + 2}{3\, n_\infty}\right)^2 \frac{N}{\pi}\left(\frac{\partial \mu}{\partial Q}\right)^2$$

where ν_{LO} and ν_{TO} are the frequencies of the LO and TO modes, N is the concentration, n_∞ the refractive index on the high frequency side of the band, and $(\partial\mu/\partial Q)$ is the dipole derivative. This type of LO–TO splitting is frequently observed in ionic crystals, but CF_4 and SiF_4 provide the only definitive examples in molecular crystals.[131] This splitting of the ν_3 band is found in the non-rotator as well as the rotator phases of CF_4 and (as shown in Figure 7.7)[128] also occurs in the liquid phase. The ν_3 spectrum in the rotator phase closely resembles that in the low temperature liquid but as the temperature is increased the two components move closer and eventually merge. This is due in part to rotational broadening, which is negligible in the plastic phase. Translational disorder in the liquid phase presumably also contributes to the broadening of the LO and TO components.

Information on molecular reorientation of CF_4 was obtained from the profiles of the other two depolarized bands $\nu_2(e)$ and $\nu_4(f_2)$. These bands were symmetric, temperature sensitive, and showed evidence of weak vibrational relaxation effects. Only a slight discontinuity in line-width (0.8 cm^{-1}) was observed at fusion, and vibrational frequency shifts were also small ($\sim$ 2 cm^{-1}) at the rotator phase—liquid transition. Angular momentum correlation times, τ_J, were obtained by fitting the theoretical spectra to the observed bandshapes using the J-diffusion model,[46] and used to compute reduced times τ_J^*. Values of τ_J^*, (the mean angles through which the molecule rotates during time intervals τ_J), of about 0.1 rad, were obtained at 90 K. This was about twice the value obtained for other tetrahedral (or quasi-tetrahedral) molecules such as CCl_4,[132] CCl_3F[133] and $SnCl_4$[134] near their melting points. However, the value of τ_J^* is smaller than that observed for SF_6 near its melting point ($\tau_J^* = 0.35$ rad)[35,136]. Unlike CF_4, SF_6 melts with much higher variations of entropy and volume than the rare gases and also has a greater plastic crystalline range. This is illustrated by the data collected in Table 7.3. A high degree of correlation is apparent between the discontinuities observed at fusion for τ_J^*, the molar volume v, and the entropy S for the rotator phase molecules CF_4, NF_3, SF_6, and MoF_6.

Carbon tetrabromide exists in the monoclinic Th^6 (Pa^3) phase at room

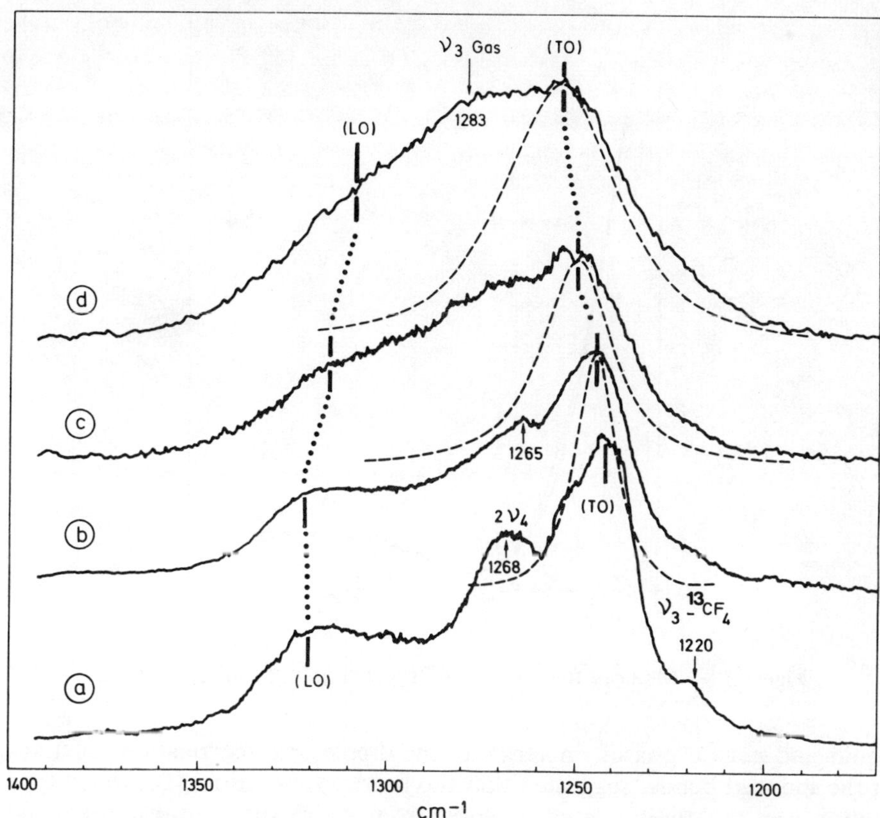

Figure 7.7 Raman profile of ν_3 in CF_4; *a*, plastic crystal, *b*, *c*, *d*, liquid at 85, 138 and 185 K, ———— experimental profile, ----- pure rotational spectra in *J*-diffusion model

temperature and undergoes a phase transition to the rotator phase at 319.86 K. The cubic rotator phase melts at 365 K. The two infrared active fundamentals of CBr_4 are found at[137] 184 (f_2) and 672 (f_2) cm^{-1}. In addition, there are four Raman active fundamentals at 267 (a_1), 123(e), 184(f_2) and 672(f_2) cm^{-1} Davies *et al.*[138] measured the far infrared induced absorption spectra of liquid, single crystal rotator phase, and monoclinic polycrystalline carbon tetrabromide. The results are shown in Figure 7.8. Broad bands were found near 32 cm^{-1} in both liquid and rotator phases and were interpreted in terms of intermolecular interactions. In the monoclinic phase at 298 K, a doublet was observed in this region, both components of which were broader than the fundamentals. In the plastic crystalline phase at 376 K, the fundamental at 124 cm^{-1} was considerably broadened compared with the monoclinic phase, and was broader still in the liquid phase. Analysis of the bandshapes of the

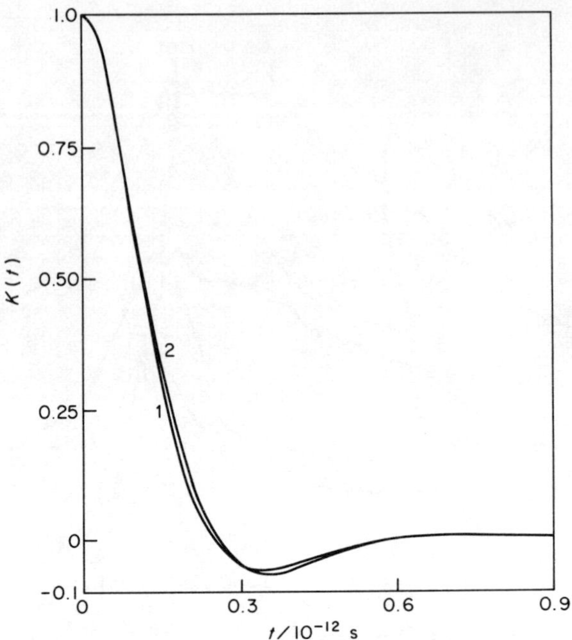

Figure 7.8 Memory function for CBr_4, (1) liquid, (2) plastic crystal

liquid and plastic phases, in terms of the dipole cross-correlation functions of the induced bands, suggested that the barrier to rotational Brownian-type motion was slightly increased on going from the plastic crystal to the liquid a few degrees above the melting point. This result was supported by the molecular dynamics calculations of O'Dell and Berne[139] who found that rotational motion is less hindered in the solid just below the melting point in an ensemble of rough hard spheres. This implies that the symmetry of packing and the resultant restriction on molecular diffusion is an important factor in determining the ease of molecular rotation in the plastic crystalline phase. At a certain density in the liquid, rotation inevitably becomes less hindered than in the rotator phase, but near the melting point, lack of packing symmetry appears to make rotation a statistically more energetic process than in the solid, resulting naturally, in the phase transition itself. The integrated absorption intensity per mole was found to be greater in the liquid phase, although the density is lower, suggesting that the spatial arrangement of the long range part of the intermolecular potentials was important in determining the magnitude of the induced molecular dipole moment. The cross-correlation functions of the liquid and rotator phase solid were compared with that of an octopole-induced absorption of a two-molecule collison of spherical top molecules, and with the autocorrelation function for a Maxwellian ensemble of freely rotating molecules of this symmetry. It was found that the mean torque was greater initially (near $t = 0$), in both condensed phases than in the bimolecular collision of octopole fields since the correlation function was found to decay faster initially. Thereafter, rotational motions are correlated in the condensed phases, and after 0.7 ps both correlation functions become exponential and decay relatively slowly compared with the octopole case.

The corresponding order–disorder transition in CCl_4[140] is found at 225.3 K. The Raman spectrum of CCl_4 is complicated by isotopic splittings arising from ^{35}Cl and ^{37}Cl and by hot bands which become increasingly apparent above 200 K.

Bartoli and Litovitz[141] carried out a Raman lineshape study of CCl_4 in the liquid and plastic crystal phases. Reorientational correlation times were calculated from the relation

$$\tau_{OR} = 1/2\pi c \omega_{OR} \tag{7.5.1}$$

where ω_{OR} is the orientational half-width of the Raman band. The E species, 217 cm^{-1} deformation was chosen and at 296 K this had a ω_{OR} of 3 cm^{-1} and τ_{OR} of 1.8 ps. This is in agreement with an NQR measurement of 1.7 ps. The temperature dependence of $\tau_2(R)$ was also obtained in the range 230 to 343 K. These results are plotted in Figure 7.9 together with the corresponding NMR data, τ_2 (NMR). This data shown no discontinuity at the liquid-plastic crystal phase transition indicating that the molecular structure (face centred cubic) is similar in the two phases. Since large volume structural fluctuations are improbable in the plastic solid, each molecule must have sufficient free volume reorient within its cage without having to wait for structural rearrangement of its neighbours. Reorientation is probably limited by potential barriers arising from angle dependent forces in the lattice.

Sunder and McClung[142] measured the Raman band contours of the ν_2 and ν_4 bands of CCl_4 in the temperature range 240–373 K. The vibrational contributions to the bandshapes were estimated from the width of the totally symmetric ν_1 band and corrections were made for the presence of different isotopic CCl_4 molecules. For rotator phase CCl_4, they were able to fit the second spherical harmonic correlation time τ_2, obtained from Raman bandshapes, to a J-diffusion model. In this model, the angular momentum correlation time τ_J which corresponds to the mean time between collisions, was used as an adjustable parameter. The value of τ_J obtained from this model was in reasonable agreement with an experimental value derived from spin-rotation coupled NMR relaxation measurements. However, Brot[143] has pointed out that this model is not necessarily correct since 'soft' interactions could lead to the same relationship between τ_2 and τ_J Hard collision models are unlikely to be correct in solid CCl_4.

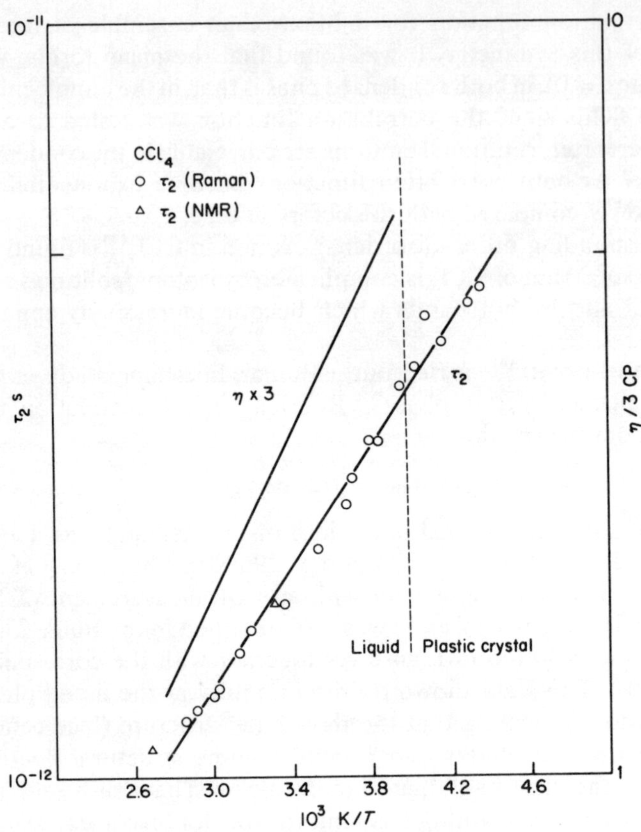

Figure 7.9 Temperature dependence of τ_2 (Raman) and τ_2 (NMR) and η in the liquid and plastic crystalline phases

7.5.7(b) *Non-symmetrical Substituted Methanes*

Methanes substituted with CH_3, NO_2 or Cl groups are all globular and show rotator phase behaviour. Sometimes, the introduction of CN or Br can also give rise to rotator phases in the solid. Most of the work on these compounds has been carried out in the far-infrared/microwave regions where the observed absorption arises from a combination of librational and relaxational behaviour.

(i) t-Butyl Chloride. This compound melts at 246 K and has a rotator type transition at 219 K. Larkin[32] was able to fit the observed far infrared/microwave spectrum to both the multi-well librator and itinerant oscillator models discussed in Section 7.3. Absorption curves were computed and matched in the frequency domain with the broad experimental bands using a number of

adjustable parameters. The barrier heights deduced from both models for the plastic phase agreed with the calculated value of Lassier and Brot[74] of 5.3 $\pm$0.9 kJ mol^{-1} and with the activation energy of Janik *et al.*[144] The librational frequency, barrier height and relaxational frequency all decreased with increasing temperature, the width of the librational absorption increasing by $\sim$ 60%. Thus, the potential wells used in these models with an angular aperture of 0.62 rad were a reasonable representation of the observed spectra.

A more searching test of these models is provided by the rotational velocity correlation function $\langle \dot{u}(0) \cdot \dot{u}(t) \rangle$ discussed previously. The time behaviour of this function illustrates directly the dipolar fluctuations and molecular interactions which give rise to the observed absorption. Evans[23] has computed experimental values of $f_2(t)$ and compared these with theoretical correlation functions $f_B(t)$ and $f_w(t)$ obtained from the Brot[35] and Wyllie[69] models. These are shown in Figure 7.10 for the liquid and rotator phases. Also shown, is the correlation function for the freely rotating molecule. At short times, in the rotator phase there is poor agreement between the experimental and predicted values. The free rotor and experimental functions show a negligible loss of corrélation in the first 0.2 ps but $f_w(t)$ falls off much too rapidly and $f_B(t)$ is virtually linear in this range; both models are thus unsatisfactory. Evans has pointed out that in the case of Wyllie's model this is due to the treatment of the molecular environment as a continuum with an opposing viscous drag and not using an explicit form of u(t), whereas in Brots model of instantaneous collisions the molecular torque tends to infinite as $t \rightarrow 0$ thus giving the wrong slope at short times. The free rotor curves emphasize the substantial torques present in the liquid and plastic phases, which are very similar to magnitude.

(ii) Methyl Chloroform CH_3 CCl_3. Methyl Chloroform has a face-centred cubic structure in the rotator phase.[145] Larkin[32] was able to obtain a good fit to both liquid and rotator phase spectra with both models. The parameters deduced were in good agreement with the molecular dynamics calculations of Lassier and Brot.[146] However, poor agreement between the rotational velocity correlation functions at short times was obtained by Evans.[23] Also, there is a marked increase in molecular torques in the rotator phase compared with the liquid.

(iii) 2,2-Dichloropropane. This material melts at 239 K and undergoes an order–disorder transition at 187 K with a $\Delta S_t > 16.7$ J mol^{-1} K^{-1}. Hoffmans and Larkin[33] measured the far infrared spectrum in both solid phases and the liquid phase. Two lattice modes were found at 61 and 78 cm^{-1} at 133 K in the ordered phase which shifted to 56 and 68 cm^{-1} at 182 K with reduced intensity. In the rotator phase, a broad band of reduced integrated intensity was found. The frequency of maximum absorption moved to lower frequencies with increasing temperature within each phase but the shift was quite small

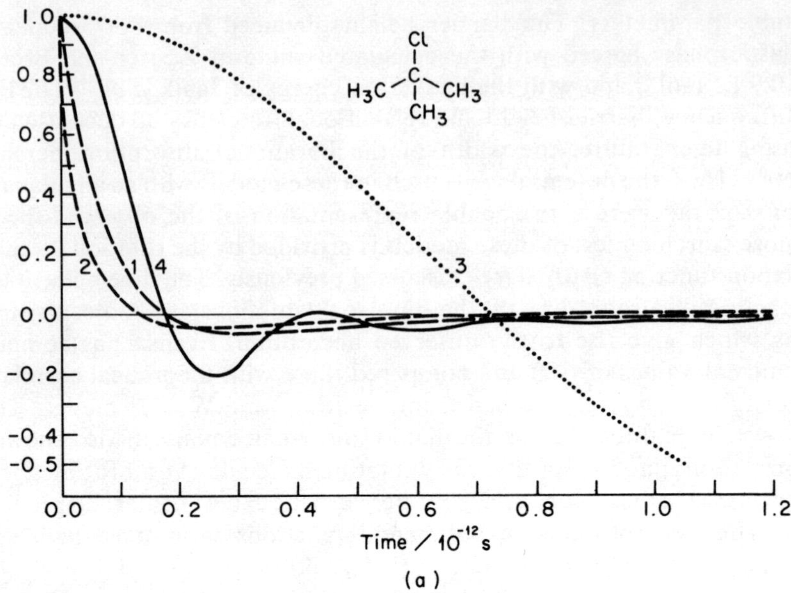

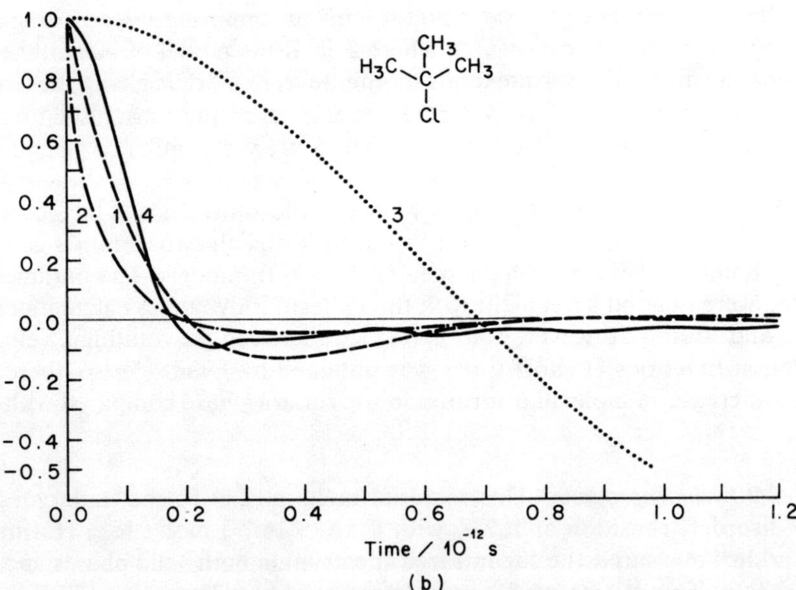

Figure 7.10 *t*-Butyl chloride, (a) rotator phase at 238 K, (b) liquid at 274 K. 1. Brot–Larkin function, $f_B(t)$; 2. Wyllie–Larkin function $f_\omega(t)$; 3. free rotor; 4. experimental curve

at the melting point. A good fit to the observed spectrum was obtained using the itinerant oscillator model. The liquid and rotator phase spectra could also be fitted with Brot's model by varying the barrier height and the shape of the potential wells at low temperatures. Again the rotational velocity correlation function suggested that these models are inadequate at short times.[23] The velocity correlation function of dichloropropane showed marked oscillations which became more significant as the temperature was lowered. These oscillations were also greater in the rotator phase than in the liquid and have been attributed to increased 'structuring' or 'order'.

(iv) 2-methyl-2-nitropropane and 2-Chloro-2-nitropropane. Both these molecules were studied by Haffmans and Larkin[33] in the far infrared region. Lattice modes which were observed in the ordered phases merged into single broad bands in the plastic phases. The spectra were found to remain constant throughout the rotator phases and past the melting point into the liquid phase, indicating no change in the degree of molecular order in this region. Good agreement was obtained between the experimental and calculated spectra for these molecules.

7.5.8 Silane and Germane, SiH_4 and GeH_4

Both SiH_4 and GeH_4 show a λ-type transition at 63.8 K and 76.5 K respectively. In addition, GeH_4 possesses three additional ordered solid phases. Fournier *et al.*[147] have reported the infrared spectra of crystalline SiH_4 is the two solid phases. The spectra show striking differences between the plastic and ordered phases.

Infrared and Raman studies of GeH_4 and GeD_4—GeH_4 mixed crystals grown from the melt have recently been published by Cabana *et al.*[148] The high temperature solid phase was shown to be plastically crystalline but the narrow linewidths of the bands in the other phases showed these to be ordered.

7.5.9 Cyclopentane

Cyclopentane exists in three crystalline modifications,[149] form 1, stable from the melting point at 179.7 K to 138.1 K, form II, from 138.1 K to 122.4 K and form III, below 122.4 K. NMR[150] and x-ray[149] work have shown that forms 1 and II are highly disordered plastic crystals. An extensive infrared study of all the condensed phases of cyclopentane was carried out by Schettino *et al.*[151] Raman spectra were recorded at 85 K (phase III) and agreed with the infrared results. The infrared spectra of the two rotator phases did not show any site or factor group splittings and were typical of the spectra of plastic phase crystals.

7.5.10 Cyclohexane

The reorientational motion in liquid and plastic crystalline cyclohexane has been studied by the infrared and Raman lineshape method.[144] The orientational linewidths and Raman reorientational times τ_2 for four fundamentals are shown in Table 7.2. Since cyclohexane, of symmetry D_{3h} is nearly spherical it is expected that reorientational times about different axes should be close in value. The two polarized bands 802 cm^{-1} and 1157 cm^{-1} refer to reorientation about the major axis of symmetry. The 1027 and 1266 cm^{-1} depolarized Raman lines are also sensitive to motion about the major axes but since the correlation function for these lines is the sum of two exponentials,

$$\left[\tau_2^{(1)} = (5D_\perp + D_{11})^{-1} \text{ and } \tau_2^{(2)} = (2D_\perp + 4D_{11})^{-1}\right]$$

they are generally difficult to analyse. However, when $D_\perp = D_{11}$ all correlation functions become $\exp(-6D_\perp t)$ and all linewidths the same. As Table 7.2 shows, this is the situation which prevails in this case.

Orientation times were computed over a wide temperature range in the liquid and rotator phases from infrared and Raman lineshapes. The infrared reorientational times τ_1(IR), were computed from the linewidth data of the 903 cm^{-1} band. The results are shown in Figure 7.11 in which τ_1(IR) and τ_2 (Raman) are plotted as a function of inverse temperature. Very little change was observed in τ_1 and τ_2 on melting confirming that the molecules are free to rotate in the plastic phase. Reorientation also takes place at a lattice site without requiring large translational motions or disruption of the lattice. This rules out models where the rotational mechanism is only via translational jumps of the molecule.

7.5.11 Norbornylene

Norbornylene (bicyclo[2.2.1]heptene–2) is unusual in that it forms a rotator phase which has a hexagonal structure.[152] This phase exists in the temperature range 129 K to the melting point 320 K.[153] The molecular reorientation in liquid

TABLE 7.2 Raman reorientational times for liquid cyclohexane at 296 K

Line (cm^{-1})	Point group, symmetry species and assignment	ω_{OR} (cm^{-1})	τ_2 (Raman) (ps)
802	A_{1g} *C–C* stretch	3.5 ±0.5	1.5 ±0.3
1027	E_u *C–C* stretch	3.1 ±0.5	1.7 ±0.3
1157	A_{1g} CH_2 rock	3.5 ±0.5	1.5 ±0.3
1266	E_g CH_2 twist	3.1 ±0.5	1.7 ±0.3

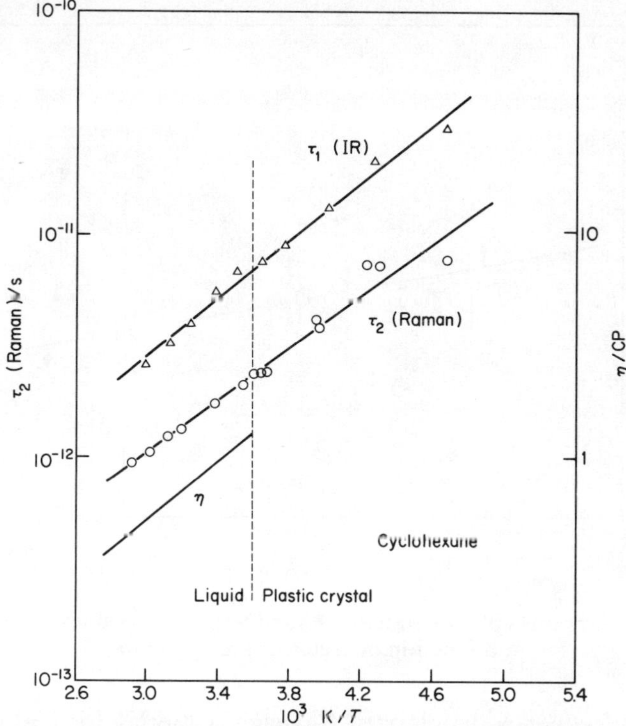

Figure 7.11 Temperature dependence of τ_2 (Raman), τ_1 (*IR*) and η for cyclohexane in liquid and rotator phases

light scattering.[154] Since plastic crystalline norbornylene is anisotropic, it was necessary to use oriented single crystals. An intense, strongly polarized Raman line at 873 cm^{-1} was chosen from the complex vibrational spectrum, and linewidth measurements carried out in the temperature range 266.5 to 363 K. The depolarization ratio also changed from 0.04 at 129 K to 0.025 at 320 K but the integrated intensity remained constant. Raman correlation times were obtained from reorientational half-widths ω_{OR} separated from the vibrational contribution. These are shown in Figure 7.12 as a function of inverse temperature together with the values obtained from NMR spin-lattice relaxation measurements of the rotator phase. Good agreement was obtained between the correlation times, which show the same Arrhenius temperature dependence in both solid and liquid phases with no obvious discontinuity at the melting point. The apparent activation energy for reorientation was 6.3 kJ mol^{-1} which probably corresponds to an essentially isotropic rotation. The vibrational linewidth remained constant at 1.62 cm^{-1} over the temperature range studied which corresponds to an upper vibrational state lifetime of 3.3 ps.

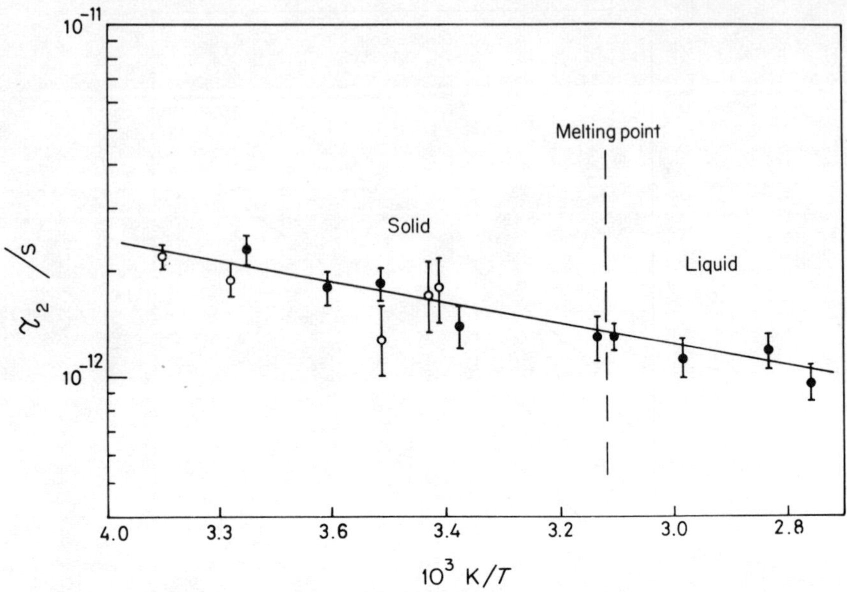

Figure 7.12 Arrhenius plot of Raman (●) and NMR (○) τ_2 values for norbornylene as a function of reciprocal temperature

The high frequency depolarized Rayleigh scattering for norbornylene just below the melting point was virtually identical to that of the liquid phase. This suggests that not only is molecular reorientation identical in both rotator and liquid phases but that the short time (<0.1 ps) molecular dynamics responsible for the high frequency intermolecular scattering are also similar.

7.5.12 Neopentane $C(CH_3)_4$

Infrared and Raman bandshape studies of neopentane in the temperature range 140 to 345 K were recently reported.[156] The order-disorder transition in neopentane is found at 140 K and the melting point is 253 K. The activation energy for reorientation was found to be about 4 kJ mol^{-1} from NMR measurements[157] and about 3.6 kJ mol^{-1} by quasi-elastic neutron scattering measurements.[158] The infrared lineshapes were recorded for the 924 cm^{-1} fundamental, the CH_3 deformation mode belonging to the F_2 species. The Raman lineshape measurements were also carried out on the 924 cm^{-1} band but in this case the band is predominantly E species. The width of the totally symmetric A_1 Raman band at 733 cm^{-1} was essentially equal to the instrumental width (~ 1 cm^{-1}) and fairly temperature insensitive in the range 180 to 300 K. This corresponds to a lower limit on the vibrational lifetime of about 10 ps, an order of magnitude longer than the reorientational correlation times. On this evidence, the vibrational contributions to the infrared and Raman reorientational correlation functions was neglected at short times. The infrared and Raman correlation functions computed on this basis indicated that the correlation behaviour of molecular rotations is the same in the liquid as in the rotator phase. There was no appreciable oscillatory behaviour of $\langle \mathbf{u}(0) \cdot \mathbf{u}(t) \rangle$ in the solid phase region and the decay of $C(t)$ with time indicated that the molecules perform almost free rotations of 31° (303 K, liquid) and 13° (178 K plastic solid) respectively about each axis of rotation.

Correlation times were also computed from the bandwidths, a technique only applicable for Lorentzian bandshapes. There was no significant change in τ in passing through the liquid–plastic crystal transition again emphasizing the similarity in dynamical behaviour near the melting point. An activation energy of 4.1 kJ mol^{-1} was computed from the temperature variation of correlation times in good agreement with neutron scattering and proton spin resonance measurements. The ratio of τ_1 to τ_2 was about 2 over the entire temperature range 170–260 K. Small angle diffusion models predict a ratio of 3 while a random distribution of jump angles predict $\tau_1/\tau_2 = 1.2$. These results, together with the relatively large free rotation angles extracted from the correlation functions indicate that the extended diffusion models of Gordon might be appropriate for the rotator phase solid.

The observed P_1 and P_2 correlation functions for the 942 cm^{-1} band at a temperature of 213 K in the plastic phase are shown in Figure 7.13 together with the corresponding functions computed from the M- and J-diffusion models. The experimental correlation functions are thus intermediate between the M- and J-model predictions. Brot[143] has however pointed out that these results do not necessarily indicate the validity of the model since vibrational relaxation was neglected. Also, in an activated process, the collision rate should increase as $T^{1/2}$, instead of decreasing with temperature, and the correlation function should thus increase with temperature in any hard collision model.

7.5.13 **Quinuclidine** (1 Aza-bicyclo-[2,2,2]-octane)

Quinuclidine belongs to the C_{3v} point group and is sterically nearly globular. It exists in a face-centred cubic plastic phase at room temperature and undergoes a transition to the ordered state at about 200 K with an entropy change of 26.53 J mol^{-1} K^{-1}.[159] A wide line NMR study has shown the disorder to be orientational and dynamical.[160] The far IR spectrum between 10 and 130 cm^{-1} was measured as a function of temperature by Lassier and Brot.[161] The shape of the absorption band, which was much broader than the librational band in rigid solids, but narrower than in liquids, was characteristic of highly anharmonic and frequently perturbed librations which take place between less frequent large-angle jumps.

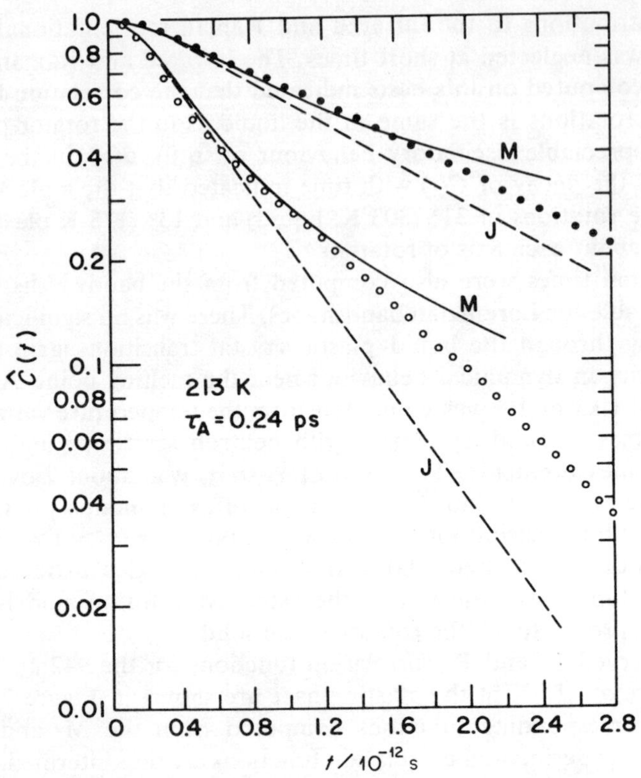

Figure 7.13 Comparison of experimental infrared (●) and Raman (○) correlation functions of the 942 cm^{-1} band of rotator phase neopentane at 213 K with the extended M and J-diffusion models. τ_A is the common angular momentum correlation time of this fit

7.5.14 Succinonitrile

Succinonitrile $(CH_2CN)_2$ is a somewhat unusual plastic crystal in that it is not a globular molecule and also has considerable internal flexibility. In the liquid state, the molecule exists in an equilibrium mixture of gauche and trans forms[162] with an energy barrier of 5 kJ mol^{-1}. The rotator phase is body-centred cubic[163] with the order–disorder transition occuring at about 233 K. This structure implies that a considerable degree of cooperative motion must occur during reorientation in the rotator phase. The rotator phase contains molecules in both gauche and trans forms but only gauche isomers are present in the monoclinic ordered phase. Infrared measurements also indicate an energy difference of about 4 kJ mol^{-1} between the trans form in the vapour and the gauche isomer in the liquid.[164,165] These conclusions were supported by a comprehensive infrared and Raman study[166] of liquid and solid succinonitrile.

Table 7.3 Physical properties of some plastic crystals near the phase transition

	Solid–solid transition T_t/K	$\Delta S_t/J\ \mathrm{mol}^{-1}\ \mathrm{K}^{-1}$	Fusion T_f/K	$\Delta S_f/J\ \mathrm{mol}^{-1}\ \mathrm{K}^{-1}$	Volume change on fusion[a] $\Delta v/v$	Reduced times τ_J^* Plastic crystal	Liquid	Reference
SF_6	94.30	17.05	222.5	22.78	0.21 ± 0.03	~ 0.2	0.35	135
MoF_6	263.48	30.97	290.7	14.88	0.127	0.23	0.27	136
CF_4	76.23	22.53	89.5	7.94	0.04	0.09	0.1	128
NF_3	56.61	26.71	66.4	5.97	—	$\parallel$ 0.014 $\perp$ —	$\parallel$ 0.15 $\perp$ 0.045	135
Ar	—	—	83.8	14.17	0.12			

[a] $\Delta v/v = (v_{\mathrm{liquid}} - v_{\mathrm{solid}})/v_{\mathrm{solid}}$.

7.5.15 Adamantane

Very little detailed spectroscopic work has been carried out on the order–disorder transition in plastic crystal. This is due in part to the accurate temperature control and careful annealing techniques necessary for a study of this nature. Dows *et al.*[167] performed such an infrared study on adamantane which undergoes an order–disorder transition 208.6 K.[168] The infrared spectra of adamantane were recorded above and below the transition temperature and also while the sample was cycled through the phase transition. The splittings of the vibrational transitions and the appearance of gas phase forbidden transitions below 208 K were in agreement with the predictions based on the site and factor group symmetries (S_4 and D_{2d} respectively). In the plastic phase, the local symmetry is tetrahedral (T_d). Thus the selection rules will be the same as for the gas phase and no splitting will occur.

The phase transition was followed by observing the disappearance of the induced bands, (e.g. 1369 cm^{-1}) and by the loss of multiplet structure (e.g. 968 cm^{-1}) in the rotator phase. The absorption spectra showed a hysteresis with a loop width of about 2 K for the 1369 cm^{-1} band. This suggests the existence of a 'smeared' transition in which the structure and surface energy terms are important in the transition mechanism, and where the phase transition involves nucleation and growth of one phase within crystals of the other.[169] Calculations of the lattice energy, frequency splittings and barriers to molecular rotation using central force functions between non-bonded atoms were also carried out (Table 7.3).

REFERENCES

1. E. R. Andrew and R. G. Eades, *Proc. Roy. Soc.*, **A218**, 537 (1953).
2. H. A. Resing, *Mol. Cryst.*, **9**, 101 (1969).
3. W. M. Lipscomb, *J. Chem. Phys.*, **60**, 5138 (1974).
4. J. O. Clayton and W. F. Giauque, *J. Amer. Chem. Soc.*, **54**, 2610 (1932).
5. C. Brot and I. Darmon, *J. Chem. Phys.*, **53**, 2271 (1970).
6. R. T. Bailey, *Specialist Periodical Reps. C.S.*, Vol. 2, *Molecular Spectroscopy*, 173 (1974).
7. H. P. Gush, W. F. Hare, E. J. Allin, and H. L. Welsh, *Can. J. Phys.*, **38**, 240 (1960).
8. R. Kopelman, *J. Chem. Phys.*, **47**, 2631 (1967).
9. J. L. Hollenberg and D. A. Dows, *J. Chem. Phys.*, **37**, 1300 (1962).
10. R. M. Hexter, *J. Chem. Phys.*, **36**, 2285 (1962); **39**, 1608 (1963).
11. D. A. Dows, *J. Chem. Phys.*, **36**, 2836 (1962).
12. D. A. Dows, *J. Chem. Phys.*, **32**, 1342 (1960).
13. A. I. Kitaigorodskii, *J. Chem. Phys.*, **63**, 6 (1966).
14. D. E. Williams, *J. Chem. Phys.*, **63**, 68 (1966).
15. Y. A. Schwartz, A. Ron, and S. Kimel, *J. Chem. Phys.*, **54**, 99 (1971).
16. S. Abragam, *The Principles of Nuclear Magnetism*, Oxford U.P., London, (1961)
17. R. G. Gordon, *Adv. Magn. Resonance*, **3**, 1 (1968).
18. L. A. Nafie and W. L. Peticolas, *J. Chem. Phys.*, **57**, 3145 (1972).

19. R. G. Gordon, *J. Chem. Phys.*, **40**, 1973 (1964).
20. R. G. Gordon, *J. Chem. Phys.*, **42**, 3658 (1965).
21. R. G. Gordon, *J. Chem. Phys.*, **38**, 1724 (1963).
22. R. G. Gordon, *J. Chem. Phys.*, **43**, 1307 (1965).
23. M. Evans, *J. C. S., Faraday II*, **71**, 2051 (1975).
24. R. G. Gordon, *J. Chem. Phys.*, **44**, 1830 (1966).
25. B. Keller, P. Ebersold, and F. Kneubuhl, *Proc. Phys. Soc.* (**B**), **3**, 688 (1970).
26. B. J. Berne, P. Pechukas, and G. D. Harp, *J. Chem. Phys.*, **49**, 3125 (1968).
27. A. A. K. Al-Mahdi and A. R. Ubbelohde, *Proc. Roy. Soc.*, **220A** 143 (1953).
28. B. J. Hunt and J. G. Powles, *Proc. Phys. Soc.*, **88**, 513 (1966).
29. J. E. Anderson and R. Ullman, *J. Chem. Phys.*, **47**, 2178 (1967).
30. R. Folland, S. M. Ross, and J. H. Strange, *Mol. Phys.*, **26**, 27 (1973).
31. N. I. Liu and J. Jonas, *Chem. Phys. Lett.*, **14**, 555 (1972).
32. I. Larkin, *J. C. S., Faraday II*, **69**, 1278 (1973).
33. R. Haffmans and I. W. Larkin, *J. C. S., Faraday II*, **68**, 1729 (1972).
34. C. Brot, *J. Physique*, **28**, 789 (1967).
35. B. Lassier and C. Brot, *Chem. Phys. Lett.*, **I**, 581 (1968).
36. A. G. St. Pierre and W. A. Steele, *Phys. Rev.*, **184**, 192 (1969).
37. P. Debye, *Polar Molecules*, Reinhold, New York (1929).
38. W. H. Furry, *Phys. Rev.*, **107**, 7 (1957).
39. L. D. Favro, *Phys. Rev.*, **119**, 53 (1960).
40. W. R. Hackelman and P. S. Hubbard, *J. Chem. Phys.*, **39**, 2688 (1963).
41. P. Rigny and J. Virlet, *J. Chem. Phys.*, **47**, 4645 (1967).
42. M. Bloom, F. Bridges, and W. N. Hardy, *Can. J. Phys.*, **45**, 3533 (1967).
43. R. C. Livingston, W. G. Rothschild, and J. J. Rush, *J. Chem. Phys.*, **59**, 2498 (1973).
44. J. D. Barnes, *J. Chem. Phys.*, **58**, 5193 (1973).
45. R. G. Gordon, *J. Chem. Phys.*, **41**, 1819 (1964).
46. R. E. D. McClung, *J. Chem. Phys.*, **51**, 3842 (1969); *J. Chem. Phys.*, **54**, 3248 (1971).
47. R. E. D. McClung and H. Versmold, *J. Chem. Phys.*, **57**, 2596 (1972); R. E. D. McClung, *J. Chem. Phys.*, **55**, 3459 (1971).
48. R. E. D. McClung, *J. Chem. Phys.*, **57**, 5478 (1972).
49. F. Perrin, *J. Phys. Radium*, **7**, 1 (1936).
50. P. Bliot and E. Constant, *Chem. Phys. Lett.*, **18**, 253 (1973).
51. B. J. Berne and G. D. Harp, *Adv. Chem. Phys.*, **17**, 63 (1970).
52. M. Fixman and K. Rider, *J. Chem. Phys.*, **51**, 2425 (1969).
53. A. G. St. Pierre and W. A. Steale, *Phys. Rev.*, **184**, 172 (1969).
54. A. G. St. Pierre and W. A. Steele, *J. Chem. Phys.*, **57**, 4638 (1972).
55. D. Frenkel, G. H. Wegdam, and J. van der Elsken, *J. Chem. Phys.*, **57**, 2691 (1972).
56. R. T. Cukier and K. Lakatos-Lindenberg, *J. Chem. Phys.*, **57**, 3427 (1972).
57. D. Kivelson and T. Keyes, *J. Chem. Phys.*, **57**, 4599 (1972).
58. H. Mori, *Progr. Theor. Phys.*, **33**, 423 (1965).
59. E. N. Ivanov, *Zhur. Eksp. Teor. Fiz. 3*, **45**, 1509 (1963).
60. K. Kometani and H. Shimizu, *J. Phys. Soc. Japan*, **30**, 1036 (1971).
61. H. Shimizu, *J. Chem. Phys.*, **43**, 2453 (1965).
62. S. Bratos and J. Rios, *Comptes Rends.*, **269**, 90 (1969).
63. Y. Guissani and S. Bratos, *Comptes Rends.*, **269**, 137 (1969).
64. S. Bratos, J. Rios, and Y. Guissani, *J. Chem. Phys.*, **52**, 439 (1970).
65. A. D. Buckingham, *Proc. Roy. Soc.*, **A255**, 32 (1960).

66. K. A. Valiev, *Optika i Spektroskopija*, **11**, 465 (1960).
67. Y. Rocard, *J. Phys. Rad.*, **4**, 247 (1933).
68. J. G. Powles, *Trans. Faraday Soc.*, **44**, 802 (1948).
69. G. Wyllie, *J. Phys. C*, **4**, 564 (1971).
70. M. Evays, M. Davies, and I. Larkin, *J. C. S., Faraday II*, **69**, 1011 (1973).
71. I. Larkin and M. Evans, *J. C. S., Farad II*, **70**, 477 (1974).
72. I. W. Larkin, *J. C. S., Faraday II*, **70**, 1457 (1974).
73. I. Darmon and C. Brot, *Mol. Phys.*, **21**, 785 (1971).
74. B. Lassier and C. Brot, *J. Chim. Phys.*, **65**, 1723 (1968).
75. W. G. Rothschild, *J. Chem. Phys.*, **53**, 990, 3265 (1970).
76. N. E. Hill, *Proc. Phys. Soc.*, **82**, 723 (1963).
77. R. P. Young and R. N. Jones, *Chem. Rev.*, **71**, 219 (1971).
78. J. P. Hawranek, P. Neelakantan, R. P. Young, and R. N. Jones, *Spectrochim Acta*, **32A**, 75 (1975); **32A**, 85 (1975).
79. J. P. Hawranek and R. N. Jones, *Spectrochim Acta*, **32A**, 99 (1975).
80. K. S. Seshadri and R. N. Jones, *Spectrochim Acta*, **19**, 1013 (1963).
81. R. N. Jones, *Pure and App. Chem.*, **18**, 303 (1969).
82. R. N. Jones, *App. Optics*, **8**, 597 (1969).
83. R. N. Jones, R. Venkataraghavan, and J. W. Hopkins, *Spectrochim Acta*, **23A**, 925, 941 (1975).
84. A. V. Rokov, *Trudy Fiz. Inst. Akad Nauk SSSR.*, **27**, 111 (1964), *Optics and Spectroscopy*, **7**, 128 (1959).
85. F. J. Bartoli and T. A. Litovitz, *J. Chem. Phys.*, **56**, 404 (1972).
86. P. H. Martin and S. H. Walmsley, *Mol. Phys.*, **27**, 49 (1974).
87. S. S. Bhatnagar, E. J. Allin, and H. L. Welsh, *Can. J. Phys.*, **40**, 9 (1962).
88. J. van Kranendonk, *Can. J. Phys.*, **38**, 240 (1960).
89. J. van Kranendonk and G. Karl, *Rev. Mod. Phys.*, **40**, 531 (1968).
90. H. P. Gush, W. F. J. Hare, E. J. Allin, and H. L. Welsh, *Can. J. Phys.*, **38**, 176 (1960).
91. M. Clovter and H. P. Gush, *Phys. Rev. Lett.*, **15**, 200 (1965).
92. W. N. Hardy, I. F. Silvera, K. N. Klemp, and O. Schrepp, *Phys. Rev. Lett.*, **21**, 291 (1968).
93. C. H. Wang and P. A. Fleury, *J. Chem. Phys.*, **53**, 2243 (1970).
94. C. H. Wang and R. B. Wright, *Mol. Phys.*, **27**, 345 (1974).
95. W. F. Giaugue and J. O. Clayton, *J. Amer. Chem. Soc.*, **55**, 4875 (1933).
96. J. O. Clayton and W. F. Giaugue, *J. Amer. Chem. Soc.*, **54**, 2610 (1932).
97. A. L. Smith, W. E. Keller, and H. L. Johnson, *Phys. Rev.*, **78**, 728 (1950).
98. J. E. Cahill and G. E. Leroi, *J. Chem. Phys.*, **51**, 1324 (1969).
99. G. E. Ewing and G. C. Pimental, *J. Chem. Phys.*, **35**, 925 (1961).
100. G. E. Ewing, *J. Chem. Phys.*, **37**, 2250 (1962).
101. A. Anderson, T. S. Sun, and M. C. A. Dunkersloot, *Can. J. Phys.*, **48**, 2265 (1970).
102. K. Clusius. *Z. Electrochem*, **39**, 598 (1933).
103. H. A. Kruis, L. Popp, and K. Clusius, *Z. Electrochem*, **43**, 664 (1937).
104. J. Harada and N. Kitamura, *J. Phys. Soc. Japan*, **19**, 328 (1964).
105. H. A. Kruis and K. Clusius, *Phys. Z.*, **38**, 510 (1937).
106. D. C. Look, I. J. Lowe, and J. A. Northby, *J. Chem. Phys.*, **44**, 3441 (1966).
107. F. P. Redding and D. F. Hornig, *J. Chem. Phys.*, **27**, 1024 (1957).
108. C. C. Stephenson and W. F. Giauque, *J. Chem. Phys.*, **5**, 149 (1937).
109. T. H. Huang, *Ph.D. Thesis*, Oregon State Univ., Corvallis, (1973).

110. A. H. Hardin and K. B. Harvey, *Can. J. Chem.*, **42**, 84 (1964).
111. A. Heinemann, *Ber. Bunsen Gesell. Phys. Chem.*, **68**, 280 (1964).
112. M. D. Francia and E. R. Nixon, *J. Chem. Phys.*, **58**, 1061 (1973); M. D. Francia, *Ph.D. Thesis*, Univ. of Pennsylvania, (1970).
113. M. Schwartz and C. H. Wang, *Chem. Phys. Lett.*, **25**, 26 (1974).
114. N. Boden and R. Folland, *Chem. Phys. Lett.*, **32**, 127 (1975).
115. J. H. Colwell, E. K. Gill, and J. A. Morrison, *J. Chem. Phys.*, **39**, 635 (1963).
116. J. H. Colwell, E. K. Gill, and J. A. Morrison, *J. Chem. Phys.*, **42**, 3144 (1965).
117. W. Press, *J. Chem. Phys.*, **56**, 2597 (1972).
118. G. E. Ewing, *J. Chem. Phys.*, **40**, 179 (1964).
119. M. F. Crawford, H. L. Welsh, and J. H. Harrold, *Canad. J. Phys.*, **30**, 81 (1952).
120. R. Sovoie and R. P. Fournier, *Chem. Phys. Lett.*, **7**, 1 (1970).
121. C. Chapados and A. Cabana, *Chem. Phys. Lett.*, **7**, 191 (1970).
122. C. Chapados and A. Cabana, *Can. J. Chem.*, **50**, 3521 (1972).
123. A. Cabana, R. Bardoux, and A. Chamberland, *Can. J. Phys.*, **47**, 2915 (1969).
124. M. Marsault-Herail, J. P. Marsault, M. Chalaye, J. L. Saulnier, and G. Levi., *Compt. Rends.*, **275B**, 307 (1972).
125. A. Eucken and E. Shroder, *Z. Phys. Chem.*, **41B**, 307 (1938).
126. R. G. Steinhardt, W. Neilsen, H. W. Morgan, and P. Staats, *J. Chim. Phys.*, **63**, 176 (1966).
127. R. P. Fournier, R. Savoie, F. Bessette, and A. Cabana, *J. Chem. Phys.*, **49**, 1159 (1968).
128. M. Gilbert and M. Drifford, *J. Chem. Phys.*, **66**, 3205 (1977).
129. P. N. Schatz and D. F. Hornig, *J. Chem. Phys.*, **21**, 1516 (1953).
130. C. Haas and D. F. Hornig, *J. Chem. Phys.*, **26**, 707 (1957).
131. R. S. Savoie, *Raman Spectra of Molecular Crystals*, in *The Raman Effect*, ed. A. Anderson, M. Dekker, N.Y. (1973), Vol. 2, p. 800.
132. D. E. O'Reilly, E. M. Paterson, and C. E. Scheie, *J. Chem. Phys.*, **60**, 1603 (1974).
133. K. T. Gillen, D. C. Douglas, M. S. Malmberg, and A. A. Maryott, *J. Chem. Phys.*, **57**, 5170 (1972).
134. R. R. Sharp, *J. Chem. Phys.*, **57**, 5321 (1972).
135. M. Gilbert and M. Drifford, *Adv. Raman Spectrosc.*, **1**, 204, (1972).
136. M. Gilbert and M. Drifford, in *Molecular Motions in Liquids*, ed. J. Lascombe, Reidel (1973), p. 279.
137. G. Herzberg, *Infrared and Raman Spectra*, van Nostrand, N.Y., 1962, p. 167.
138. G. J. Davies, G. J. Evans, and M. Evans, *J.C.S., Faraday II*, **72**, 2147 (1976).
139. J. O'Dell and B. J. Berne, *J. Chem. Phys.*, **63**, 2376, (1975).
140. J. F. G. Hicks, J. G. Hooley, and C. C. Stephenson, *J. Amer. Chem. Soc.*, **66**, 1064 (1944).
141. F. J. Bartoli and T. A. Litovitz, *J. Chem. Phys.*, **56**, 413 (1972).
142. S. Sunder and R. E. D. McClung, *Chem. Phys.*, **2**, 467 (1973).
143. C. Brot and B. Lassier-Govers, *Ber. Bunsen Gesell. Phys. Chem.*, **80**, 31 (1976).
144. S. Urban, J. A. Janik, J. Lenik, J. Mayer, T. Walnga, and S. Wrobel, *Phys. Stat. Solid*, **10A**, 271, 1972.
145. R. Rudman, *Mol. Cryst.*, **6**, 427 (1970).
146. B. Lassier and C. Brot, *J. Mol. Phys.*, **21**, 785 (1971).
147. R. P. Fournier, R. Savoie, N. D. Thé, R. Belzile, and A. Cabana, *Can. J. Chem.*, **50**, 35 (1972).
148. N.D. Thé, J. M. Gagnon, R. Belzile, and A. Cabana, *Can. J. Chem.*, **52**, 327 (1974).

149. B. Post, R. S. Schwartz, and I. Fankuken, *J. Amer. Chem. Soc.*, **73**, 5113 (1951).
150. F. A. Rushworth, *Proc. Roy. Soc.*, **A222**, 526 (1954).
151. V. Schettino, M. P. Marzocchi, and S. Califano, *J. Chem. Phys.*, **51**, 5264 (1969).
152. R. L. Jackson and J. H. Strange, *Acta Cryst.*, **B28**, 1645 (1972).
153. E. F. Westrum in *Molecular Dynamics and Structure of Solids*, edited by R. S. Carter and J. J. Rush, 1969 (NBS Publication 301).
154. R. Folland, D. A. Jackson, and S. Rajagopal, *Mol. Phys.*, **30**, 1063 (1975).
155. R. Folland, R. L. Jackson, and J. H. Strange, private communication quoted in Reference 154.
156. R. C. Livingston, W. G. Rothschild, and J. J. Rush, *J. Chem. Phys.*, **59**, 2498 (1973).
157. E. O. Stejskal, D. E. Woessner, T. C. Farrar, and H. S. Gutowsky, *J. Chem. Phys.*, **31**, 55 (1959).
158. R. E. Lechner, J. M. Rowe, K. Skold, and J. J. Rush, *Chem. Phys. Lett.*, **4**, 444 (1969).
159. P. Bruesch, *Spectrochim Acta*, **22**, 861 (1966).
160. I. Darmon and C. Brot, *Mol. Cryst.*, **1**, 417 (1966).
161. B. Lassier, C. Brot, and N. Dat-Xuong, *Mol. Phys.*, **27**, 1697 (1974).
162. G. L. Lewis and C. P. Smyth, *J. Chem. Phys.*, **7**, 1085 (1939).
163. C. Finback, *Arch. Math. Naturv*, **42B**, 71 (1938).
164. G. J. Janz and W. E. Fitzgerald, *J. Chem. Phys.*, **23**, 1973 (1955).
165. W. E. Fitzgerald and G. J. Janz, *J. Mol. Spect.*, **1**, 49 (1957).
166. T. Fujiyama, K. Tokumaru, and T. Shimanouchi, *Spectrochim Acta*, **20**, 415 (1964).
167. P. J. Wu, L. Hsu, and D. A. Dows, *J. Chem. Phys.*, **54**, 2714 (1971).
168. S. S. Chang and E. F. Westrum, *J. Phys. Chem.*, **64**, 1547 (1960).
169. A. R. Ubbelohde, *Quart. Rev.*, **11**, 246 (1957).

8

Neutron Scattering Studies

A. J. Leadbetter and R. E. Lechner

8.1 INTRODUCTION

It is well established that the orientational disorder of the molecules in the plastically crystalline phases is dynamic. In only a few cases however are the rotational motions connected with this disorder thoroughly understood. Neutron scattering, which simultaneously yields both geometrical and dynamic information on the re-orientation process offers the most promise of all techniques to define these motions. Up to the present, neutron scattering studies of rotational motions of molecules in crystals are few. Consequently the present discussion will not be confined to plastic crystals alone. We shall review more generally, studies of orientational disorder in translationally ordered crystals including rotational motions in ionic crystals and liquid crystals. Most of the work reported concerns incoherent, quasi-elastic scattering, which gives information about the motion of individual atoms and hence molecules. We shall concentrate on a description of these experiments.

Three extreme types of rotational motions may be envisaged:

(1) Free rotation
(2) Rotational diffusion
(3) Jump reorientation

Free rotation is a definite periodic motion with classical rotation frequencies Ω of the order of $(k_B T/I)^{1/2}$, where k_B is the Boltzmann constant, T the temperature and I the moment of inertia. Progressive damping of this motion would lead to a rotational diffusion type of motion. This motion involves random small angle displacements in the classical diffusion sense, where the motion of the diffusing particle is limited to the surface of a sphere. Jump reorientation requires a set of preferred orientations in which the molecule undergoes librational motion for an average time τ between jumps, the jumps being instantaneous. The jump rate is then τ^{-1} and if the jump distance becomes small enough this motion goes over to rotational diffusion. In the condensed phase, except perhaps for very special cases, the relatively strong interactions between neighbouring molecules precludes free rotational motion. The free rotation frequency Ω defined above is useful however in estimating the lower limit of the reorientation time between preferred sites for the jump reorientation mechanism. This and rotational diffusion may then be regarded as the simple extremes of possible molecular rotational motions in plastic crystals, although in general the real motion may well be more complex. In all cases so far investigated, an adequate description of the motion has been possible in terms of some kind of large angle jump reorientations, rotational diffusion or a combination of the two. Incoherent neutron scattering experiments give information about the time evolution of the molecular orientation, which includes geometrical information about the molecular orientations which may be adopted.

In discussing models for the rotational motion, usually the coupling between different molecules is only taken into account indirectly, but for a full discussion of the rotator phases it is necessary to consider explicitly the time dependent correlations in the orientations of different molecules. The static aspects of these correlations may be studied by diffuse neutron or x-ray scattering which is the subject of a previous chapter, and the time dependent part by coherent inelastic neutron scattering. Little work has so far been done in this area but we will briefly comment on the situation.

In the following section we give a brief review of the theoretical background in which we shall attempt to give a physical insight into the possible approaches and models used rather than to elaborate the details. These may be found in the original papers. The importance of the experimental conditions will be emphasized and it will be shown that the recent availability of high resolution instruments enables us to make much more detailed interpretations than were possible even a few years ago. A review will then be given of the results which have been obtained on a number of different materials with particular emphasis on recent high resolution studies.

8.2 NEUTRON SCATTERING

8.2.1 General Aspects

A number of detailed discussions of the basic aspects of neutron scattering have appeared in recent years[1,2]. We shall here only summarize briefly those results which are essential to this article.

In the present context the most important property of neutrons is that they combine a wavelength of the order of atomic or molecular dimensions 0.05–1.2 nm with a low energy (300–0.8 meV).† This is why both the structure and dyna-

†1 meV $\equiv$ 8.07cm^{-1} $\equiv$ 0.242 THz $\equiv$ 1.52 $\times$ 10^{12} rad s^{-1}.

mics of a system may be studied simultaneously in the same experiment. The basic quantity which is measured is the double differential cross section $\partial^2\sigma(E_0, E_1, \phi)/\partial\Omega\partial E_1$ which gives the fraction of neutrons of incident energy E_0 scattered through an angle ϕ into an element of solid angle $d\Omega$ with an energy between E_1 and $E_1 + dE_1$. The natural variables, as functions of which the scattered neutron intensity is measured are the energy exchanged between neutron and system

$$\hbar\omega = E_0 - E \tag{8.2.1}$$

and the scattering vector $\mathbf{Q}$ (or momentum exchange $\hbar\mathbf{Q}$) where

$$\mathbf{Q} = \mathbf{k}_0 - \mathbf{k} \tag{8.2.2}$$

and $\mathbf{k}_0$ and $\mathbf{k}$ are respectively the incident and scattered wave vectors.

The scattering cross section is determined by the properties of the scattering system. It is proportional to a flux factor k/k_0 and is related to the scattering lengths b of the different atoms in the system. Nuclei of the same element may have different values of b because of the existence of different isotopic species and different nuclear spin states. Since these are in general not correlated with nuclear positions this gives rise to an incoherent component

$$\sigma_{\text{inc}} = 4\pi(\langle b^2\rangle - \langle b\rangle^2) \tag{8.2.3}$$

as well as to a coherent component

$$\sigma_{\text{coh}} = 4\pi\langle b\rangle^2 \tag{8.2.4}$$

of the total bound scattering cross section

$$\sigma_s = 4\pi\langle b^2\rangle = \sigma_{\text{inc}} + \sigma_{\text{coh}} \tag{8.2.5}$$

Here the angular brackets denote an average over all nuclei of the same element. Whereas σ_{coh} is responsible for the effects of interference between waves scattered from all the nuclei, σ_{inc} does not contribute to such effects except for those interference phenomena which are due to the scattering from one and the same nucleus.

If we assume for simplicity that the scattering system contains only one kind of atom, then the double differential scattering cross-section per atom is given by

$$\frac{\partial^2\sigma}{\partial\Omega\partial E} = \frac{1}{\hbar}\frac{k}{k_0}[\sigma_{\text{coh}}\, S_{\text{coh}}(\mathbf{Q}, \omega) + \sigma_{\text{inc}}\, S_{\text{inc}}(\mathbf{Q}, \omega)] \tag{8.2.6}$$

where S_{coh} and S_{inc} are called the coherent and incoherent scattering functions (or scattering laws) respectively. The scattering laws depend only on the properties of the system and are independent of the experiment.

The bound scattering cross-sections of most of the atoms commonly occuring in plastic crystals are of the order of a few barns. The very large incoherent cross-section for hydrogen (~ 80 barns) due to the different nuclear spin states means that in most hydrogenous organic molecules (including plastic crystals) the incoherent scattering from the protons is dominant. Provided that the fraction of protons is not too small then the only coherent scattering which is important arises from the (elastic) Bragg reflections. Since these occur at particular values of **Q**, whereas the incoherent scattering is a continuous function of **Q**, they may be experimentally separated. Hence the incoherent scattering i.e. the second term in (8.2.6) may be considered in isolation. This greatly simplifies the problem and we now concentrate our discussion on this area.

8.2.2 General theory of Incoherent Quasielastic Neutron Scattering from Rotational Motions of Molecules in Crystals

In this section we will consider the incoherent scattering function $S_{\text{inc}}(\mathbf{Q}, \omega)$, or shorter $S_s(\mathbf{Q}, \omega)$ of a proton; (if the system contains different kinds of protons which are dynamically not equivalent, then $S_s(\mathbf{Q}, \omega)$ must be averaged over them). The scattering function will be calculated classically. The resulting symmetrical function $S_s(\mathbf{Q}, \omega)$ does not fulfil the detailed balance condition.[1] For the purpose of a comparison with experimental data it must therefore be corrected with the detailed balance factor $\exp(\hbar\omega/2k_B T)$, i.e. $S_s(\mathbf{Q}, \omega)$ must be replaced by $\exp(\hbar\omega/2k_B T) \cdot S_s(\mathbf{Q}, \omega)$.

The scattering function is the time Fourier transform

$$S_s(\mathbf{Q}, \omega) = \frac{1}{2\pi}\int_{-\infty}^{\infty} e^{-i\omega t} I_s(\mathbf{Q}, t)dt \tag{8.2.7}$$

of the intermediate scattering function $I_s(\mathbf{Q}, t)$, which is given by

$$I_s(\mathbf{Q}, t) = \langle \exp i\, \mathbf{Q}[\mathbf{r}(t) - \mathbf{r}_0] \rangle \tag{8.2.8}$$

The angle brackets denote a thermal average of the exponential with the conditional probability function $G_s(\mathbf{r}, \mathbf{r}_0, t)$ and the probability distribution $g(\mathbf{r}_0)$ of initial positions $\mathbf{r}_0$ as weighting factors:

$$I_s(\mathbf{Q}, t) = \iint \exp i\, \mathbf{Q}[\mathbf{r} - \mathbf{r}_0] \cdot G_s(\mathbf{r}, \mathbf{r}_0, t) \cdot g(\mathbf{r}_0)\, d\mathbf{r}\, d\mathbf{r}_0 \tag{8.2.9}$$

In other words one may say that $I_s(\mathbf{Q}, t)$ is the spatial Fourier transform of $G_s(\mathbf{r}, \mathbf{r}_0, t)$ averaged over all initial positions. For the case of a polycrystal $I_s(\mathbf{Q}, t)$ must be averaged over all directions of **Q**.

For classical systems the function $G_s(\mathbf{r}, \mathbf{r}_0, t)$ is the probability that a given atom will be at position **r** at time t if it was at $\mathbf{r}_0$ at time zero. For such systems **r** is the position vector of the scattering nucleus, and it may be split into different components:

$$\mathbf{r}(t) = \mathbf{a}(t) + \mathbf{R}(t) + \mathbf{u}(t) \tag{8.2.10}$$

where $\mathbf{a}(t)$ is the position vector of the (molecular) centre of rotation (centre of mass); $\mathbf{R}(t)$ represents the 'equilibrium' position of the proton relative to the centre of mass $\mathbf{a}(t)$ and at the same time the variable orientation of the molecule; $\mathbf{u}(t)$ is the displacement vector of the proton from the endpoint of $\mathbf{R}(t)$, due to intramolecular and torsional vibrations. The latter may alternatively be included in $\mathbf{R}(t)$ if they are treated explicitly in the theory.

Since in studying reorientational motions one is mainly interested in the quasielastic part of the cross section it is of interest to separate $S_s(\mathbf{Q}, \omega)$ into a 'quasi-elastic' and an 'inelastic' term. Such a separation seems empirically justified if the experimental spectra are well separated into a quasielastic and an inelastic part. Theoretically the separation can be made under the following assumptions:

(i) The observed quasielastic scattering is due to the (diffusive) rotational motion of the molecule alone.

(ii) The dynamical coupling between the centre of mass vibration of the molecule, the torsional and intramolecular vibrations, and the diffusive rotational motion may be neglected. This assumption is necessary in order to make calculations feasible. It becomes probably less satisfactory in the large-$\mathbf{Q}$ region where the low energy part of the inelastic spectrum begins to blend into the quasielastic peak. In other words at large $\mathbf{Q}$ the experiment explores mainly small displacements and the distinction between rotational and vibrational displacements becomes more difficult.

In spite of the assumed dynamic independence, the contributions of different kinds of motion to $I_s(\mathbf{Q}, t)$ do not separate completely. This is because the components of the displacement vector $\mathbf{u}$, which are due to intramolecular motions (and possibly librations), rotate together with $\mathbf{R}$: therefore they are not really independent of rotation. Thus an exact calculation of expression (8.2.9) would be very tedious (if feasible at all), so that in practice one is led to approximate methods. One usually assumes that the thermal averaging in equation (8.2.9) can be done separately for the different components of the motion. This permits a separation of $I_s(\mathbf{Q}, t)$ into factors corresponding to different kinds of motion:

$$I_s(\mathbf{Q}, t) = I_s^a(\mathbf{Q}, t) \cdot I_s^R(\mathbf{Q}, t) \cdot I_s^u(\mathbf{Q}, t) \tag{8.2.11}$$

where each factor is of a form analogous to (8.2.8), e.g.

$$I_s^R(\mathbf{Q}, t) = \langle \exp i\mathbf{Q}\cdot[\mathbf{R}(t) - \mathbf{R}_0]\rangle \tag{8.2.12}$$

It is then assumed that the intermediate scattering functions $I_s^a(\mathbf{Q}, t)$ and $I_s^u(\mathbf{Q}, t)$ may be written like phonon expansions,[1] e.g.

$$I_s^u(\mathbf{Q}, t) = \exp - \langle(\mathbf{U}\cdot\mathbf{Q})^2\rangle\cdot[1 + I_1(\mathbf{Q}, t) + \ldots] \tag{8.2.13}$$

where $\exp -\langle[\mathbf{U}\cdot\mathbf{Q}]^2\rangle$ is a Debye–Waller factor. If $2W = [\langle(\mathbf{U}_a\cdot\mathbf{Q})^2\rangle +$

$\langle(\mathbf{U}_u \cdot \mathbf{Q})^2\rangle]$ we can now write

$$I_s(\mathbf{Q}, t) = \exp[-2W] \cdot I_s^R(\mathbf{Q}, t) \cdot [1 + I_s^{inel}(\mathbf{Q}, t)] \quad (8.2.14)$$

Thus all inelastic effects are concentrated in the term $I_s^{inel}(\mathbf{Q}, t)$. The exponent of the Debye–Waller factor is the sum of the contributions from translational (and possibly torsional) lattice vibrations and intramolecular vibrations. Although the 'thermal clouds' of a proton are not expected to have cubic symmetry the Debye–Waller factor may be simplified to $\exp[-\langle U_Q^2\rangle Q^2]$ (where U_Q is the component of the displacement along **Q**) to a good approximation for polycrystals because of the orientational averaging required in this case.[3] In the general case we obtain from equation (8.2.7) and (8.2.14):

$$S_s(\mathbf{Q}, \omega) = \exp[-2W] \cdot [S_s^R(\mathbf{Q}, \omega) + S_s^I(\mathbf{Q}, \omega)] \quad (8.2.15)$$

The first term on the right hand side of equation (2.15) corresponds to quasi-elastic scattering, whereas the second term is a convolution of the quasi-elastic scattering function with the time Fourier transform of $I_s^{inel}(\mathbf{Q}, t)$ and is thus related to the frequency distribution of all the molecular vibrations. Due to its smallness in the quasielastic region (at not too high Q) this 'inelastic' term is usually approximated near $\hbar\omega = 0$ by a slowly varying function of Q and $\hbar$ to give ω. Particularly for very high energy resolution (FWHM $\leqq 10\,\mu$eV) and not too large momentum transfer ($Q < 2\text{Å}^{-1}$) this is of negligible importance. After these general considerations we may now limit the discussion to the rotational scattering function $S_s{}^R(\mathbf{Q}, \omega)$.

8.2.3 Rotational Scattering Functions

8.2.3(a) General

For simplicity we now assume that all scattering nuclei are dynamically equivalent. When this is not the case, however, then the incoherent scattering function is simply the average of the different functions of the individual nuclei. In order to derive rotational scattering functions we write equation (8.2.12) explicitly (compare equation (8.2.9)):

$$I_s^R(\mathbf{Q}, t) = \iint \exp i\mathbf{Q} \cdot [\mathbf{R} - \mathbf{R}_0]\, G_s^R(\mathbf{R}, \mathbf{R}_0, t) \cdot g^R(\mathbf{R}_0) \mathrm{d}\mathbf{R}\, \mathrm{d}\mathbf{R}_0 \quad (8.2.16)$$

where $G_s^R(\mathbf{R}, \mathbf{R}_0, t)$ is the conditional probability function of the scattering nucleus at time t, i.e. the probability that the nucleus is at **R** at time t if it was at an origin $\mathbf{R}_0$ at zero time. $g^R(\mathbf{R}_0)$ is the distribution function for the initial position. The integrals are taken over the space available for the (rotational) motion of the nucleus. This space is restricted to part or all of the surface of a sphere and this restriction has a very important general consequence for the nature of the experimental scattering law. Thus if we consider the long time limit of $G_s^R(\mathbf{R}, \mathbf{R}_0, t)$ namely $G_s^R(\mathbf{R}, \infty)$ which is the infinite time (or time averaged) distribution, then since by the ergodic theorem

$$g^R(\mathbf{R}_0) = G_s^R(\mathbf{R}, \infty)$$

we have

$$I_s^R(\mathbf{Q}, \infty) = \left| \int [\exp i\mathbf{Q} \cdot \mathbf{R}]\, G_s^R(\mathbf{R}, \infty)\, d\mathbf{R} \right|^2 \tag{8.2.17}$$

$$= |\langle \exp i\mathbf{Q} \cdot \mathbf{R} \rangle|^2 \qquad \text{(compare (2.12)).}$$

When the motion is such that the proton is confined to a restricted volume of space, $G_s^R(\mathbf{R}, \infty)$ and thus $I_s^R(\mathbf{Q}, \infty)$ are finite and it is therefore useful to split the intermediate scattering function into this time-independent and a time-dependent term:

$$I_s^R(\mathbf{Q}, \mathrm{t}) = I_s^R(\mathbf{Q}, \infty) + I_s'^R(\mathbf{Q}, t) \tag{8.2.18}$$

Then, since the Fourier transform of a constant gives a δ-function, one obtains from equation (8.2.7):

$$S_s^R(\mathbf{Q}, \omega) = A_0(\mathbf{Q})\, \delta(\omega) + S_s^{qe}(\mathbf{Q}, \omega) \tag{8.2.19}$$

where $A_0(\mathbf{Q}) = I_s^R(\mathbf{Q}, \infty)$. This means that $S_s^R(\mathbf{Q}, \omega)$ always contains a purely elastic component, $A_0(\mathbf{Q})\delta(\omega)$, superimposed on the quasielastic component $S_s^{qe}(\mathbf{Q}, \omega)$ (cf. Figures 8.2, 8.4, 8.6). $A_0(\mathbf{Q})$ is called the Elastic Incoherent Structure Factor (EISF). It gives a direct measure of the time-averaged spatial distribution of the proton which is very important information about the geometry of the rotational motions of the molecules, whereas the time evolution of the proton position is contained in $S_s^{qe}(\mathbf{Q}, \omega)$. Note that $S_s^{qe}(\mathbf{Q}, \omega) \to 0$ and $A_0(\mathbf{Q}) \to 1$ as $Q \to 0$. The separation of the elastic and quasielastic components of the scattering in general requires a high resolution experiment.

We now consider various models† which have been applied to specific cases and show examples of the predicted forms of $A_0(Q)$ for some of these models in Figure 8.1. Further examples are included in the discussion of experimental results.

8.2.3(b) Rotational Jumps between a Finite Number of Allowed Orientations[2,5,6,7]

In this model it is assumed that the jump time‡ is very much shorter than the time interval between jumps τ (instantaneous jumps). Then, the conditional probability function is given by

$$G_s^R(\mathbf{R}, \mathbf{R}_0, t) = \sum_{j=1}^{m} W_{ji}(t)\, \delta(\mathbf{R}_i - \mathbf{R}_j) \tag{8.2.20}$$

† For a review of rotation models see also reference 4.

‡ This is the time which an atom spends jumping between sites. It can be estimated from the free rotation frequency Ω and the jump angle.

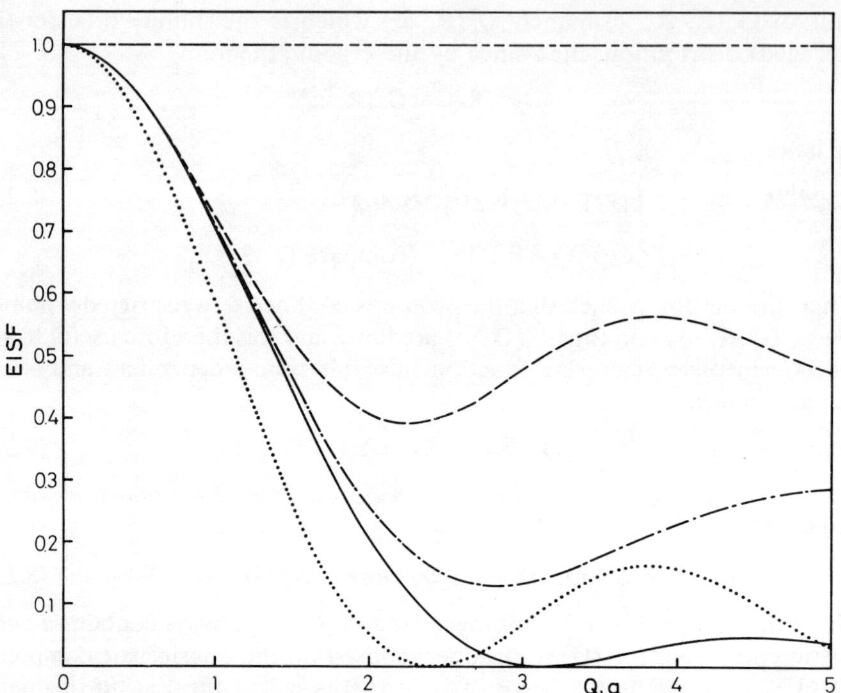

Figure 8.1 The elastic incoherent structure factor for various models of rotational motion (EISF ≡ A_0(Q); cf. equations (8.2.19) and (.3.1): ––·––·- No component of motion along **Q**, e.g. 2-fold jump with Q⊥a (equation (8.2.23)) or motion on circle with **Q** perpendicular to the plane of the circle (cf. equation (8.2.35)); --- Random jump diffusion between two sites separated by a distance $2a$, for a powder sample (cf. equation (8.2.24) and equations (8.2.26) and (8.2.27); –·–·–. Random jump diffusion between 4 sites situated on a circle of radius a, for a powder sample (cf. equations (8.2.26) and (8.2.27)); ..... Continuous diffusion on a circle of radius a, for **Q** in the plane of the circle ($J_0^2(Qa)$ cf. equation (8.2.35)); —— Any isotropic motion on the surface of a sphere of radius a, ($j_0^2(Qa)$, equation (8.2.33))

In terms of the motion of a single proton $W_{ji}(t)$ is the probability of finding the proton at the allowed sites $\mathbf{R}_j$ if it was at the origin $\mathbf{R}_0 = \mathbf{R}_i$ at $t = 0$. The probabilities W_{ji} are calculated from the system of simultaneous differential equations ('rate equations'):

$$\frac{\mathrm{d}W_{ji}(t)}{\mathrm{d}t} = \frac{1}{\tau}\left[\frac{1}{n}\sum_{l=1}^{n} W_{li}(t) - W_{ji}(t)\right] \tag{8.2.21}$$

where τ is the average time between two successive jumps of a proton, n is the number of proton sites that can be reached by a single rotational jump and equal probabilities are assumed for all n sites. The normalization conditions are

$$\sum_{j=1}^{m} W_{ji}(t) = 1 \qquad i = 1, 2, \ldots m.$$

and the initial conditions are

$$W_{ii}(0) = 1; \quad W_{ji}(0) = 0 \quad \text{for } j \neq i$$

m is the total number of available sites for a particular proton.

For illustration we will now demonstrate the calculation of $G_s^R(\mathbf{R}, \mathbf{R}_0, t)$ and $S_s^R(\mathbf{Q}, \omega)$ for the most simple case of rotational jump diffusion. This is when there are only two sites and correspond e.g. to a rotation by 180° such that a nucleus is transferred from one site to the other by this rotation.

The sites are at $\mathbf{R}_1$ and $\mathbf{R}_2$ and the probabilities of finding the proton at each if it started at R_1 at $t = 0$ are $W_1(t)$ and $W_2(t)$. According to equation (8.2.21) we have

$$\frac{dW_1}{dt} = \frac{1}{\tau}(W_2(t) - W_1(t)).$$

In this simple case $W_2(t) = 1 - W_1(t)$, and for the initial conditions $W_1(0) = 1$, $W_2(0) = 0$, the solution is

$$W_1(t) = \tfrac{1}{2}(1 + \exp(-2t/\tau)) \equiv W(t)$$

With the initial site $\mathbf{R}_1 = 0$ and with $\mathbf{R}_2 = 2\mathbf{a}$, equation (8.2.20) gives

$$G_s^R(\mathbf{R}, \mathbf{R}_0, t) = \delta(\mathbf{R})W(t) + \delta(\mathbf{R} - 2\mathbf{a})(1 - W(t))$$

A similar result is obtained if the second site ($\mathbf{R}_2$) is chosen as initial site. Furthermore $G_s^R(\mathbf{R}, \infty) = g^R(\mathbf{R}_0) = \tfrac{1}{2}\delta(\mathbf{R}) + \tfrac{1}{2}\delta(\mathbf{R} - 2\mathbf{a})$
The evaluation of equation (8.2.16) then gives

$$I_s^R(\mathbf{Q}, t) = W(t) + \cos 2\,\mathbf{Q}\cdot\mathbf{a}\,(1 - W(t))$$

and, finally, time Fourier transformation leads to the incoherent scattering law

$$S_s^R(\mathbf{Q}, \omega) = \frac{1}{2}\delta(\omega)\,(1 + \cos 2\,\mathbf{Q}\cdot\mathbf{a}) + \frac{1}{2}(1 - \cos 2\,\mathbf{Q}\cdot\mathbf{a})\frac{1}{\pi}\mathscr{L}\frac{2}{\tau} \qquad (8.2.22)$$

where $\mathscr{L}(2/\tau)$ is the Lorentzian function $(2/\tau)/(\omega^2 + (2/\tau)^2)$ (this has a half-width at half-maximum of $2/\tau$ rad s^{-1}).

The scattering law clearly depends on the relative orientations of **Q** and **a**, and in particular, for **Q** perpendicular to **a**

$$S_s^R(\mathbf{Q}, \omega) = \delta(\omega) \qquad (8.2.23)$$

and for **Q** parallel to **a** then

$$S_s^R(\mathbf{Q}, \omega) = \frac{1}{2}\delta(w)(1 + \cos 2|\mathbf{Q} \parallel \mathbf{a}|) + \frac{1}{2}(1 - \cos 2|\mathbf{Q} \parallel \mathbf{a}|)\frac{1}{\pi}\mathscr{L}\frac{2}{\tau}$$

For the case of random orientation (powder average) the result is

$$S_s^R(\mathrm{Q}, \omega) = \frac{1}{2}\,\delta(\omega)\left(1 + \frac{\sin 2\,\mathrm{Q}a}{2\,\mathrm{Q}a}\right) + \frac{1}{2}\left(1 - \frac{\sin 2\,\mathrm{Q}a}{2\,\mathrm{Q}a}\right)\frac{1}{\pi}\mathscr{L}\left(\frac{2}{\tau}\right) \quad (8.2.24)$$

The elastic incoherent structure factor (EISF) for this case, $A_0(\mathbf{Q}) = (1/2 + 1/2\,(\sin 2\,\mathrm{Q}a)/(2\,\mathrm{Q}a)$, is shown in Figure 8.1 as a function of Qa.

Returning now to the general case we note that *a priori*, the number of equations in the system (8.2.21) is equal to the number of sites; however this can often be appreciably reduced due to the symmetry of the sites. For example in the case of reorientations by 120°-jumps of plastic neopentane[7] one has $m = 12$ and $n = 8$. In this case the system of 12 equations reduces to 3 coupled equations only. An alternative method to solve the system of equations (8.2.21) uses group theoretical concepts.[8] This applies when the various orientations of the molecules belong to a simple point group (e.g. the cubic point group) and is particularly useful, when the number of possible orientations is large and/or when several different kinds of rotational jumps with different jump probabilities are occurring. In this latter case equation (8.2.21) must be generalized to include the different jump probabilities.

One obtains for the incoherent rotational scattering function the general expression:[9]

$$S_s^R(\mathbf{Q}, \omega) = \sum_{\alpha=1}^{N_\alpha} f_\alpha(\mathbf{Q}) \cdot S_\alpha(\omega) \quad (8.2.25)$$

where

$$S_\alpha(\omega) = \frac{1}{\pi}\,\mathscr{L}\left(\frac{1}{\tau_\alpha}\right) \quad \text{if} \quad \frac{1}{\tau_\alpha} \neq 0$$

and

$$S_\alpha\,(\omega) = \delta(\omega) \quad \text{if} \quad \frac{1}{\tau_\alpha} = 0.$$

The Lorentzian widths τ_α^{-1} and the coefficients $f_\alpha(\mathbf{Q})$ are determined by the subgroup of rotations to which the allowed reorientations belong. The sum in equation (8.2.25) is taken over the N_α irreducible representations of this group. For the cubic point group the quantities τ_α^{-1} and the coefficients $f_\alpha(\mathbf{Q})$ are tabulated in Reference 9.

When the m allowed sites are equidistant points on a circle of radius a then, starting from equation (8.2.21) the following general formula is obtained for the incoherent scattering function in a powder sample[10]

$$S_s^R(Q, \omega) = A_0(Q)\,\delta(\omega) + \frac{1}{\pi}\sum_{l=1}^{m-1} A_l(Q)\mathscr{L}(\tau_l^{-1}) \tag{8.2.26}$$

where

$$A_l(Q) = m^{-1}\sum_{j=1}^{m} j_0(2\,Qa\sin\pi j/m)\cos(2\pi lj/m) \tag{8.2.27}$$

$j_0(x)$ is the zero order spherical Bessel function ($j_0(x) = \sin x/x$) and

$$\tau_l = \frac{\tau}{2\sin^2(\pi l/m)} \tag{8.2.28}$$

For the definition of the residence time τ see equation (8.2.21). The EISF, $A_0(Q)$ of expression (8.2.26), is shown in Figure 8.1 as a function of Qa for $m = 2$ and for $m = 4$. It should be mentioned that equations (8.2.26) to (8.2.28) can also easily be derived using the group theoretical technique mentioned above. For $m \to \infty$ they are equivalent to continuous rotational diffusion on the circle[11] with $D_r = 1/\tau_1$ (see also equation (8.2.35)).

8.2.3(c) Continuous Rotational Diffusion

(i) On a Sphere. This is most usefully introduced as a special case of 'isotropic' rotational motion in which on a time average a molecule has no preferred orientations in space. For such a case there is of course no difference in the scattering law between a single crystal and a powder since the motion of each individual molecule is already spatially isotropic. Sears[12] has shown that the intermediate scattering function for a proton moving on the surface of a sphere of radius a may be written

$$I_s^R(Q, t) = \sum_{l=0}^{\infty} (2l + 1)\,j_l^2(Qa)\,F_l(t) \tag{8.2.29}$$

where j_l is a spherical Bessel function and

$$F_l(t) = \langle P_l[\cos\beta(t)]\rangle \tag{8.2.30}$$

is a rotational correlation function, P_l being a Legendre polynomial and $\beta(t)$ the angle through which some vector fixed in the molecule rotates in a time t. The $F_l(t)$ contain the details of the motion and for isotropic motions (e.g. rotational diffusion but not free rotation) all $F_l(t)$ decay to zero at long times except $F_0(=1)$. The scattering law quite generally reads

$$S_s^R(Q, \omega) = j_0^2(Qa)\,\delta(\omega) + \sum_{l=1}^{\infty} (2l + 1)\,j_l^2(Qa)\,S_l(\omega) \tag{8.2.31}$$

where the $S_l(\omega)$ are the time Fourier transforms of the $F_l(t)$.

For the case of rotational diffusion

$$F_l(t) = \exp(-l(l+1)D_r t) \tag{8.2.32}$$

and thus $S_l(\omega)$ is the Lorentzian function $\mathscr{L}(l(l+1)D_r)$, where D_r is a rotational diffusion constant. It should be noted that in this case—as for any isotropic rotational motion, whatever the details of the motion at finite times—the EISF is given by

$$A_0(Q) = j_0^2(Qa) \tag{8.2.33}$$

This function is shown in Figure 8.1.

(ii) On a Circle. If the diffusional motion is confined to a circle of radius a, then it is necessary to consider the relative orientation of circle and scattering vector. If the angle between **Q** and the rotation axis is θ then[11,13]

$$I_s^R(\mathbf{Q}, t) = \sum_{m=-\infty}^{-\infty} J_m^2(Qa \sin\theta) \exp(-m^2 D_r t) \tag{8.2.34}$$

and

$$S_s^R(\mathbf{Q}, \omega) = J_0^2(Qa \sin\theta)\,\delta(\omega) + \frac{2}{\pi} \sum_{m=1}^{\infty} J_m^2(Qa \sin\theta)\mathscr{L}(m^2 D_r) \tag{8.2.35}$$

where J_m is a Bessel function of the first kind.

For a powder sample equation (8.2.35) must be averaged over all values of θ and the result is not simple. It is preferable then to use equation (8.2.26) for the powder averaged jump model with m sufficiently large for the scattering law to be independent of further increase in m. At any given Q this occurs for finite values of m. (Remember that $1/\tau_1 \to D_r$ for $m \to \infty$[11]). The behaviour of $J_0^2(Qa \sin\theta)$ as a function of Qa is shown in Figure 8.1 for $\theta = \pi/2$, i.e. for **Q** in the plane of the circle.

8.2.3(d) More Complicated Special Cases

It cannot be expected that either of the above simple models should accurately describe the reorientational motion of molecules in plastic phases generally, although in a number of simple cases studied recently the instantaneous jump model works remarkably well (see below). However, a number of more complicated models has been used in which some kind of mixed motions have been assumed.

(i) Isotropic Jump Diffusion with a Distribution of Jump Angles[14]

The basic idea here is that at long times the distribution of orientations is spherical, but this is achieved by a random distribution $W(\beta)$ of finite jump angles β around an average value $\langle\beta\rangle$. Equations (8.2.29) and (8.2.31) then apply with

$$F_l(t) = \exp(-\tau^{-1}(1 - \lambda_l)t) \tag{8.2.36}$$

and

$$S_l(\omega) = \frac{1}{\pi}\mathscr{L}\left(\frac{1 - \lambda_l}{\tau}\right) \tag{8.2.37}$$

where τ is the average time between jumps

$$\lambda_l = (2l + 1)^{-1}\int_0^\pi \mathrm{d}\beta\, W(\beta)\,\frac{\sin\left[(l + \frac{1}{2})\beta\right]}{\sin(\beta/2)}$$

with the normalization

$$\int_0^\pi W(\beta)\mathrm{d}\beta = 1$$

For low Q the scattering law for this model resembles that of the simpler models treated above. At higher Q, where terms for higher l become important, the Lorentzian widths saturate at τ^{-1} as long as $\langle\beta\rangle \ll 2\pi$. This is similar to the instantaneous rotational jump model; the widths progressively increase for the continuous rotational diffusion case.

(ii) Two-step Diffusion Involving Rotational Motions and Librations about a Fixed Orientation.

The idea here is to describe the overall rotational motions in terms of consecutive steps in which a diffusive (rotational) motion occurs for an average time τ_R followed by an oscillatory (librational) motion for an average time τ_L.

The evaluation of the scattering law for such a step model was first given for translational motions by Singwi and Sjölander[15] and has been derived for the rotational case by Larsson.[16] The result may be written in the form of equation (8.2.31) but now

$$S_l(\omega) = \frac{1}{\tau_R + \tau_L}\frac{1}{\pi}\,\mathrm{Re}\,\frac{\tau_L a_l(\omega) + \tau_R b_l(\omega) + 2a_l(\omega)b_l(\omega)}{1 - a_l(\omega)b_l(\omega)/\tau_R\tau_L} \tag{8.2.38}$$

where

$$a_l(\omega) = \int_0^\infty \mathrm{d}t\, \exp(-\mathrm{i}\omega t - t/\tau_L)\, g_l(t)$$

and

$$b_l(\omega) = \int_0^\infty \mathrm{d}t\, \exp(-\mathrm{i}\omega t - t/\tau_R)\, h_l(t)$$

The functions $g_l(t)$ and $h_l(t)$ are the coefficients of the expansions in spherical harmonics of the correlation functions $g(\Omega_0, \Omega, t)$ and $h(\Omega_0, \Omega, t)$ of the librational and rotational motions, respectively; Ω_0 and Ω are the solid angles characterizing the molecular orientation at zero time and at time t. Note that, even if the model of rotational motion on which $h(\Omega_0, \Omega, t)$ is based is not isotropic (e.g. free rotation), the time averaged distribution of molecular orientations remains spherical as long as the life time τ_R is finite (damped free rotation).

Therefore again—as for all isotropic rotational models—the EISF is given by equation (8.2.33). The full scattering law is fairly complex and in general involves a relatively large number of unknown parameters.

(iii) Simultaneous Motions of Different Kinds. The basic assumptions are that a molecule may reorient by several different mechanisms and that these motions occur quite independently of each other and therefore 'simultaneously'. In consequence, if for instance, there are two different types of rotational motion with intermediate scattering functions $I_s^{R_1}(\mathbf{Q}, t)$ and $I_s^{R_2}(\mathbf{Q}, t)$, then (cf. equation (8.2.11)) the combined function is the product

$$I_s^{R}(\mathbf{Q}, t) = I_s^{R_1}(\mathbf{Q}, t) \cdot I_s^{R_2}(\mathbf{Q}, t) \tag{8.2.39}$$

The incoherent rotational scattering function $S_s^R(\mathbf{Q}, \omega)$ is then the convolution of the individual scattering functions:

$$S_s^R(\mathbf{Q}, \omega) = \int S_s^{R_1}(\mathbf{Q}, \omega - \omega') S_s^{R_2}(\mathbf{Q}, \omega')\, d\omega' \tag{8.2.40}$$

Such a calculation was carried out[17] for reorientational jumps occurring simultaneously with rotational diffusion. The resulting rotational scattering function for a polycrystal is

$$S_s^R(Q, \omega) = \frac{1}{\pi} \sum_{i=1}^{m} \sum_{l=0}^{\infty} (2l + 1) j_l^2(Qa) P_l(\cos \xi_i)\, M_{il}(\omega) \tag{8.2.41}$$

where a is the rotation radius and ξ_i are the angles between the position vectors of the different proton sites, if the center of rotation is taken as the origin.

The functions $M_{il}(\omega)$ are convolutions of the functions $S_i(\omega)$ (which are the time Fourier transforms of the solutions $W_i(t)$ of equations (8.2.21)) and $S_l(\omega)$ (see equation (8.2.31)) including the $\delta(\omega)$ function. The physical picture underlying this particular model is the following:

(a) A molecule may have any orientation and on a time average all orientations of an individual molecule are equally probable.

(b) Because of relatively tight packing (of molecules that are not strictly spherical) a molecule may normally perform a collective rotational motion in small random angular steps together with its neighbours. This motion may be described by the simple rotational diffusion equation.[12]

(c) From time to time the molecule is activated beyond the instantaneous angular potential barrier created by its neighbours. This leads to large angle jumps with a tendency to end in configurations which are indistinguishable from the original ones, because the neighbouring molecules essentially stay at rest during the rotation. The geometry of these jumps will therefore largely be determined by the symmetry of the molecule itself. This is the reason for the appearance of the angles ξ_i in equation (8.2.41).

8.3 EXPERIMENTAL RESULTS

8.3.1 General Comments

One of the most important results of Section 8.2.3 is the fact that, whatever the detailed nature of the rotational motion, the incoherent rotational scattering law may be written in the form

$$S_s^R(\mathbf{Q}, \omega) = A_0(\mathbf{Q})\delta(\omega) + S_s^{qe}(\mathbf{Q}, \omega) \tag{8.3.1}$$

where $S_s^{qe}(\mathbf{Q}, \omega)$ is the form of the quasielastic scattering underlying the elastic component $A_0(\mathbf{Q})\delta(\omega)$.

The form of the spectra actually observed experimentally will depend on the experimental energy resolution $\Delta(\hbar\omega)$ and only when this is good enough (i.e. roughly when $\Delta(\hbar\omega) < \tau^{-1}$, where τ is the rotational correlation time which essentially determines the width of $S_s^{qe}(\mathbf{Q}, \omega)$, will a direct determination of $A_0(\mathbf{Q})$ be possible. This has been the case so far only for a relatively small number of recent experiments and it puts the analysis of the data on a much more certain footing. The direct experimental determination of $A_0(\mathbf{Q})$ immediately permits the establishment of whether rotational motion occurs on the time scale of the experiments and gives information about the rotation radius and type of rotational motion (see Figure 8.1). Neutron experiments should therefore, where this is feasible, aim at the determination of the EISF. Of course, geometrical information about the possible orientations of the molecules may also be obtained from x-ray or neutron diffraction but only when the different orientations are distinguishable. This will not always be so, and furthermore structure factors from diffraction are only sampled at values of $\mathbf{Q}$ corresponding to the Bragg reflections. Therefore important regions of $\mathbf{Q}$, which can be studied by incoherent neutron scattering (INS), may not be accessible to diffraction. The quasielastic part of the INS-spectrum can be analysed to obtain detailed information about the time dependence of the orientations. This has sometimes been attempted by direct Fourier transformation but more usually by calculation of $S_s^R(\mathbf{Q}, \omega)$ for various models, convolution with the resolution function and comparison with experiment. It should be emphasized that with present limitations of experimental resolution only motions with correlation times shorter than about 10^{-9} can be studied.

For experiments with poor resolution it is naturally much more difficult to obtain geometrical and time information about molecular rotations and sometimes all that can then be derived is an approximate rotational correlation time.

We shall now review, in the light of the above discussion, the experimental results which have been obtained on a variety of plastic crystalline materials.

8.3.2 Uniaxial Rotations

In this section we consider a variety of examples in which rotational motion of all or part of a molecule occurs about a single axis.

8.3.2(a) End Group Rotation in Solid Phases of Liquid Crystalline Systems

Liquid crystals usually exhibit several different mesophases differing by the degree of randomness of the molecular motion. Neutron data on such systems are difficult to interpret because in general many different kinds of molecular motion are observed simultaneously and thus are superimposed on oneanother in the experimental result. In particular this may include effects from random rotational motion of whole molecules and/or parts of them. In some cases internal molecular random rotation can already be observed in the solid phase, where no other rapid random motions are present. Such results may then be extrapolated to other phases existing at higher temperatures and facilitate the analysis of the more complex dynamical phenomena in these phases. It was recently demonstrated in a series of experiments[18–21] how, by combining high resolution neutron quasi-elastic scattering techniques with partial deuteration of the sample, one can separate different kinds of motion and thereby achieve a simple and clear interpretation of the experimental results. In the case of solid PAA (para-azoxyanisole)[21] two different partially deuterated derivatives of PAA were studied:

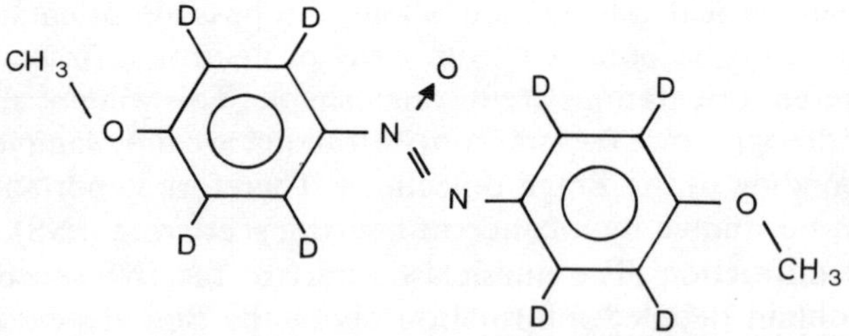

In this experiment use was made of the fact that–due to the dominant incoherent scattering cross-section of the proton–spectra of PAA-ϕD4 mainly contain information on the motion of the two methyl groups. Thus a comparison with the spectra from PAA-CD_3 permits the separation of the methyl group motion from the motion of the whole molecule. This is clearly demonstrated in Figure 8.2 (reproduced from Reference 21) which shows time-of-flight spectra of the two partially deuterated PAA-derivatives. The PAA-ϕD4 spectrum contains two dominant features which disappear completely when the methyl groups are deuterated: the quasielastic scattering (underneath a well resolved elastic peak) and the strong inelastic peak near $h\omega = 31$ meV. This immediately shows that the quasielastic scattering is caused by a random rotational motion involving the methyl groups, but not the phenyl groups. One may imagine the following three different models: rotation of the methoxy group or of the methyl group alone or rotation of both simultaneously. A careful analysis of the EISF can help to decide.

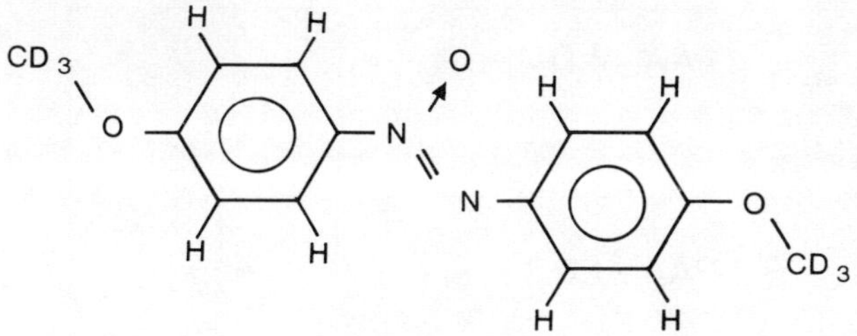

In Reference 21 it is clearly shown by a comparison of the EISFs for different models (Figure 8.3) that a rotation of the methoxy group must be excluded in favour of a three-fold rotation of the methyl group alone (model **B**). The small difference between model **B** and the experimental data is taken care of by a correction for incoherent elastic scattering due to (non-rotating) deuterons and non perfect deuteration of the phenyl rings. The large difference between the model curves **B** and **C** is due to the appreciable difference in the rotation radii of the protons characterizing the different types of rotation. The three-fold rotational jump model was fitted to the quasielastic data taken at different temperatures and examples of the fit are reproduced in Figure 8.4. The width of the quasielastic peak is closely related to the rotational correlation time τ. This was measured as a function of temperature and found to follow an Arrhenius law as follows:

$$\tau = \tau_0 \exp(E_a/k_B T)$$

with

$$\tau_0 = (0.35 \pm 0.07) \times 10^{-12}\ \text{s}$$

and an activation energy

$$E_a = 10.6 \pm 0.5\ \text{kJ mol}^{-1}$$

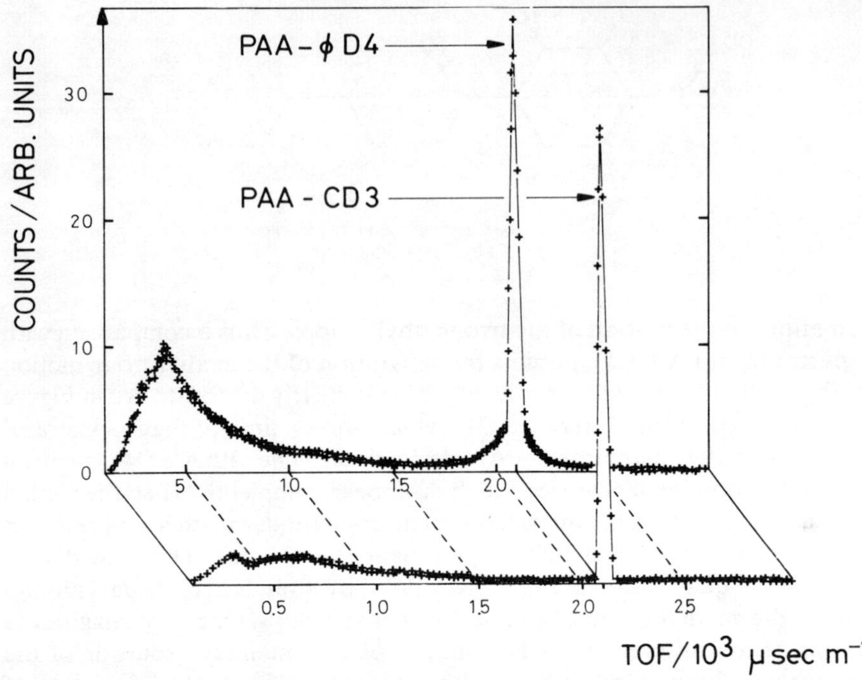

Figure 8.2 Time of flight spectra for two partially deuterated derivatives of PAA at 100 °C. Incident wavelength $\lambda_0 = 0.825$ nm and elastic momentum transfer $Q = 1.35\ \text{Å}^{-1}$

It should be mentioned that the value of the activation energy obtained in this way may be used to check on calculations of the barrier height, assuming simple rotational potentials and using the position of inelastic peaks which are due to torsional excitations. For instance the assumption of a three-fold cosine potential leads to an activation energy which is definitely too large and which shows that the true potential shape deviates appreciably from the cosine form.

A similar study was made by the same authors[19] on solid TBBA (terephthal-*bis*-butyl aniline, C_4H_9–⟨⊙⟩–NCH–⟨⊙⟩–CHN–⟨⊙⟩–C_4H_9 where the molecule is somewhat more complicated. Again it was of interest to separate the body motion from the end group motion, and the same technique of partial deuteration was used together with high resolution neutron scattering. Although from the experimental result it is immediately clear that in the solid

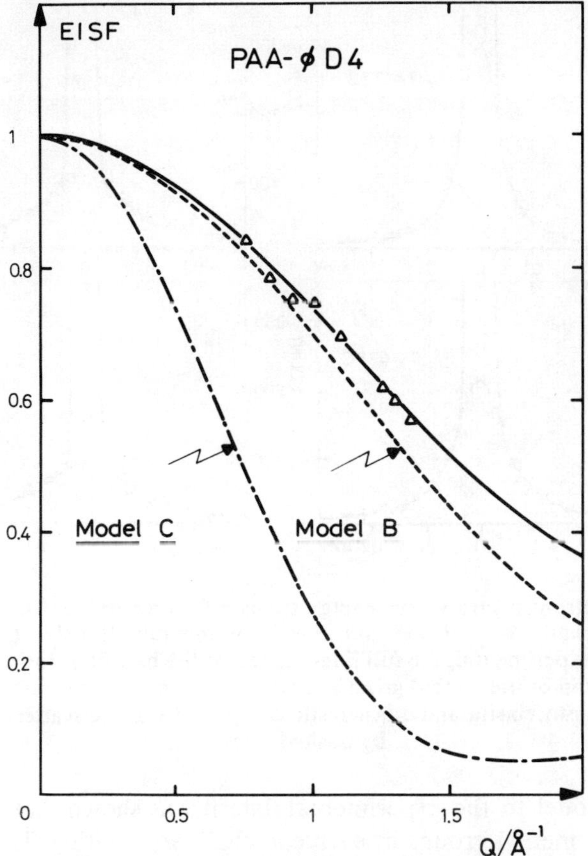

Figure 8.3 The elastic incoherent structure factor as a function of Q for PAA-φD4 at 100 °C: Δ Experimental points determined by graphical integration; –·–· Model C, simultaneous rotation of O–CH_3 and of CH_3 groups; --- Model B, rotation of methyl groups only; ——— Model B corrected for incoherent elastic scattering due to deuterium and due to remaining hydrogen (incomplete deuteration) on the phenyl rings. Radius of gyration of methyl protons was taken as a = 0.1032 nm

phase the chains are involved in a random rotational motion whereas the phenyl groups are fixed, the interpretation is complicated by the fact that the different protons in the end chain are not dynamically equivalent. The spectra, which reflect an average over all the individual proton motions, indicate that different chain protons have different rotational correlation times. A simple model was used in which protons closer to the phenyl rings are assumed to be fixed and therefore contribute only to the purely elastic part of the low energy spectrum, while protons closer to the chain ends are assumed to rotate. From

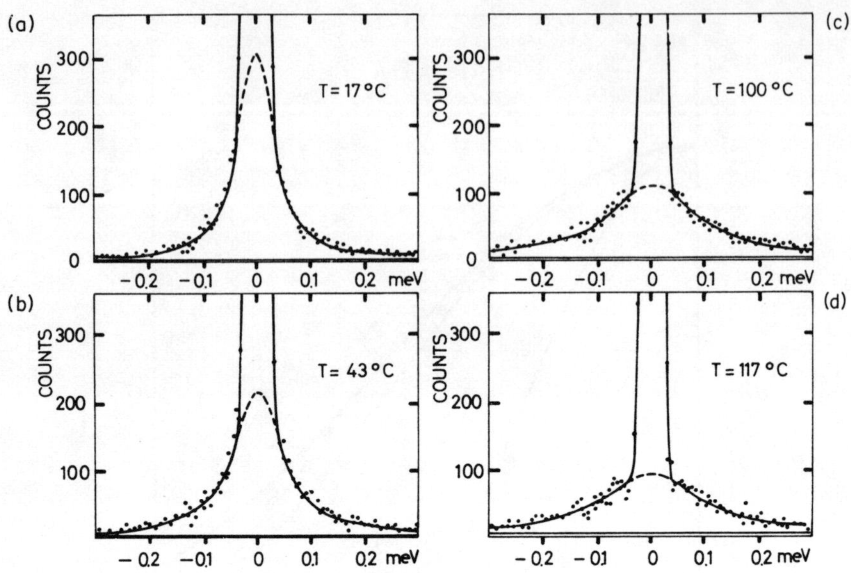

Figure 8.4 Neutron spectra versus energy transfer for PAA-φD4 at four temperatures. Incident wavelength $\lambda_0 = 0.945$ nm elastic momentum transfer $Q = 1.15$ Å^{-1}. The points are experimental, the full lines represent the best fit of the model of 3-fold jump reorientation of the methyl groups to the broadened components of the spectra. The separation into elastic and quasielastic components of the scattering is indicated by dashed lines

a fit of this model to the experimental data it was shown that only the last methylene and methyl groups in each end chain are rotating. The correlation time is about 1.4×10^{-11} s at room temperature and the activation energy about 7 kJ mol^{-1}.

8.3.2(b) Trimethyl Ammonium Chloride (TMA-Cl), $(CH_3)_3NHCl$

In this example of uniaxial rotation, the method of combining results obtained with different experimental resolutions was used in order to separate the effects of internal molecular and whole molecule rotation and to deduce the mechanism of the rotational motion.[22,23] The combination of such different results is necessary because $S(Q, \omega)$ must be determined in a sufficiently large range of Q and ω in order to achieve these goals, and for a given experimental resolution each technique is restricted to a certain region in (W, ω)-space.

In the case of TMA-Cl different techniques were used with the following elastic energy resolutions (FWHM [meV]]) and (Q, ω)-restrictions, respectively: neutron time-of-flight (TOF) with $\Delta(\hbar\omega) = 0.036$ meV and $Q \leqq 1.25$

Å^{-1}; TOF with $\Delta(\hbar\omega) = 0.27$ meV and $Q \leqq 2.5$ Å^{-1}; neutron back-scattering (BS) with $\Delta(\hbar\omega) = 0.0006$ to 0.0017 meV, $\hbar\omega \leqq 0.0075$ meV and $Q \leqq 1.9$ Å^{-1}.

It is seen that the higher resolution experiments are generally limited to the lower regions of Q or ω. They permit however an exact determination of the apparent correlation times via the measurement of the quasielastic widths. This result is then used as an input for the analysis of lower resolution data which cover a larger range of energy and/or momentum transfer. In this way it was possible to determine important details of the EISF although the energy resolution in the larger Q-range would not have been sufficient to separate directly elastic from quasielastic scattering. It was found that in the low temperature phase ($T < 308$ K) of TMA-Cl the cation is reorienting by 120°-jumps around the long axis of the molecule, while the methyl groups are performing (internal) 120°-rotational jumps about their threefold symmetry axes, both with a correlation time τ of about 4×10^{-10} s. At the phase transition (308 K) there is a sudden increase of the quasielastic width due to the cation motion which changes over to uniaxial rotational diffusion about the long molecular axis with $1/D_r \sim 2 \times 10^{-12}$ s at 333 K. The methyl group's correlation time decreases more slowly and is of the order of 5×10^{-11} s at 317 K.

8.3.2(c) Pivalic Acid, $(CH_3)_3CCOOH$

This is an example of a low resolution neutron study[24] ($\Delta(\hbar\omega) \gtrsim 0.4$ meV but extending up to $Q > 2$ Å^{-1}, on a substance for which a good deal is known about the molecular motions from extensive previous work using other techniques. The high temperature phase exists between 279.9 K ($\Delta S_f = 25.4$ J mol^{-1} K^{-1}) and the melting point at 309.5 K ($\Delta S_f = 6.5$ J mol^{-1}K^{-1}). The structure is face-centred-cubic and a combination of dielectric, NMR, light scattering and tracer diffusion studies have established that the molecules are randomly dimerised with one of the twelve nearest neighbours, the dimer axis being along a [110] direction. The molecules reorient by making and breaking hydrogen bonds with a correlation time of $\sim 10^{-8}$ s and an activation energy of ~ 36 kJ mol^{-1}. By exchange of acid protons between molecules this reorientation also provides the mechanism for proton diffusion through the lattice, but is too slow to be observed in the neutron experiments.

From inelastic neutron scattering measurements the methyl group torsional frequencies were identified and the potential barriers to methyl group reorientation obtained. This leads to an estimate of the correlation time for methyl reorientations of 10^{-10} s in pivalic acid at 290 K. This is close to a value extrapolated from low temperature NMR measurements, lies between the values measured directly for PAA and TMA-Cl above, and is too slow to contribute significantly to the observed quasielastic broadening (corresponding to $\tau \sim 10^{-12}$ s) in these experiments. Although the resolution was not good enough to permit the EISF to be determined, calculation of $S_s(Q, \omega)$ for various models showed clearly that rotation of the tertiary butyl groups about the long (dimer) axis is occurring and that the mechanism is a small step rotational diffusion rather than a 3-fold jump process. The motion is characterised by a rotational diffusion constant $D_r \cong 1.1 \times 10^{11}$ rad s^{-1} at $T \sim 290$ K. No information was obtained on the temperature dependence nor on whether the whole dimers or only the tertiary butyl groups are rotating.

8.3.2(d) Rotation in Liquid Crystalline Phases

There exist many different liquid crystalline phases and neutron scattering studies are playing a major role in elucidating their nature. In particular, the more ordered smectic phases are similar to plastic crystals in that there is dynamic orientational disorder of the molecules, while—on the time scale of the rotational motion—their centres of mass remain fixed in equilibrium positions. In this connection the measurement of the EISF has been extremely useful.

The first series of such experiments was undertaken on TBBA. The results obtained on the solid phase have already been discussed above.[19] Measurements on the smectic H phase[18,20] in which the molecules are hexagonally packed in layers but tilted with respect to the layers to give a monoclinic symmetry, have revealed that the molecules perform a random rotational motion about their long axes. The EISF shows that this rotation may be a sixfold jump motion (with a correlation time of about 1.8×10^{-11} s at 119°C), but cannot eliminate the possibility that the jumps may be smaller. However, diffuse x-ray scattering studies of the local order in the S_H phase[25] suggest that the motion is indeed a sixfold jump reorientation. The neutron scattering results also suggest a much faster random motion of the aliphatic end chains which is superimposed on the rotation of the body. It is interesting to note that the rotational motion of the body in the H-phase is slower than the end group rotation observed in the solid phase at much lower temperature, (see 8.3.2(a)). An explanation is of course provided by the smaller diameter and smaller steric hindrance of the end chains. The picture of a rotating molecule is at variance with the model of orientational ordering around the long molecular axis suggested for this phase.[26] Such an ordering may exist, however, in the metastable smectic VI (or tilted E) phase.[27,28] In addition to the rotation of the molecule around its long axis there are in the H-phase, fluctuations of this axis about its equilibrium orientation. The amplitude of these fluctuations increases with temperature and appears to become very large in the smectic C phase (50 to 60 °). This is suggested as a possible reason for the breakdown of the crystalline order in the smectic planes.[29]

A similar series of experiments has been carried out on BPBAC.[30,31]

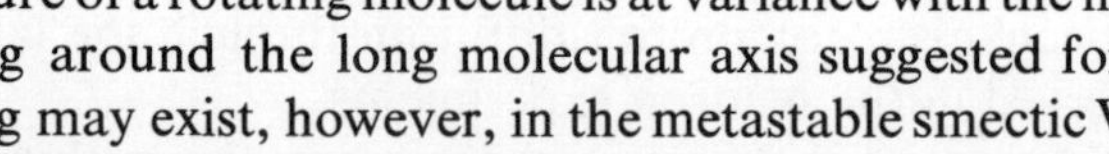

C_6H_5–C_6H_4–CH = N–C_6H_4–CH=CHCOOC$_4$D$_9$

with deuteration of the *n*-butyl chain used to remove the effect of fast motions of the tail. This substance has smectic E, B and A phases in which the molecules are normal to the smectic planes. Measurements of the EISF have shown that there is some restricted rotational motion about the long molecular axis in the orthorhombic *E* phase, with τ about 3×10^{-11} s. In the hexagonal B phase this motion appears to be similar to that observed in TBBA-S_H, with an average jump step of $\pi/3$ or smaller and a correlation time for the six-fold reorientation of $\sim 2 \times 10^{-11}$ s. In addition there are also in the B phase some other bound molecular motions on a similar time scale which increase in amplitude with increasing temperature. These may be fluctuations of, or screw motions along, the long molecular axes but their nature has yet to be established.

8.3.3 Rotation of Globular Molecules

In this section we discuss examples where the rotating unit is a molecule of globular (close to spherical) shape. Due to the spherical symmetry the onset of rotation is usually connected with a phase transition to a cubic crystalline system. The most interesting question is perhaps to what extent the rotational motion is isotropic. The different models presented in Section 8.2.3 provide different degrees of orientational anisotropy. Several of them have been tested by comparison to experiment.

8.3.3(a) Adamantane

Adamantane, $C_{10}H_{16}$, was probably the first plastic crystal studied with the incoherent quasielastic neutron scattering technique[5,32]. It has a face-centred cubic plastic phase extending from 208.6 K ($\Delta S_t = 16.2$ J mol^{-1}K^{-1}) to the melting point (540 K; $\Delta S_f = 20.9$ J mol^{-1} K^{-1}). In the analysis of these early measurements the basic mechanism of the rotational motion was already recognized although the experimental resolution had not been sufficient to prove the conclusion. In a more recent study high resolution neutron techniques were employed and have been able to reveal a certain number of details of the rotational motion by determination of the EISF.[9,33] It is known from x-ray diffraction[34] that the molecules are randomly distributed among two distinguishable orientations, related to each other by 90°-rotations about their two-fold symmetry axes, which are parallel to the cube edges of the f.c.c. crystal lattice. All isotropic rotation models can therefore be ruled out in favour of those rotational jump models which are consistent with the above-described orientational disorder. The x-ray structure factors can only reveal the distinguishable orientations and the EISF of a powder sample was determined[33] in order to choose the model for the reorientational motions which comes closest to reality. It turns out that this procedure is limited by the fact that several different jump models have the same EISF. Nevertheless, important new information is obtained. In particular it is concluded that either C_4–or C_2'–rotations or both are dominant on the time scale of the neutron experiments and that C_2- and/or C_3-rotations (if they exist) certainly do not occur alone.† For symmetry reasons preference should probably be given to C_4-rotational jumps. These conclusions are illustrated by Figure 8.5 taken from Reference 33. It shows theoretical EISF's for C_4– (full line), C_3- (dashed line) rotations and for rotational diffusion (dotted line) together with the experimental points.

† C_4: 90°-reorientations around the three [100]-axis of the lattice.
C_2': 180°-reorientations around the six [110]-axis of the lattice.
C_3: 120°-reorientations around the four [111]-axes of the lattice.
C_2: 180°-reorientations around the three [100]-axes of the lattice.

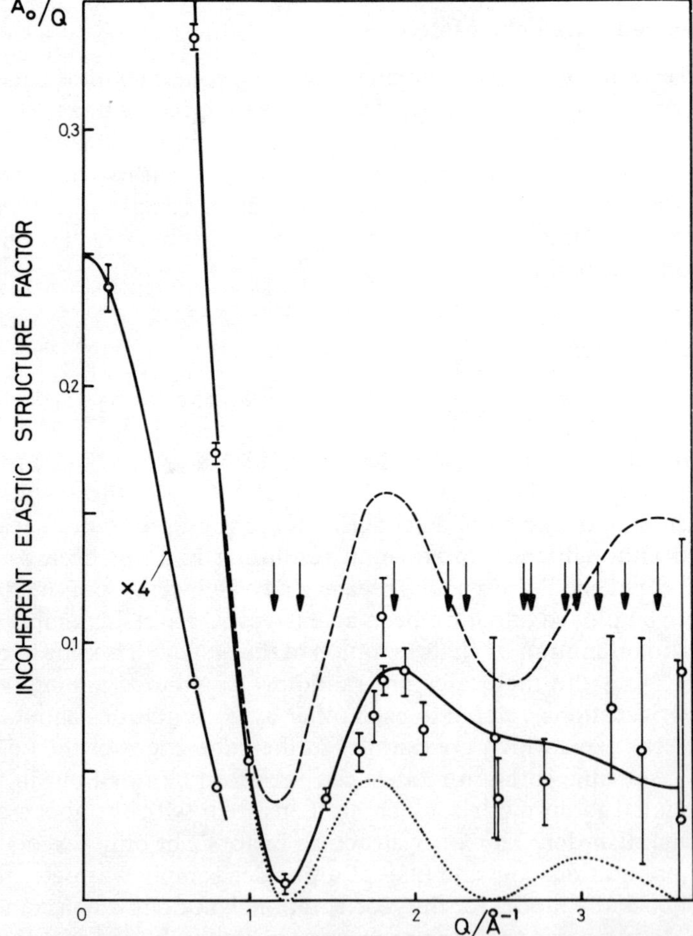

Figure 8.5 The EISF for adamantane. Experimental data (o) compared with Models for (1) C_4 rotations ———; (2) C_3 rotations - - - - -; (3) Rotational diffusion · · · · · ·. The arrows denote the positions of Bragg reflections

The quasielastic part of the scattering (Figure 8.6) reveals a correlation time of $\tau_{C_4} = 1.7 \times 10^{-11}$ s at 300 K. Its temperature dependence can be described by an Arrhenius law[9]:

$$\tau_{C_4} = 0.192 \times 10^{-12} \exp(1350/T)\text{s}$$

corresponding to a barrier height of 11.2 kJ mol^{-1}.

Some coherent neutron scattering experiments have also been performed[35] on a single crystal of a fully deuterated specimen. Translational phonons were observed in a number of symmetry directions and the dispersion curves fitted very well a simple model based on atom-atom potential functions. No librational modes or other torsional excitations were clearly detected although considerable near-elastic intensity was observed whose origin is not explained. The jump model discussed above implies the existence of librons but these may well be highly anharmonic and heavily damped so that the failure to observe discrete excitations is not too surprising, but it would be of interest to do experiments at higher resolution.

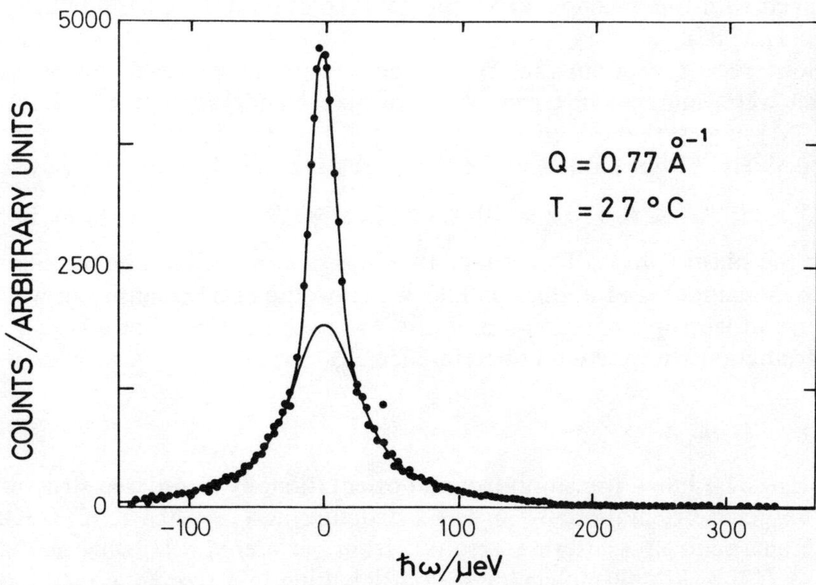

Figure 8.6 Typical neutron spectrum of adamantane showing fit of C_4 rotation model to experimental data and separation into elastic and quasielastic components of the scattering

8.3.3(b) Neopentane

Neopentane has an orientationally disordered phase extending from 140 K (ΔS = 18.4 J mol^{-1} K^{-1}) to the melting point (253 K; ΔS_f = 12.7 J mol^{-1} K^{-1})

and there have been several neutron scattering studies.[7,17,36–39] While early measurements[7,36] were of relatively poor resolution, sufficiently good resolution was used in one of the later experiments[37] ($\Delta(\hbar\omega) \sim 75\ \mu$eV) to give some information about the EISF. Two different approaches have been used to try to explain these data. First, simple rotational jump models involving threefold or threefold and fourfold reorientation were compared to the experiment and they are everywhere smaller than the experimental EISF. A model for twofold rotational jumps of the molecule was not tested, although this would give in general a higher EISF.

A second approach, which gives agreement with the experimental EISF, is based on an earlier stage of development[38] of the model represented by equation (8.2.38). This approximation can be interpreted as treating the sample as a mixture of molecules in two different states: one undergoing rotational motions and the other librating about (temporarily) fixed orientations. The underlying physical picture would be reasonable if the time scale on which one kind of motion alternates with the other were such that the phenomenon of damping connected with this alternation cannot be resolved with the available experimental resolution $\Delta(\hbar\omega)$, i.e. if both τ_L and $\tau_R \gg 1/\Delta(\hbar\omega)$.

More recent experimental results on the plastic phase of neopentane[39] which were analysed in terms of the two-step model (equations 8.2.31) and (8.2.38)) suggest that this condition is not fulfilled. The mean life times τ_L and τ_R, respectively, of librations and free rotations are found to be relatively short:

$$0.25 \times 10^{-12}\,\text{s} < \tau_L < 3.5 \times 10^{-12}\,\text{s and } 0.4 \times 10^{-12}\,\text{s} < \tau_R < 0.8 \times 10^{-12}\,\text{s}$$

over the plastic phase. This means that librations as well as free rotations are strongly damped and implies an EISF which would also be consistent with the picture of isotropic rotational motion;[17] we note however that this EISF is in contradiction with the data of Reference 37).

8.3.3(c) Methane

$CH_4(CD_4)$ has a transition from an orientationally disordered structure to an 'antiferro-orientationally' ordered structure at $T_t = 20.4$ K (27.0 K). Incoherent neutron scattering results[40] from powdered CH_4 suggest that in phase $I(T > T_t)$ all molecules are participating in a reorientational motion of the jump type.[6] The correlation time is of the order of 10^{-12} s. The details of the motion are, however, not known. Phase II of CH_4 is believed to have the same structure as CD_4-II, i.e. face-centred cubic with two different symmetry sites for the molecules: six of eight molecules in the primitive cell are orientationally ordered, two are orientationally disordered. Experimentally one would expect to find some kind of rotational motion for the disordered molecules and collective torsional excitations (librons) for the ordered ones.

The presence of 'almost free' rotations in CH_4 has indeed been observed with neutrons between 4.2 K[41] and 18 K.[42] At 4.2 K energy transfers of 1.06 meV and 1.8 meV were found and attributed to $J = 0 \rightarrow J = 1$ and $J = 1 \rightarrow J = 2$ rotational transitions of the disordered molecules (cf. for free molecules: $E_{0\rightarrow 1} = 1.3$ meV; $E_{1\rightarrow 2} = 2.6$ meV). No well-defined librational excitations could be detected; however, some information on the dynamics of ordered molecules in phase II was gained from a measurement of the tunnel splitting of the librational ground state.[43]

8.3.3(d) Cyclohexane and Related Compounds

In this section we consider the results obtained by a number of different investigators on the approximately globular compounds listed in the table below.

	T_t/K	ΔS_t/J mol^{-1} K^{-1}	T_f/K	ΔS_f/J mol^{-1} K^{-1}
Cyclohexane C_6H_{12}	186.1	36.0	279.8	9.3
Cyclopentane C_5H_{10}	122.0	39.8	179.0	3.3
Cyclohexanol $C_6H_{11}OH$	263.0	31.2	297.0	5.9
Perfluorocyclohexane[a] C_6F_{12}	185.1	37.8	332.6	18.8
Undecafluorocyclohexane[a] $C_6F_{11}H$	152.1	13.7	317.1	14.4
Nonafluorocyclohexane[a] $C_6F_9H_3$	163.2	13.2	310.2	12.1

[a]Thermal data from A. J. Leadbetter and P. Wrighton (unpublished)

Cyclohexane and cyclopentane[44] were among the first plastic crystals to be investigated by neutron quasielastic scattering. This work, in common with that on the other materials considered in this section, was of relatively low resolution, (typically $\Delta(\hbar\omega) \sim 0.4$ meV) and no values for the EISF were obtained, thus limiting the amount of detailed information obtainable.

In the first study of cyclohexane and cyclopentane the results were interpreted in terms of a model which did not properly take account of the fact that the protons are confined to the surface of a sphere and hence the model scattering law was not of the correct form (cf. equation (8.3.1). The geometrical information derived about the rotations must therefore be suspect but the correlation times ($\sim 10^{-12}$ s) will nevertheless be approximately correct. This is confirmed by a later discussion[45] of the data for cyclohexane using models of the correct form (assuming isotropic rotation) which give $\tau \sim 2 \times 10^{-12}$ s at 220 K. No conclusions were reached about the nature of the rotational motion but the author was concerned rather with comparisons of correlation times from neutron experiments with those from Raman and Rayleigh scattering.

The correlation times from the light scattering experiments were found to be about an order of magnitude longer than those from the neutron data which was taken as evidence about the time scale of the cooperative reorientation of a group of molecules as compared with that for a single molecule.

The fluorinated cyclohexane derivative $C_6F_9H_3$ has been studied over a wide temperature range.[46,47] In the low temperature phase no quasielastic broadening was observed showing rotational correlation times to be longer than about 10^{-10} s. In the orientationally disordered phase data were obtained at $T \tilde{<} 295$ K and $Q \tilde{<} 1.85$ A^{-1}. Except for the higher Q results at 295 K the data were shown to be consistent with an isotropic jump model of the type discussed in Section 8.2.3.d(i) with an average jump angle of $\pi/3$ and an activation energy of 13 kJ mol^{-1}. Similar results were obtained for $C_6F_{11}H$ but suggested a barrier height of only 2 kJ mol^{-1}. In all three fluorinated cyclohexanes the inelastic scattering was interpreted as showing librational modes near 30 cm^{-1}, which is consistent with the presence of preferred orientations, and for $C_6F_9H_3$ is quantitatively in agreement with a librational frequency calculated from the experimental barrier height with the assumption of a simple cosine potential. The presence of a significant librational contribution to the scattering was also confirmed by an analysis of the Debye–Waller factors. However, the higher Q results at the highest temperature (295 K) were not wholly explained using the large angle jump model and further data were obtained to still higher Q($Q \lesssim$ 3.2 A^{-1}).[47] These were in much better accord with a rotational diffusion model ($D_r = 3 \times 10^{10}$ s^{-1}) than with a $\pi/3$ jump model showing that as the melting point is approached the reorientational motion becomes freer, although the molecules must still spend an appreciable fraction of time librating about fixed orientations. The motion should perhaps be described as an itinerant oscillator but higher resolution data are required to provide more definite answers and to establish the extent to which the motion is truly isotropic.

Some preliminary studies of coherent scattering from C_6F_{12} in the near elastic region have also been reported.[47] These showed the presence of excitations propagating in the [110] direction with very small velocity (240 m s^{-1}) and an energy at the zone boundary of only ~1.3 meV. It was suggested that the modes involved might be coupled molecular reorientations but further work at higher resolution is required to confirm this.

The investigation on cyclohexanol is noteworthy in that the scattering law data were interpreted using Fourier transform techniques to obtain the intermediate scattering function.[48–50] In principle the elastic scattering component should then show up as a constant value of $I(Q, t)$ at long time and there is indeed some indication of this in the results.[48] However $I(Q, t)$ does not extend to t-values sufficiently large for a determination of the EISF.

In conclusion it should be emphasized that from all of this work, no positive information is obtained regarding the extent to which the average distribution function for the molecular orientations is anisotropic.

8.3.4 Succinonitrile $(CH_2CN)_2$ and Dichloroethane $(CH_2Cl)_2$

These are two substituted ethanes which have high temperature orientationally disordered phases. Both substances have been extensively studied by a variety of techniques in addition to neutron scattering.

Dichloroethane is the simpler example and will be considered first. The monclinic low temperature phase transforms to a high temperature phase of the same symmetry at a broad transition centred around ~177 K (ΔS_t = 11.7 J mol^{-1} K^{-1}) the melting point being at ~237.8 K (ΔS_f = 37.2 J mol^{-1} K^{-1}). x-ray work[51,52] shows that the molecules are always in the trans configuration and there is rotational disorder in the high temperature phase about the axis of least moment of inertia which is almost coincident with the Cl–Cl axis. Both structural[51] and theoretical[53] work suggest a single weakly-favoured orientation, which implies that the orientational distribution about the Cl–Cl axis is not random and that the molecules spend a significant time undergoing more or less damped librational motions about this axis. This picture has found support from inelastic neutron scattering experiments in which a well-defined librational frequency about the Cl–Cl axis has been identified[54,24] in the low temperature phase at about 105 cm^{-1} and found to broaden and shift to lower frequencies (~90 cm^{-1}) in the disordered phase. The reorientational motion has been investigated by relatively low resolution ($\Delta\hbar\omega \sim 0.4$ meV) incoherent quasielastic experiments in which broadening was observed only in the high temperature phase. Since the EISF could not be obtained no definitive information about the geometry of the rotation could be established but the results were consistent with a uniaxial rotation with jump steps of ~ $\pi/3$ or less, D_r ($\cong \tau_1^{-1}$ from equation (8.2.28) $= 8 \times 10^{10}$ rad s^{-1} at 227 K and an activation energy of 5 ± 4 kJ mol^{-1}. These results are in reasonable agreement with those of an NQR study[55] which gave an activation energy of ~6 kJ mol^{-1} and a 're-orientation frequency' (τ_c^{-1}) about a factor of six greater than D_r.

Succinonitrile is more complicated because the molecules exist in both trans and gauche configurations in the plastic phase. Nevertheless, some detailed information on the geometry and dynamics of the reorientational motions is now available. The high temperature phase extends from 233 K (ΔS_t = 26.6 J mol^{-1} K^{-1}) to the melting point (331 K, ΔS_f = 11.2 J mol^{-1} K^{-1}). It is body-centred cubic with the central C–C bond of the molecule always located along a body diagonal.[56,57] Both gauche (G) (~80%) and trans (T) (~20%) forms are present and the proportions are weakly temperature dependent.[58] The N atoms are always located near the middle of a cube face so that in the T-form the N–N axis is nearly on a 4-fold axis of the cubic unit cell. Rotation of a (cyanide) group by $2\pi/3$ leads to G $\leftrightarrow$ T interchange (isomerization) and rotation of the T-form by $\pi/2$ rotates the central C–C bond from one 3-fold axis to another. It follows that each isomer has available to it 12 distinguishable orientations in the lattice, and there are altogether 24 possible sites for each proton.

In describing the dynamics of the molecular motions two correlation times are important

(1) τ_i which describes the isomerisation, where[58]

$$\tau_i^{-1} = \tau_{GT}^{-1} + 2\,\tau_{TG}^{-1}$$

and τ_{GT}^{-1} and τ_{TG}^{-1} are jump probabilities per unit time. Here it is assumed. that τ_{GG}^{-1} is small enough to be neglected.

(2) τ_4 which describes the fourfold rotation of a trans molecule about the N–N (100) axis.

τ_4 is the mean residence time (or τ_4^{-1} the jump probability per unit time).

The first neutron study[24] was of relatively low resolution and the EISF was not determined. However, the results suggested that both correlation times were necessary to explain the neutron data. The analysis in terms of an approximate model gave values for τ_4 and τ_i near the mp of $\sim 2 \times 10^{-12}$ s and $\sim 10^{-11}$ s respectively at the transition temperature while τ_i was too long to be determined in this experiment for $T < 275$ K. A later[59,60] high resolution study was made for $T \gtrsim 300$ K and the EISF determined. This has confirmed the geometrical description of the disorder given above. Furthermore, analysis of the quasielastic scattering was made using a very detailed model of the molecular motions and gave $\tau_4 \cong 5 \times 10^{-12}$ s and $\tau_i \cong 2.5 \times 10^{-11}$ s at room temperature. The agreement between the two neutron studies is satisfactory in view of the relatively poor resolution and approximate model used in the earlier work.

A number of Rayleigh scattering[61,62] and dielectric relaxation studies[63] have been made. These techniques are sensitive to the isomerization. The resultant correlation times are about 6 times longer than those deduced from the neutron experiments and this might be due to the cooperative nature of the reorientation.[45]

The correlation times obtained from nmr[64] and Brillouin scattering[65] on the other hand are rather similar near the melting point to τ_i deduced from neutron experiments, although there may be some divergence at lower temperatures. The barrier heights for T–G isomerisation and four-fold rotation of the T molecules are respectively about 25 and 15 kJ mol^{-1}.

8.3.5 Rotation in Ionic Crystals

Traditionally the name 'plastic' crystals has been reserved, because of their macroscopic properties, to molecular crystals with more or less globular molecules. There are however numerous ionic crystals, notably ammonium salts, hydrosulphides, and cyanides which exhibit crystalline phases with orientational disorder. On a microscopic scale they are very similar to plastic crystals, with the essential difference that here more or less globular (polyatomic) ions are rotating within the (monatomic) crystal lattice of the second ion. Consequently on a macroscale they behave quite differently from plastic molecular crystals, i.e. they cannot be as easily deformed or extruded, have higher melting entropies and lower vapour pressures. So far as the rotational motion is concerned it may well be more anisotropic than in the average plastic molecular crystal, although the rotational correlation times may be of the same order of magnitude.

8.3.5(a) Ammonium Halides

Three of the ammonium halides, namely NH_4Cl, NH_4Br and NH_4I, exhibit orientationally disordered crystalline phases. If only the positions of nitrogen atoms and halide ions are considered, then phases I (just below the melting point) have the NaCl-structure. At lower temperatures they transform to the CsCl-type structure (phase II). In both phases, due to the rotational motion of the ammonium ions, the hydrogen atoms do not have fixed lattice sites. Early neutron diffraction results have shown, e.g. for NH_4Br,[66] that in phase II the ammonium ions are distributed at random over two orientations differing from each other by a 90°–rotation around a twofold axis of the NH_4–tetrahedron. Each of these orientations corresponds to a closest approach of the four protons to four halide ions (NH-bond in [111]–direction). In phase I a simultaneous 'closest approach' of all four protons to four halide ions cannot be achieved. Therefore convincing physical models describing the NH_4 ion orientation are more difficult to establish. It is more convenient to express the distribution of nuclear density of the hydrogen atoms around the central nitrogen atom in terms of cubic harmonics. This was done for NH_4I[67] and more recently also for ND_4Br.[68] The result is that the four hydrogen atoms are smeared out in sixfold octahedral positions around the central nitrogen atom. Their ensemble-averaged density distributions have maxima in the directions of the halide ions with full widths at half maximum of about 40° (ND_4Br, 473 K) and 70° (NH_4I, 298 K), respectively.

The first neutron studies of the rotational dynamics of ammonium ions were performed with relatively modest resolution.[69] These concerned mainly the torsional oscillations, which the ammonium ions perform within the potential produced by the surrounding ions at low temperature. Later some detailed information on the random rotational motion of the $(NH_4)^+$ groups in phase II was obtained by incoherent quasielastic neutron experiments on powders of NH_4Cl[70] and NH_4Br[71] and on single crystals of both substances.[72,73] The results of the powder experiments on phases II are consistent with the picture of jump rotation; however, mainly because of the restricted Q-range covered in the experiments, it was not possible to arrive at definite conclusions regarding the specific mode of reorientation. More conclusive results were obtained when single crystals† were used. In the case of NH_4Br the results are similar to those

† Single crystals permit a more rigorous test of theoretical models than polycrystalline samples since in the latter many details of the models are lost because of the orientational averaging.

for the plastic phase of adamantane (see 8.3.3(a)): the reorientations of the $(NH_4)^+$ ions are dominated by 90°-jumps about the C_4-axes of the crystal lattice, although the possibility of a small admixture of C_3-jumps cannot be excluded. This is in qualitative agreement with calculations[69] of the interionic potential which predict a barrier about 80% higher for rotations around the threefold axes than around the twofold axes of the NH_4-tetrahedron in phase II. The average time between rotational jumps (residence time) was found to be $3.2 \pm 0.4 \times 10^{-12}$ s at 373 K.

In the case of NH_4Cl a separate determination of the two different residence times for three-fold (τ_{C3}) and four-fold (τ_{C4}) rotational jumps was achieved by measurements at carefully chosen locations in the reciprocal space of the sample crystal.[72] At the III/II-phase transition (243 K) both times are of the order of 10^{-9} s. However, τ_{C4} shows a discontinuous increase through the transition of a factor of 4 whereas τ_{C3} does not change within the experimental error.

As far as phase I is concerned inelastic neutron data are available only on ND_4I for which dispersion curves of translational phonons propagating in the high-symmetry directions have been determined.[74] These measurements provide information on the degree to which well-defined phonons exist in this highly disordered phase. In fact the successful determination of translational phonon branches suggests that the high molecular reorientation rates present in phase I have very little effect on translational phonon lifetimes. On the other hand well-defined librations apparently do not exist in this system probably because of the low barrier to rotation. As for the alkali halides the tanslational phonon branches can be well reproduced by shell models. This indicates that rotation–translation coupling in phase I of ND_4I is at most a second-order effect. These results should be compared with the findings of a neutron study on the orientationally disordered cubic phases of KCN and NaCN (where the cyanide ions are undergoing rapid reorientational motion).[75] The failure to observe well-defined optic and zone-boundary acoustic phonons in the cyanides may be attributed to the rotation–translation coupling and/or to the different nature of the reorienting ions (linear versus quasi-spherical). In neither case, however, do these studies provide information on the nature of the rotational motions of ammonium and cyanide ions, respectively.

8.3.5(b) Other Ammonium Salts

Here we discuss two incoherent quasielastic neutron studies on ammonium compounds with orthorhombic structure. In $(NH_4)_2SO_4$[76] the results are consistent with instantaneous reorientations of the ammonium ions about their four C_3 axes with $\tau = 2 \times 10^{-11}$ s at 300 K. 90°-Rotations about the twofold axes of the NH_4-group were not considered in Reference 76. They can, however, be discarded since the corresponding EISF differs appreciably from the experimental points (compare Figure 2 of Reference 76 and Figure 2(A) of Reference

70). It should be mentioned that at room temperature the $(NH_4)^+$ ions are highly distorted with the H–N–H angles varying from 104.7° to 118.5° in one ammonium ion and from 100.2° to 116.2° in the other ammonium ion.[77] Strong coupling between rotational and intramolecular motions is therefore to be expected and should be included in the theory. In NH_4ClO_4 the ammonium ions have essentially ideal tetrahedral structure; however, one hydrogen has a stronger hydrogen bond than the other three, two of which have identical libration amplitudes.[78] Therefore the rotational motion is probably more complex than described by the simple rotational jump models. Nevertheless the model of instantaneous reorientation about the four C_3-axes of the NH_4-group gives good agreement with the measured quasielastic neutron spectra[79] although this was of relatively poor resolution and the EISF was not determined. Furthermore in this case also 90°-reorientations can be excluded in agreement with the diffraction results. The residence times τ derived for the C_3-rotation model vary from 18×10^{-12} s to 1.8×10^{-12} s between 66 K and 150 K. From this, assuming an Arrhenius law, an activation energy of 2.3 ± 0.3 kJ mol^{-1} was derived. It is interesting to note that, although the barrier to rotation is extremely low, the (time averaged) orientation distribution of the ammonium ion is definitely anisotropic, and there is no 'quasifree' rotation or small-step rotational diffusion.

8.3.5(c) Alkali Hydrosulphides

The alkali hydrosulphides NaSH, RbSH and CsSH, like many other ionic crystals with orientational disorder, have cubic symmetry in the solid phase just below the melting point. At lower temperatures they transform to crystal phases with lower symmetry. In a series of quasielastic neutron studies[80–82] it was shown that in the cubic phases the hydrosulphide ions reorient rapidly between a set of equilibrium orientations, but because of the limited resolution and restricted Q-range of these experiments, the geometry of these equilibrium orientations could not be established. The available data are equally well described by three different models in which the $(SH)^-$ ions are aligned at random in the [111], [110] or [100] directions, respectively, of the unit cell. The residence times found in the fitting procedure are somewhat model dependent; they are about 1×10^{-12} s for CsSH (373 K) and about 0.4×10^{-12} s for RbSH (414 K) and NaSH (376 K). In the trigonal phase of RbSH, just below the transition from the cubic phase, it could be established that the ions are reorienting between two equilibrium sites situated in opposite directions along the trigonal axis. Such rotational jumps correspond to 180° flips and are occurring with $\tau = (4.0 \pm 0.4) \times 10^{-12}$ s at 393 K. This is an order of magnitude slower than in the cubic phase. The high temperature phase transition thus involves the onset of much faster reorientation among at least six equilibrium orientations in the cubic phase. It is interesting to note that on the other hand the mean square amplitude of the hydrogen about the equilibrium sites is larger in the trigonal phase (~0.18 Å^2) than in the cubic phase (~0.14 Å^2) of RbSH. This is consistent with the observation of a relatively low lying librational frequency band with a maximum around 320 cm^{-1}. The activation energy for the rotational jumps in the trigonal phase of RbSH is estimated to 16 ±4 kJ mol^{-1}.

ACKNOWLEDGEMENT

We thank Dr. F. Volino for critically reading the manuscript.

REFERENCES

1. See for example: *Thermal Neutron Scattering*, ed. P. A. Egelstaff, Academic Press, New York, 1965; *Chemical Applications of Thermal Neutron Scattering*, ed. B. T. M. Willis, Oxford U. P., London 1973; Theory of Thermal Neutron Scattering, W. Marshall and S. W. Lovesey Oxford U. P., London 1971; I. I. Gurevich and L. V. Tarsov, *Low Energy Neutron Physics*, North Holland, Amsterdam, 1968.
2. See for example: T. Springer, *Quasielastic Neutron Scattering for the Investigation of Diffusive Motions in Solids and Liquids, Springer Tracts in Modern Physics*, **64**, (1972).
3. H. Hahn, *Neutron Inelastic Scattering*, IAEA, Vienna, (1965). Vol. II, p. 279.
4. C. Brot and B. Lassier-Govers, *Ber Bunsenges Phys. Chem*, **80**, 31 (1976).
5. R. Stockmeyer and H. Stiller, *Phys. Stat. Sol*; **27**, 269 (1968).
6. K. Sköld, *J. Chem. Phys.*, **49**, 2443 (1968).
7. R. E. Lechner, J. M. Rowe, K. Sköld, and J. J. Rush, *Chem. Phys. Letts.*, **4**, 444 (1969).
8. C. Thibaudier and F. Volino, *Molec. Phys.*, **26**, 1281 (1973); see also: C. Thibaudier and F. Volino, *Molec. Phys.*, **30**, 1159 (1975); P. Rigny, *Physica*, **59**, 707 (1972).
9. R. E. Lechner, in *Proceedings of the Conference on Neutron Scattering*, Gatlinburg, 1976, Vol. 1, p. 310. Natl. Techn. Inform. Service, U.S. Dept. of Commerce, Springfield, Virginia 22161.
10. J. D. Barnes, *Neutron Inelastic Scattering*, IAEA, Vienna, 1972, p. 287; *J. Chem. Phys.*, **58**, 5193 (1973).
11. A. J. Dianoux, F. Volino, and H. Hervet, *Mol. Phys.*, **30**, 1181 (1975).
12. V. F. Sears, *Can. J. Phys*; **45**, 237 (1967).
13. A. J. Leadbetter, F. P. Temme, A. Heidemann, and W. S. Howells, *Chem. Phys. Letts.*, **34**, 363 (1975).
14. E. N. Ivanov, *Soviet Phys. JETP*, **18**, 1041 (1964).
15. K. S. Singwi and A. Sjölander, *Phys. Rev.*, **119**, 863 (1960).
16. K. E. Larsson, *J. Chem. Phys.*, **59**, 4612 (1973).
17. R. E. Lechner, *Solid State Comm.*, **10**, 1247 (1972).
18. H. Hervet, F. Volino, A. J. Dianoux, and R. E. Lechner, *J. Physique Lett.*, **35**, L-151 (1974).
19. F. Volino, A. J. Dianoux, R. E. Lechner, and H. Hervet, *J. Physique. Colloq.* **36**, Cl–83 (1975).
20. H. Hervet, F. Volino, A. J. Dianoux, and R. E. Lechner, *Phys. Rev. Lett.*, **34**, 451 (1975).
21. H. Hervet, A. J. Dianoux, R. E. Lechner, and F. Volino, *J. Physique.* **37**, 587 (1976).
22. A. Heidemann, J. C. Lassegues, R. E. Lechner, and M. Schlaak in *Molecular Spectroscopy of Dense Phases*. Proc. 12th European Congress on Mol. Spectroscopy, Strasbourg, July 1975; Elsevier Sci. Publ. Comp., Amsterdam 1976, p. 327.
23. M. Schlaak, J. C. Lassegues, A. Heidemann, and R. E. Lechner, *Mol. Phys.*, **33**, 111 (1977).
24. A. J. Leadbetter and A. Turnbull, *J. C. S. Faraday II*, **73**, 1788 (1977).

25. A. M. Levelut, *J. Physique Colloq.*, **37**, C3–51, (1976).
26. R. J. Meyer and W. L. McMillan, *Phys. Rev.*, **A9**, 899 (1974).
27. F. Volino, A. J. Dianoux, and H. Hervet, *Solid State Comm.*, **18**, 453 (1976).
28. F. Volino, A. J. Dianoux, and H. Hervet, *J. Physique Colloq.*, **37**, C3–55 (1976).
29. F. Volino, A. J. Dianoux, and H. Hervet, Proc. 4th Int. Liq. Cryst. Conf., Kent (1976), *Mol. Cryst. Liq. Cryst.*, **38**, 483 (1977).
30. A. J. Leadbetter, R. M. Richardson, and C. J. Carlile, *J. Physique Colloq.*, **37**, C3–65 (1976).
31. R. M. Richardson, *Thesis, University of Bristol*, 1977. R. M. Richardson, A. J. Leadbetter, C. J. Carlile and W. S. Howells, *Mol. Phys.*, **35**, 1697 (1978). R. M. Richardson, A. J. Leadbetter and J. C. Frost *Ann. Phys.*, **3**, 177 (1978).
32. R. Stockmeyer. *Disc. Faraday Soc.*, **48**, 156 (1969).
33. R. E. Lechner and A. Heidemann, *Communications on Physics* **1**, 213 (1976).
34. C. E. Nordman and D. L. Schmitkons, *Acta Cryst.*, **18**, 764 (1965).
35. J. N. Sherwood, D. Taylor, G. S. Pawley, D. H. Saunderson, and C. G. Windsor, *NBRC Report*, p. 60 (1973).
36. L. A. de Graaf and J. Sciesinski, *Physica*, **48**, 79 (1970).
37. U. Dahlborg, C. Gräslund, and K. E. Larsson, *Physica*, **59**, 672 (1972).
38. U. Dahlborg and K. E. Larsson, private communication.
39. T. Månsson, L. G. Olsson, and K. E. Larsson, *J. Chem. Phys.*, **66**, 5817 (1977).
40. Y. D. Harker and R. M. Brugger, *J. Chem. Phys.*, **46**, 2201 (1967).
41. H. Kapulla and W. Gläser, *Neutron Inelastic Scattering* (IAEA, Vienna, 1972), p. 841.
42. W. Press and A. Kollmar, *Solid State Comm.*, **17**, 405 (1975).
43. W. Press, A. Hüller, H. Stiller, W. Stirling, and R. Currat, *Phys. Rev. Lett.*, **32**, 1354 (1974).
44. L. A. de Graaf, *Physica*, **40**, 497 (1969).
45. P. A. Egelstaff, *J. Chem. Phys.*, **53**, 2590 (1970).
46. A. J. Leadbetter, D. Litchinsky, and A. Turnbull, *Neutron Inelastic Scattering*, IAEA, Vienna, 1972, p. 231.
47. A. J. Leadbetter, A. Turnbull, and P. M. Smith, *J. C. S. Faraday II*, **72**, 2205 (1976).
48. F. F. M. de Mul and J. D. Bregman, *Phys. Letts.*, **33A**, 87 (1970).
49. J. D. Bregman and F. F. M. de Mul, *Nucl. Instr. and Meth.*, **93**, 109 (1971).
50. F. F. M. de Mul and J. D. Bregman, *Nucl. Instr. and Meth.*, **98**, 53 (1972).
51. M. E. Milberg and W. N. Lipscomb, *Acta Cryst.*, **4**, 369 (1951).
52. T. B. Reed and W. N. Lipscomb, *Acta Cryst.*, **6**, 45 (1953).
53. P. Manzelli and G. Taddei, *Chem. Phys. Letts.*, **13**, 132 (1972).
54. A. Ozora, M. Ito, N. Niimura, and N. Watanabe, *Chem. Phys Letts.*, **18**, 306 (1973).
55. T. Tokuhiro, *J. Chem. Phys.*, **41**, 438 (1964).
56. H. Fontaine, W. Longueville, and F. Wallart, *J. Chim. Phys.*, **68**, 1593 (1971).
57. H. Fontaine, *Thèse Université de Lille* (1973).
58. H. Fontaine and R. Fouret, *Adv. Mol. Relax. Proc.*, **5**, 391 (1973).
59. J. P. Amoureux, *Thèse Université de Lille* (1976).
60. R. E. Lechner, J. P. Amoureux, M. Bée, and R. Fouret, *Comm. on Physics*, **2**, 207 (1977).
61. M. Adam, L. Boyer, R. Vacher, P. Bergé, and L. Cecchi, *J. Physique Colloq.*, **32**, C5a–233 (1971).
62. D. A. Jackson, M. J. Bird, H. T. A. Pentecost, and J. G. Powles, *Phys. Letts.*, **35A**, 1 (1971).
63. W. Longueville, H. Fontaine, and A. Chapoton, *J. Chim. Phys.*, **68**, 436 (1971).
64. J. G. Powles, A. Begum, and M. O. Norris, *Mol. Phys.*, **17**, 489 (1969).
65. L. Boyer, R. Vacher, M. Adam, and L. Cecchi, *Proc. 2nd. Int. Conf. on Light Scat-*

tering in Solids, Ed. M. Balkanski, Flammarion, Paris, 1971, p. 498.
66. H. A. Levy and S. W. Peterson, *J. Amer. Chem. Soc.*, **75**, 1536 (1953).
67. R. S. Seymour and A. W. Pryor, *Acta Cryst.*, **B26**, 1487 (1970).
68. W. Press, *Acta Cryst.*, **A29**, 257 (1973).
69. G. Venkataraman, K. Usha Deniz, P. K. Iyengar, A. P. Roy, and P. R. Vijayaraghavan, *J. Phys. Chem. Solids*, **27**, 1103 (1966).
70. K. Sköld and U. Dahlborg, *Sol. State Comm.*, **13**, 543 (1973).
71. R. E. Lechner, F. Volino, A. J. Dianoux, and H. Hervet in *Neutron Research Facilities at the HFR of the ILL*, p. 25, Jan. 1975, Grenoble.
72. J. Töpler, *Thesis*, Technische Hochschule Aachen, 1976, and J. Töpler, D. Richter, and T. Springer, to be published.
73. R. C. Livingston, J. M. Rowe, and J. J. Rush, *J. Chem. Phys.*, **60**, 4541 (1974).
74. N. Vagelatos, J. M. Rowe, and J. J. Rush, *Phys. Rev.*, **B12**, 4522 (1975).
75. J. M. Rowe, J. J. Rush, N. Vagelatos, D. L. Price, D. G. Hinks, and S. Susman, *J. Chem. Phys.*, **62**, 4551 (1975).
76. H. J. Kim, P. S. Goyal, G. Venkataraman, B. A. Dasannacharya, and C. L. Thaper, *Sol. State Comm.*, **8**, 889 (1970).
77. E. O. Schlemper and W. C. Hamilton, *J. Chem. Phys.*, **44**, 4498 (1966).
78. C. S. Choi, H. J. Prask, and E. Prince, *J. Chem. Phys.*, **61**, 3523 (1974).
79. H. J. Prask, S. F. Trevino, and J. J. Rush, *J. Chem. Phys.*, **62**, 4156 (1975).
80. J. J. Rush, L. A. de Graaf, and R. C. Livingston, *J. Chem. Phys.*, **58**, 3439 (1973).
81. J. M. Rowe, R. C. Livingston, and J. J. Rush, *J. Chem. Phys.*, **58**, 5469 (1973).
82. R. M. Rowe, R. C. Livingston, and J. J. Rush, *J. Chem. Phys.*, **59**, 6652 (1973).

The following is a selection of papers relevant to the subject matter of this article which have appeared since its completion in March 1977:

1. Incoherent Neutron Scattering Study of Molecular Reorientation and Rotational Isomerization in Plastic Succinonitrile. J. P. Amoureux, M. Bée, R. Fouret, and R. E. Lechner, *Proc. of the Int. Sympos. on Neutron Inelastic Scattering*, Oct. 17–21 (1977), IAEA, Vienna. Vol. 1, p. 397. Molecular order and Dynamics in liquid hydrates, A. J. Dianoux and F. Volino, *ibid.* p. 533. Also other papers in this proceedings.
2. Random motion of a uniaxial rotator in an N-fold cosine potential: Correlation functions and neutron incoherent scattering law. A. J. Dianoux and F. Volino, *Molecular Physics*, **34**, 1263 (1977).
3. The Problem of Orientational Order in Tilted Smectic Phases: A High Resolution Neutron Quasielastic Scattering Study. A. J. Dianoux, H. Hervet, and F. Volino, *J. Physique*, **38**, 809 (1977).
4. Complement to 'Rotational Diffusion and Reorientations in Molecular Crystals'. C. Brot and B. Lassier-Govers, *Ber. Buns. Ges. Physikal Chemie*, **81**, 444 (1977).
5. J. D. Axe, L. M. Corliss, J. M. Hastings, W. L. Roth, and O. Muller. Neutron scattering study of NH_4^+ motion in NH_4^+ β-alumina, *J. Phys. Chem. Solids*, **39**, 155 (1978).
6. Orientational Order in Biaxial Liquid Crystals: The smectic VI and -H phases. F. Volino and A. J. Dianoux, *Phys. Rev. Letts.*, **39**, 763 (1977).
7. Local Orientation in the Mesophases of TBBA: Results of a Consistent Analysis of Neutron, NQR and NMR Data. A. J. Dianoux and F. Volino, *J. de Physique* (1978). In press.
8. The Dynamics of the Crystal, S_E, S_B and S_A Phases of IBPBAC. A. J. Leadbetter, R. M. Richardson and J. C. Frost, *J. de Physique* (1978). In press.
9. Molecular Motions in a Smectic A Phase by Incoherent Quasi-elastic Neutron Scattering. A. J. Leadbetter and R. M. Richardson, *Mol. Phys.*, **35**, 1191 (1978).

9

Correlation of Dynamic Observations on Molecular Motion in Plastic Crystals

R. A. Pethrick

9.1 INTRODUCTION

The existence of orientational freedom is the property which distinguishes plastic crystals from other molecular solids.[1,2] In comparison with atomic and ionic crystals, molecular solids are infinitely more complex, primarily due to the angular dependence of the intermolecular interactions. In attempting to characterize the motion of a particular entity the following factors have to be considered:

(i) the extent to which the motion of an individual molecular species can be considered free or coupled to that of its neighbours.

(ii) the importance of local disorder and free volume in defining the extent to which motion occurs.

(iii) the importance of long range morphological factors, line defects, faults etc., in the discussion of long range diffusion.

The properties of an atomic solid can be described adequately by a simple positional correlation function $g(R)$, whereas those of a molecular crystal require the addition of angular dependent terms through two Euler angles, Ω_1, Ω_2, i.e. $g(R, \Omega_1, \Omega_2)$. The distribution function for orientation is thus a function of nine variables, although the number of independent variables is often considerably reduced as a consequence of symmetry. In a plastic crystal the additional complication of orientational freedom introduces a time dependence into the function which now has the form $g(R, \Omega_1, \Omega_2, t)$. Even this form of function is not adequate to describe the form of interaction for a particular type of dynamic observation.

9.2 CORRELATION FUNCTIONS

The work of Green[3] and Kubo[4] in the 1950's began the development of a non-equilibrium statistical description of transport processes and other time dependent phenomena. Using this approach, phenomenological coefficients called time correlation functions can be formulated, the integrals of which over time describe the transport and dynamics of the ensemble. These correlation functions play a similar role in non-equilibrium statistical mechanics to that played by partition functions in equilibrium statistical mechanics.

Before considering the application of this approach to specific dynamic observations it is desirable to discuss briefly the concept of a time correlation function in its most general form.

A particle can have its energy and position defined in terms of two functions, momentum coordinate $\mathbf{p}(t)$ and the spatial coordinate $\mathbf{q}(t)$. At some initial time $t = 0$ the value of the phase space coordinate will have values $\mathbf{p} = \mathbf{p}(0)$ and $\mathbf{q} = \mathbf{q}(0)$. The values of $\mathbf{p}(t)$ and $\mathbf{q}(t)$ are related to the values of $\mathbf{p}$ and $\mathbf{q}$ through the equations of motion of the system. To emphasize this, we write

$$\begin{aligned} \mathbf{p}(t) &= \mathbf{p}(\mathbf{p}, \mathbf{q}; t) \\ \mathbf{q}(t) &= \mathbf{q}(\mathbf{p}, \mathbf{q}; t) \end{aligned} \tag{9.2.1}$$

Let $A\{\mathbf{p}(t), \mathbf{q}(t)\}$ be some function of the phase space coordinates, then

$$A\{\mathbf{p}(t), \mathbf{q}(t)\} = A(\mathbf{p}, \mathbf{q}; t) = A(t)$$

The classical time correlation function of $A(t)$ is defined by

$$C(t) = \langle A(0) \cdot A(t) \rangle = \int \cdots \int \mathbf{dpdq}\, A(\mathbf{p}, \mathbf{q}; 0) A(\mathbf{p}, \mathbf{q}; t) f(\mathbf{p}, \mathbf{q}) \tag{9.2.3}$$

where $f(\mathbf{p}, \mathbf{q})$ is the equilibrium phase space distribution function and where $\mathbf{dpdq}$ stands for $\mathbf{dp}_1, \mathbf{dp}_2, \ldots \mathbf{dp}_N, \mathbf{dq}_1, \mathbf{dq}_2, \ldots \mathbf{dq}_N$. If $A(t)$ is a function such as velocity or momentum then

$$C(t) = \langle A(0) \cdot A(t) \rangle = \int \cdots \int \mathbf{dpdq}\, A(\mathbf{p}, \mathbf{q}; 0) A(\mathbf{p}, \mathbf{q}; t) f(\mathbf{p}, \mathbf{q}) \tag{9.2.4}$$

For simplicity, it is usual to consider the motion of a particular molecule relative to its position—self-correlation, or by its motion relative to its neighbours—auto-correlation. The appropriate function for a single molecule would have the form.

$$C(t) = \langle \mathbf{v}(0) \cdot \mathbf{v}(t) \rangle \tag{9.2.5}$$

where $\mathbf{v}(0)$ and $\mathbf{v}(t)$ are respectively the initial and instantaneous values of the velocity after time t. Since $\mathbf{v}(t)$ depends upon the momentum and position of many other interacting particles in the system an explicit calculation of this function is obviously very difficult. What then is the advantage of the time correlation function formulism?

The principle advantage of this approach is that the function is completely general. It does not depend on the details of any particular model nor is it limited to any particular density region. Because the relationships are completely general they provide a convenient means of comparing various types of dynamic observation. The concept of a correlation function is one which has been used earlier in this book to describe the specific form of molecular interaction responsible for the randomization of some molecular vector. In practice, accurate modelling of an experiment is very difficult and it is only through a comparison of data from a variety of different types of observation that any definite conclusions can be drawn about the nature of molecular motion in a particular situation.

9.3 CLASSIFICATION OF MOTION IN MOLECULAR SYSTEMS

Molecular motion can be broadly sub-divided into two extreme situations—completely free or highly cooperative motions. Between these two limiting cases will lie a number of possible sub-classes, certain of which relate to the motion in plastic crystals. Since the motion and position of any one molecule depends on that of its neighbours, then all motions are in some sense ‘cooperative’. However, some types of motion will involve only essentially a few neighbours and approximate to the free rotation limit, whereas collective motions may involve virtually the whole crystal. As we shall see, certain experiments are dominated by the short time motion of the molecule whereas others are dominated by the longer time behaviour and hence reflect a different aspect of the overall dynamics. Although the motion of a molecule may not strictly be divided into translational and reorientation, this approach provides a convenient sub-division of the correlation functions obtained experimentally and will be used to classify the motions discussed in the next section.

9.4 EXTENT OF MOTION IN SIMPLE MOLECULAR SOLIDS

The motions which are observed in a simple molecular system at low temperature are rather limited and are usually restricted to those of phonons and librons, rotational and translation motion being almost non-existent. An increase in temperature allows ‘free’ volume to be created due to expansion of the lattice and the increased energy of the system is manifested as motion. In most cases the motion is still rather restricted due to strong orientational correlations. Easy motion is usually a precursor of a phase change. There are numerous exceptions to this rule e.g. benzene,[5] reorientational motion occurring in the solid phase. It could be argued that the observation of motion is, even in this case, a manifestation of a phase change, the lattice symmetry changing from tetragonal to cubic with an appreciable expansion of the lattice. The expansion of the crystal lattice is sufficient to unlock the molecules and allow reorientational motion without a significant increase in translational motion. It also decreases the repulsion forces encountered when the molecule attempts to reorientate and subsequently increases the probability of a molecule acquiring sufficient energy to move from its lattice site. Self-diffusion may be observed to change from virtually zero to something which is observable (10^{-17} $m^2 s^{-1}$). The process by which librational freedom is acquired may not be simple and several distinct stages have been identified in molecular crystals,[6] e.g. HBr and PH_3. The complexity of the rotor phases in the hydrogen halides have been described in detail in Chapter 4, and will not be considered further here. The reader will recall that changes in the ease of motion correlate well with the modification of the crystal structure of the different phases.

In an attempt to understand the mechanism whereby the motions of a particular type are influenced by disorder, crystal structure and coupled motion between specific molecule groupings, it is necessary to investigate the form of the correlation functions defining such motions.

Although such functions are generated theoretically it is by comparison with the experimentally determined functions that the detail of the motion is understood. A schematic representation of the ways in which coupling between rotation and translation motion can occur is presented in Figure 9.1. Unless the experiment differentiates between these forms of motion there is an increased uncertainty in the analysis of the data. Before considering the complexity of differentiating between the various possibilities presented by Figure 9.1 we shall consider the various techniques available for the experimental observation of translational and rotational motion.

9.5 TRANSLATIONAL MOTION

9.5.1 Radiotracer and NMR Experiments

The simplest form of translational motion is that of self-diffusion. For long times and large volumes the probability or concentration $c(R, t)$ always obeys the equation,[7]

$$\mathrm{d}c/\mathrm{d}t = D\Delta^2 c. \tag{9.5.1}$$

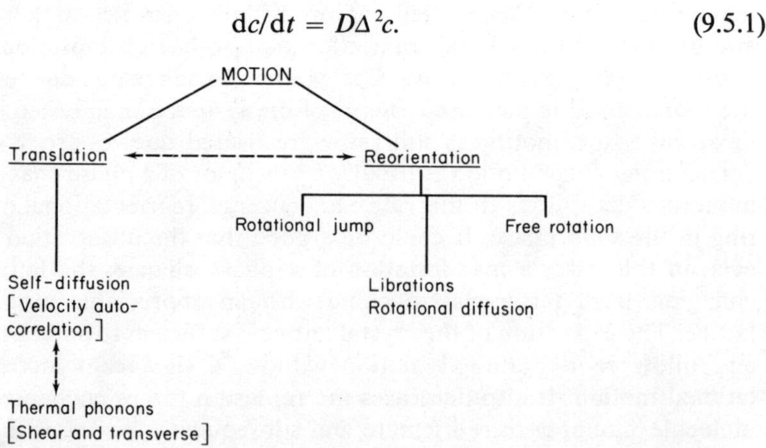

Figure 9.1 The interrelationship of molecular motions in molecular crystals

D the diffusion coefficient is a tensor for most crystals, although the anisotropy is usually very small. The diffusion constant D, can be measured by 'tagging' molecules with radioactive tracers (Chapter 2) or by assessing the re-orientation of their spins as they move in a magnetic field gradient (Chapter 5). The former gives D from the relation

$$\langle R^2 \rangle_t = 6Dt \tag{9.5.2}$$

Such a measurement of D should be clearly distinguished from that obtained from NMR relaxation experiments which gives essentially a translational correlation time. With the aid of a model it is possible to guess $\langle R^2 \rangle_t$ and it is often found that $\langle r^2 \rangle \sim a^2$ applies, where a is a nearest neighbour distance. Hence[8]

$$D_{NMR} \sim a^2/6t$$

which is not the same as D but is very close to it in magnitude.

The diffusion constant clearly depends on the magnitude of the molecular velocity and more importantly on the persistence of velocity which in the Kubo formulism[4] and can be written as

$$D = \frac{1}{3} \int_0^\infty \langle \mathbf{v}_i(t) \cdot \mathbf{v}_i(t_0 + t) \rangle \mathrm{d}t \tag{9.5.4}$$

where $\langle \ldots \rangle$ means the average over t_0 and i or both and this is readily generalized if D is tensorial.

Equation (9.5.4) is very important because it provides a specific relationship whereby comparison can be made between a macroscopic property D and a microscopic property, the molecular velocity, $v_i(t)$. The form of equation (9.5.4) implies that severe averaging due to $\langle \ldots \rangle$ and $\int \mathrm{d}t$ removes the detail of the velocity distribution from measurement of D. Nevertheless D is a very important property since any model or theory for $\mathbf{v}(t)$ should be correct for both it and $\mathbf{v}(t)$.

Dynamic experiments such as NMR are dominated by the short time form of the velocity correlation function and 'see' a simple decay function. The velocity correlation function is usually described as the normalized integrand of equation (9.5.4), which has a value of unity when t = 0;

$$\phi_v(t) = \langle \mathbf{v}(t_0) \cdot \mathbf{v}(t_0 + t) \rangle / \langle \mathbf{v}^2 \rangle \tag{9.5.5}$$

It is a decaying function, Figure 9.2, which may oscillate and often goes negative, with the constraint that $|\phi_t(t)| \leqq 1$. This function itself is usually used to define a correlation time[9]

$$\tau = \int_0^\infty \phi_v(t) \mathrm{d}t \tag{9.5.6}$$

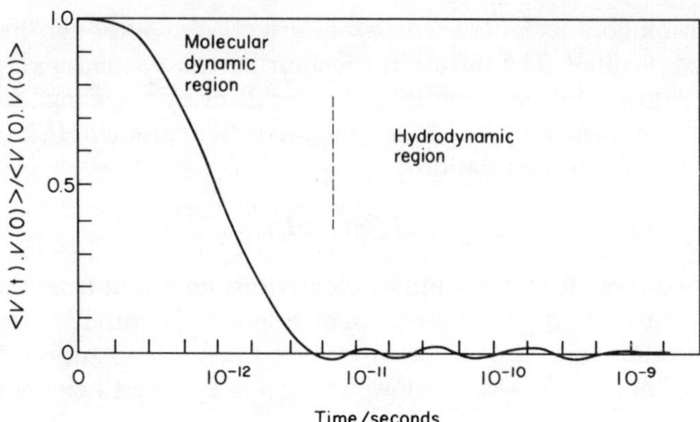

Figure 9.2 The velocity correlation function for a typical fluid

where, ϕ_v is a function of the constraints on the system such as temperature and pressure. In many cases we cannot measure $\phi_v(t)$ directly but only τ; this is true for instance of the NMR observations. If as in the case of a fluid, the net overall translational motion of a particle is significantly influenced by the collective behaviour of small oscillations in the tail of the function $\phi_v(t)$ then significant differences are to be expected between dynamic microscopic and macroscopic observations of diffusional motion.[10] In this situation tracer studies are expected to be far more sensitive to the precise form of the tail of the function than are dynamic studies.

The tracer and NMR techniques have been used for by far the largest number of experimental determinations of self-diffusion coefficients in plastic crystals. As might be expected from the above outline there are examples for which there is excellent agreement between the two types of measurement. (Within the errors of experiment and choice of model essential to the interpretation of the nmr experiment). This is in accord with the premise that both techniques should reflect a closely similar process on the two times scales of the velocity correlation function. Presumably the influence of the tail of the correlation function is inconsequential. The measure of agreement is shown in Figure 9.3 for adamantane and hexamethylethane. In other cases however, e.g. *dl*-camphene Figure 9.3 (some other examples are shown in Chapter 2), although the diffusion coefficients determined by both techniques near the melting point are similar, the temperature variations and hence the diffusion coefficients at lower temperatures are distinctly different.

This disagreement is surprising. It has been proposed that it might result from the possibly more defective nature of the material concerned (see Sherwood Chapter 2). Whatever the cause, an explanation of the difference is not possible without the detailed definition of the form of the tail of the correlation function in both the pure and defective materials.

Although NMR measurements lead directly to an effective diffusion constant

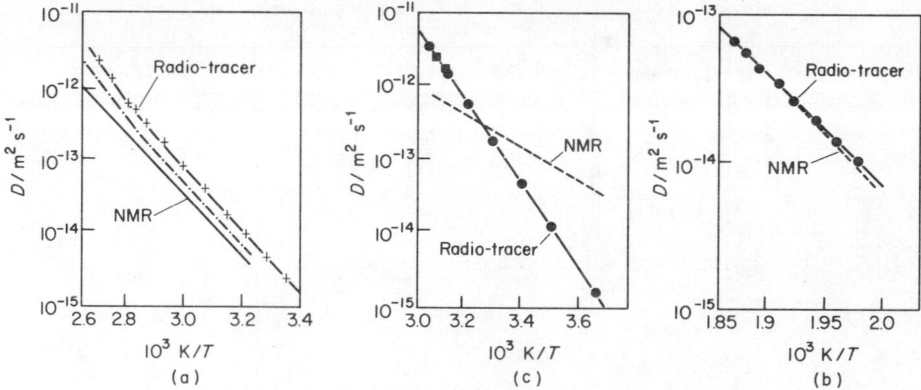

Figure 9.3 The relationship between radiotracer and NMR studies of self-diffusion in: (a) Hexamethylethane; (b) Adamantane; (c) *dl*-Camphene

a number of experiments sense the Fourier transform of $\phi_v(t)$ through the fluctuations of the density as characterized by the density of states distribution in the frequency domain[11] $-\rho(w)$:

$$\rho(w) \propto \int_{-\infty}^{\infty} \phi(t) e^{it} dt \tag{9.5.7}$$

Direct observation of the density of states is possible through scattering techniques-photons or neutrons.

9.5.2 Low Energy or Thermal Neutron Scattering

The neutron experiment observes changes in the centre of mass of the molecule through scattering of the neutron as a function of energy (frequency) and wave vector. This is incorporated in the scattering cross-section $S(\mathbf{Q}, \omega)$, where $\mathbf{Q} = \mathbf{k}_0 - \mathbf{k}$, the initial and final wave vectors of the scattering particle and ω is the angular frequency. As $Q \to 0$ the experiment looks at large regions of the sample and the scattering function has the form[12]

$$S(\mathbf{Q}, \omega) \propto e^{-i\omega t}\, dt\, e^{-\mathbf{Q}\cdot\mathbf{r}}{}_{dr} \langle n(\mathbf{r}_0, t_0) n(\mathbf{r}_0 + \mathbf{r}, t_0 + t) \rangle \tag{9.5.8}$$

where n is the density of particles. The incoherent scattering is a composite of rotational and translational motion. The width of the scattering will in part be related to the diffusion coefficient through an effective Doppler shift. If the motion of the particle is entirely described by equation (9.5.1) then[13]

$$S_s(\mathbf{Q}, \omega) = -\frac{\mathbf{DQ}}{\pi\,(\omega^2 + (D\mathbf{Q}^2)^2)} \tag{9.5.9}$$

which for fixed $\mathbf{Q}$, is a Lorentzian in ω of width, $\Delta\omega = 2D\mathbf{Q}^2$. It is usual to find that $\Delta\omega$ in fact varies rapidly with $\mathbf{Q}^2$, if $\mathbf{Q}^2$ is not small, Figure 9.4. The

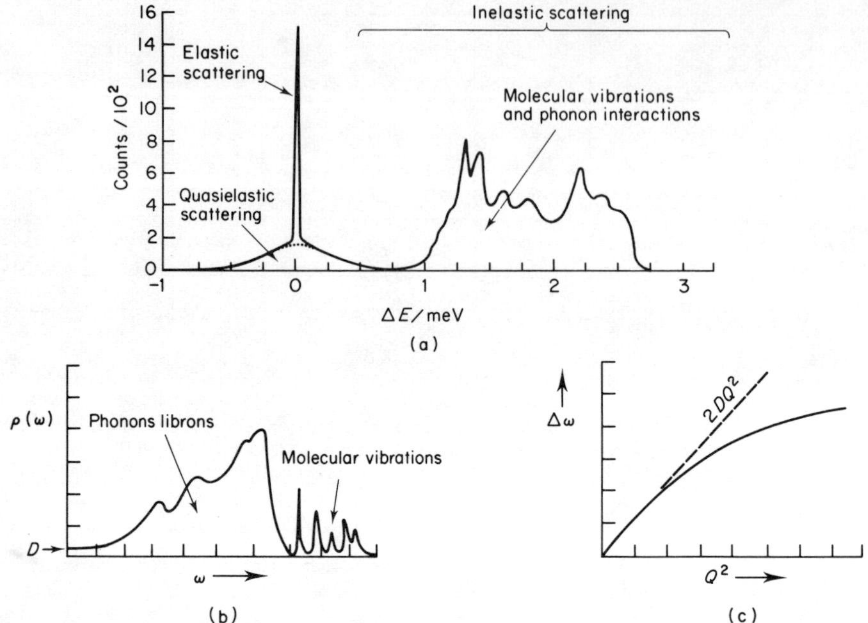

Figure 9.4 (a) A typical neutron scattering spectrum; (b) Density of states; (c) The vairation of the width ($\Delta\omega$) of the quasi-elastic peak with $\mathbf{Q}^2$

considerable disparity[10] between the values of D obtained from this approach and that of tracer studies indicates that the scattering is dominated by rotational diffusion, in which case equation (9.5.8) has the form

$$S(\mathbf{Q}, \omega) = \iint G_s(\mathbf{R}, t) \exp(\mathrm{i}\mathbf{Q}r - \mathrm{i}\omega t)\, \mathrm{d}\mathbf{Q}, \mathrm{d}\omega \tag{9.5.10}$$

where $G_s(\mathbf{R}, t)$ is the self-rotational correlation function for a scattering centre. The inelastic part of the scattering, Figure 9.3 may be decomposed into an inelastic and an elastic part. The inelastic part is determined by the rotational motion and the elastic part reflects the translational motion. The form of $G_s(\mathbf{R}, t)$ will be discussed in the next section.

9.5.3 Photon Scattering

For collective motions to be observed, coherent scattering experiments have to be performed and the advent of lasers has made this type of experiment relatively easy.[14] If the polarized light scattered with Q of the order of 10^{-3} Å^{-1}, observation over a domain of 1000 Å is possible. In practice, Brillouin scattering allows the simultaneous observation of both longitudinal and transverse thermal phonons which are similar in magnitude to the sound velocities as observed in ultrasonics at lower frequencies ($\sim 10^8$ Hz). Coherent slow neutron scattering at much higher values of Q can in principle be used to study phonon dispersion curves.

9.6 REORIENTATIONAL MOTION

The plastic phase is dominated by reorientational motion and as indicated in the chapters on infrared, Raman, and NMR relaxation, the principal question which needs to be asked is whether or not the rotational motion is free, partially hindered or corresponds to a jump diffusion mechanism. A variety of experiments can be used to probe this type of motion, however the form of the correlation function changes dramatically and allows various selection rules to be applied to the description of the motion being observed.

9.6.1 Dipole and Photon Scattering Interactions

The most direct method of observation of reorientational motion is the dielectric or far infrared absorption experiment.[4,16] If a molecular dipole is partially aligned by the application of an electric field, removal of the force allows the system to return to equilibrium under the influence of Brownian motion. The polarization produced by the application of an electric field decays to a zero value with a time constant which is characteristic of the motion of the constituent molecules. The correlation function which describes this process has the form

$$\langle \mu_i(t_0) \cdot \mu_i(t_0 + t) \rangle \qquad (9.6.1)$$

and allows the absorption coefficient to be expressed as:

$$\alpha(\omega) \propto \omega^2 \int \phi_{P_1}(t)\, e^{-i\omega t} dt \qquad (9.6.2)$$

where ϕ_{P_1} is obtained from equation (9.6.1) by writing $\mu^2 \langle \cos\theta(t_0) \cos\theta (t_0 + t)\rangle$ and expressing the modified correlation function in the form of the associated Legendre polynomial i.e. $P_1(\theta) = \cos\theta$, $P_2 = \frac{1}{2}(3\cos^2\theta - 1)$, etc. In practice this simplified formulism does not allow for the effects of the field produced by the surrounding molecular dipoles and other long range interactions of an induced dipolar nature. However, if these are assumed to add a constant contribution to the force field then direct correlation of the observed spectrum with the rotational motion is possible. The electromagnetic spectrum can be split into two distinct regions, the first corresponding to the change in the permittivity due to reorientational motion in which case

$$\langle P_1(t_0) \cdot P_1(t_0 + y) \rangle \propto \exp(-t/\tau_D) \qquad (9.6.3)$$

and

$$\alpha(\omega)/\omega^2 \propto \tau_D(1 + \omega^2 \tau^2_D) \tag{9.6.4}$$

Alternatively, if the observation is coupled to the vibration of the molecule through the transition moment then

$$\langle \ldots \rangle \alpha_e - i(\omega_v t + \psi)\, e^{-t/\tau_{IR}} \tag{9.6.5}$$

and the absorption can be described by

$$\alpha(\omega)/\omega^2 \propto I_{IR}(\omega) \propto \tau_{IR}/(1 + (\omega - \omega_v)^2 \tau^2_{IR}) \tag{9.6.6}$$

which corresponds to a Lorenzian infrared line. If the effects of vibrational relaxation processes are excluded then $\tau_{IR} \sim \tau_D$. This type of observation allows the correlation function to be obtained for motion about a selected molecular axis defined relative to the direction of vibration and the principal axes of rotation. The dielectric and infrared techniques provide an example of the application of selection rules to the observation of molecular motion. The experiment does not 'see' molecular translational motion and observes only molecular reorientational *via* the dipole vector i.e. $\cos\theta$. The major uncertainty about the interpretation of the infrared data is the correct model to be used to fit the absorption curve. This problem is quite general to the discussion of molecular motion in plastic and molecular crystals and will be considered in detail at the end of this section.

9.6.2 Coherent Light Scattering

An alternative method of sensing the motion in a molecular crystal is the observation of coherent light scattering.[17] The isotropic part of the electric polarization tensor senses the interactions of phonons with the scattered photons. The scattering due to the anisotropic polarizability will depend on the orientation of the molecule and will produce a partially depolarized scattered light which is a function of $P_2(\theta(t))$ i.e. $\frac{1}{2}(2\cos^2\theta - 1)$, for symmetric tops or higher symmetry. The Rayleigh line is a function of

$$\sum_{i,j} \langle P_2(\theta_i(t_0)) \cdot P_2(\theta_j(t_0 + t)) \rangle \tag{9.6.7}$$

Translational motion is not sensed in this experiment because Q is small and correlation of orientation is usually short range. It should be noted that for $i \neq j$, the information obtained is related to the correlation of the orientation, however this data is mixed up with the self terms $i = j$. The implication which this has is simply that the observations do not yield a reorientational correlation time directly but rather a composite with the self motion.

Raman molecular vibrational light scattering is related to Rayleigh light scattering in much the same manner as dielectric absorption is related to infrared spectroscopy.[16] The process of Raman scattering is related to the Fourier transform of ϕ_t of $P_2(\theta(t))$ and hence is sensitive to a slightly different aspect of the reorientational motion. The interpretation of experimental data has been discussed adequately and will not be considered further here.

9.6.3 Thermal Neutron Scattering

As indicated earlier in this Section neutron scattering can provide information on both rotational and translational diffusion of the scattering centre. Thermal neutron scattering is usually discussed in terms of two distinct types of interaction: coherent and incoherent scattering.[18–20] In the process of scattering the interactions are usually considered to be between the neutron and nuclear point centres (Fermi) and in the strict sense of spectroscopic transitions discussed above contain no selection rules. The neutron will be influenced by all the motions executed by the scattering centre; translational, reorientation and vibrational. In particular the function for reorientation will be of the form:

$$\langle P_n(\theta_i(t_0)) \cdot P_n(\theta_j(t_0 + t)) \rangle \tag{9.6.8}$$

and contains contributions from all values of n. This is in contrast to the situation for light scattering where n has a value of 2 or dielectric relaxation where n equals 1. As a consequence the neutron scattering data is very rich in information on the motion of a particle however, the separation of the various components often presents an impossible task. Studies performed at low values of Q often enable elastic and quasi-elastic peaks to be resolved allowing contributions from various types of motion to be identified, Figure 9.4 above. In the case of plastic crystals, translational motion is usually significantly slower than rotational motion and the observed scattering reflects the rate of reorientational motion in the crystal. If the reorientational motion is assumed diffusive then the orientational concentration is given by

$$\mathrm{d}c(\Omega)/\mathrm{d}t = D_\mathrm{r} \Delta^2 r^\mathrm{c}(\Omega) \tag{9.6.9}$$

This or an appropriate generalization describing anisotropic diffusion looks to the neutron just like translational diffusion with $D_{\mathrm{eff}} = D_R/a^2$ where a is the distance of the nucleus from the centre of the molecule. The rotational diffusion is obtained from the width of the quasi-elastic peak which has a width $2(D_r/a^2)Q^2$ and $\langle \theta^2 \rangle \sim 4D_r\tau_r$. In molecular crystals it is unlikely that the reorientational motion is completely free and the trial functions often used in the analysis incorporate the concept of rotational motion *via* jumps in angle of the order of a radian. The analysis of the elastic scattering has until recently suffered from problems of instrument resolution and deconvolution of the shape of the irradiating pulse, similarly the quasi-elastic scattering peak, which of necessity is broad is often difficult to resolve from the background scattering in molecular crystals, Figure 9.4 above.

9.6.4 Magnetic Interactions

The correlation function of the magnetization of the nucleus at resonance $\langle M_n(t_0) \cdot M_n(t_0 + t) \rangle$ is accessible directly from experiment by choice of a suitable radio frequency pulse sequence.[8] In practice the magnetization approximates closely to an exponential decay and depending upon the selection rules placed on the experiment in terms of the magnetization vector relative to the laboratory axis, direct observation is possible of T_2 (spin–spin), T_1 (spin-lattice) and $T_{1\rho}$ (rotating frame) relaxation times. Analysis of these parameters in terms of the detailed motions of the molecule requires certain assumptions to be made with regard to the nature of the magnetic interactions responsible for the relaxation process. The correlation function will rarely be simply related to only one of the possible motions of the molecule but will usually correspond to a composite of all possible interactions. Nuclear spin relaxation will be caused by any process which leads to a fluctuation of the magnetic field at the nucleus, $H(t)$. The nucleus usually resonates with a Larmor frequency ω_L, of the order of 10^8 Hz and it is fluctuations of the density of states with this frequency which lead to relaxation. In the rotating frame experiment it is possible to extend the spectral range contributing to the relaxation and perturbations with frequency components of the order of $10^4 - 1 \times 10^8$ Hz have been studied. In general the relaxation equations have the form

$$1/T_1 \propto \tau/(1 + \omega_L^2 \tau^2) \tag{9.6.10}$$

where τ and the constant of proportionally depend on the nucleus and the various interactions with the nuclear magnetic moment. In certain instances selection rules apply and as indicated in Chapter 5 the observations in molecular solids tend to be dominated by the nuclear electric quadrupole relaxation which is a tensorial intramolecular interaction and hence directly related to the reorientational motion. It is clear from other experiments that the values obtained from NMR either do not correspond purely to reorientational motion or may give different values if the defined molecular axes are different from those sensed by the other technique. If as in certain instances the relaxation is primarily determined by intermolecular dipole–dipole interactions then the relaxation reflects the rotational diffusion process and provides an estimate of the rotational diffusion coefficient. Clearly a greater uncertainty exists in the analysis of the NMR data when compared with either dielectric or infrared observations.

9.7 COMPARISON OF CORRELATION TIMES

The comparison of the correlation functions or more simply the correlation times, for the various techniques provides a means of:

(i) testing the validity of a particular theoretical model and
(ii) the correctness of a particular form of analysis when applied to a particular set of experimental data. In order to understand the problems associated with the correct modelling of a particular form of motion it is desirable to analyse the way in which various forces influence the overall motion of a molecule in the crystal.

9.7.1 Influence of the Strength of the Intramolecular Interactions

The potential experienced by a particular molecule in a plastic crystal will be the sum of two components:

(i) a static contribution to the overall potential which is angular dependent.
(ii) a time dependent contribution arising from the motion of the surrounding environment. It is this latter contribution which dominates the relaxation process although the form of the function is determined by the magnitude of both.

Four extreme situations can be identified, Table 9.1. The effect of change of either of the components clearly leads to a different class of motion and in real systems it is to be expected that the neat division into these specific groupings will not occur. It is possible to illustrate the effects of changes in the factors influencing the interactions between the molecule and its environment using the formulism of the coherent scattering experiment. The scattering cross

Table 9.1 Effects of variation of interaction potential on the reorientational motion in molecular crystals

Static potential \ Fluctuating potential	Strong	Weak
Strong	Rotational jumps	Librational and rotational tunneling
Weak	Rotational diffusion	Free rotation

section

$$S(Q, \omega) = \sum (2k + 1) j^2(Q_p) S_k(\omega) \tag{9.7.1}$$

where $S_k(\omega)$ is the Fourier transform of the relaxation function, $F_k(t)$ with the initial condition $F_k(t = 0) = 1$ for all k. If the molecule is assumed to be spherical then the macroscopic diffusion equation will have the form

$$D_r \Delta G_s(\omega - \omega_0; t) = G_s(\omega - \omega_0; t)\, t \tag{9.7.2}$$

which for the orientational self correlation function $-G_s\ (\omega - \omega_0; t)$ has the form

$$F_k(t) = \exp(-k(k + 1)\, D_r t) \tag{9.7.3}$$

where D_r is a rotational diffusion constant. If the molecule experiences a strong interaction from the surrounding field, the motion of the particle can be considered to occur in the presence of internal friction (ζ) which modifies the correlation function and leads to

$$F_k(t) = \exp\left[-k(k + 1)\, D_r t - (1 - \exp(\zeta(t)/\zeta))\right] \tag{9.7.4}$$

The form of the equations will appear familiar to the reader and it can readily be seen that application of selection rules leads to those presented in the previous section. For instance dielectric and NMR experiments contain contributions from $k = 1$ and $k = 2$ respectively. Clearly the use of a simple adjustable parameter, such as ζ, has a considerable advantage where comparison of the correlation function with experiment is concerned. This simple extension of the analysis is unfortunately not always valid and more specific forms of the functions have to be considered. In particular, the dielectric observations have recently been analysed in terms of a variety of correlation functions and enabled specific differences between the various theories to be specifically examined.

9.7.2 Models for Rotational Motion in Molecular Crystals

The large majority of the models have their origins in studies of liquids.[21] Since plastic crystals are intrinsically rotator phase systems, free volume models may be discarded from our consideration in most situations. This statement may not be strictly true for some brittle rotator phase materials where reorientational motion is relatively slow. Indeed if the presence of a neighbouring vacancy were to control the reorientational motion then the activation energy for rotation should be equal to that for translation diffusion. This is not so, Table 9.2. Near the melting point of adamantane[22,23] the translational diffusion jump rate is vacancy controlled and has a value of $10^6\ s^{-1}$, whereas the rotational rate is about $5 \times 10^{11}\ s^{-1}$. It is very improbable that the motion of a single vacancy can produce the relaxation of 5×10^5 molecules. Two further pieces of experimental evidence support the contention that rotational motion is an intrinsic property of the 'ideal' plastic crystal: variable pressure NMR experiments indicate that the activation volume is small for the reorientational process[24–26] and a vacancy triggered mechanism should produce a more or less asymmetric distribution of reorientational correlation times, a prediction which is not verified in practice.[27]

It is clear that three strong interaction situations in Table 9.1 are possible models for plastic crystals. In going from the weakest to the largest degree of interaction, the rotational motion becomes increasingly more hindered. For each model the motion can be either about one single axis (plane rotation) or about several axes (endospherical or isotropic).

9.7.2(a) Free Rotation

The extreme of weak interactions is the free rotational state,[28,29] a situation found in practice in a number of low density liquids but never in plastic crystals. Calculations of the correlation functions for a variety of geometries indicate that in all cases the correlation time should be of the order of $(I/kT)^{1/2}$. In the case of adamantane[23] the correlation time for the second spherical harmonic is 2×10^{-12} s, whereas $(I/kT)^{1/2}$ is 0.8×10^{-12} s. The situation in which the strength of the interaction is slightly increased has been discussed above.

9.7.2(b) Collision-interrupted Rotations

This situation may be considered as an example of a weak interaction case in which the fluctuations of the field have a finite strength. This model has been formulated by Gordan[28] under the name of 'J-diffusion' and applied successfully to liquids.[30] It has been argued that, assuming a reasonable form of the intermolecular potential, the collision duration is of the same order of magnitude as the collision interval. It is very unlikely that a collisional model would be consistent with the existence of an activation energy of the order of 4 kJ mol^{-1}. At constant volume the collision rate should increase as $T^{1/2}$ instead of decreasing with temperature, and the orientational correlation times should correlatively increase with temperature in any hard collision model, a prediction which is in contradiction with experiment. Other similar models do not appear to be able to explain observations more adequately.

9.7.2(c) Random Wells

As with the previous situations this model was developed for the description of molecular motion in a strongly interacting fluid.[31–33] The molecules are assumed trapped in potential wells where they perform more or less perturbed angular oscillations or 'librations' under the restoring torque produced by

Table 9.2 Summary of data on reorientational Motion in the plastic crystals

Molecule	T_t	T_m	$\Delta S_f/R$	$\Delta E_a/\mathrm{k}J\ \mathrm{mol}^{-1}$	Form of potential surface and comment	Reference
Adamantane	209	542	2.51	12.9	$C_4 - 1.7 \times 10^{-11}$ s	36, 64–66
			1.99	11.2	Free rotation is often assumed	
Hexamethylethane	153	374	2.41	9.2		67, 68
Perfluorocyclohexane	168	336	2.30	5.0 (168–270)		69
				1.8 (270–336)		
Methyl chloroform	224	243	2.26	18.8		70
Norbornane	131	360	1.53	Non-Arrhenius		71
Neopentane	140	257	1.46	4.2	Libration–rotator Model	70
					$\tau_L/\tau_L \sim 2$	
d-Camphor	238	449	1.40	11.7		72
Methane	20.4	—	—	—	Almost free rotor	73
Succinonitrile	236	331	1.35	25.1 (NM)	$C_4 \sim 2 \times 10^{-12}$ s	74, 76
			3.27	25.0 isomerism	neutron	
				15.0 reorientation		

Hexamethyldisilane	222	288	1.25	6.3		68
Norbornene	129	320	1.22	11.2 (129–146) 7.1 (146–250)		77
dl-Camphene	175	323	1.15	4.9 (175–220)		77
Cyclohexane	186	280	1.10	6.40	Complex surface $\tau \approx 2 \times 10^{-12}$ s	78
t-Butylchloride	183	245	0.95	3.0		79
Trimethylammonium chloride	308	—	—	— C_{3v}	$\tau = 4 \times 10^{-10}$ s	79
White phosphorus	196	317	0.95	5.7	Uncorrelated jumps −90° discrete wells	82
Norbornadiene	202	254	0.79	7.9		71
Phosphine	88	139	0.90	1.30		83
Pivalic acid	280	310	0.78	36	τ – C–C 0.12×10^{-12} τ – C–CH_3 1×10^{-10} τ – O–H 0.1×10^{-8} s Reorientates by breaking H bonds	84, 85
Dichloroethane	177	237.8	1.4	5	Librational motion, jumps of 60° to discrete wells.	85, 86

their neighbours. The slower large amplitude angular motion can be brought about by either of the following mechanisms:

(i) diffusion of the well on the surface of a sphere leading to a reorientational motion which corresponds to an initially damped oscillation followed by a motion similar to that of a classical rotational diffusion process.

(ii) fluctuations in orientation of the neighbouring molecules change the shape of the potential well and the trapped molecule follows these perturbations. In this latter model it is assumed that the wells are randomly orientated and that they bear no relationship with the crystal space group. This sort of situation appears to be rather less probable than one in which the potential surface is more intimately tied to the 'bumps' on the neighbouring molecules.

9.7.2(d) Site Models

This type of model is the logical extension of the second case described above in which the potential surface is intimately connected with the space group of the crystal; molecular motion or reorientation being preceeded by damped librational motion.[34] The detailed description of the potential surface and the related dynamics will be different for virtually every case, however the approach has several common features. As indicated in the introduction, the distribution function for a molecule in space will involve the definition of three Eulerian angles (two for a linear molecule) usually with respect to a set of crystal axes. The proper rotations of the crystal space group will generate other allowed potential wells. The initial direction can be chosen to lie along a crystallographic axis or not and can be unique or not. Clearly the choice of the initial structure can in certain cases simplify the subsequent calculations and allows the use of 'chemical' intuition.

(i) Symmetric Molecules in Potential Surfaces with Reduced Multiplicity. If we define the crystal site rotation group as the proper rotations of the crystal at the molecular centre, then it is possible for a molecular point group to have in common a subgroup with crystal site rotation group. If one or more symmetry elements in the molecule and in the crystal coincide for a potential surface, the wells will have a reduced multiplicity. This situation is to be found in molecules which correspond to elongated symmetric tops and have potential surfaces which are sensitive to repulsive interactions. The preferred molecular direction is determined by the long symmetry axis. This will tend to lie along the interstitial direction, hence minimizing the repulsive interactions and often corresponds to a direction of crystallographic symmetry.

Alternatively if the anisotropic interaction is mainly attractive, the molecule will point preferentially towards its neighbours, which often corresponds to a symmetry axis of the crystal. In both cases the multiplicity of the potential wells are considerably reduced. More generally, the reorientations can be classified into two kinds: those which bring the molecule into an undistinguishable orientation and those which produce dynamic disorder. If the jumps are fast and rare, the experimental techniques which are based upon permanent molecular properties cannot readily detect the former, whereas observation of symmetry breaking properties such as changes in vibrational line shapes or spin-labelled transitions can detect both kinds. In particular NMR relaxation can detect the orientational jumps in ordered phases,[35] such as the C_3—reorientational motion in the low temperature phase of NH_4Cl.

The obvious extension of this model is to a situation where the crystal site rotation group generates a potential surface with full multiplicity. In this situation the orientational disorder is not isotropic but corresponds to a site model with group O. Such a situation has been suggested to occur in plastic cyclohexane.[36] Obviously this situation presents a virtually impossible one to model successfully.

(ii) Potential Surface with Several Different Multiplicities. It is possible to envisage a situation in which the site rotation group generates a series of wells which are non-equivalent. The differences in energy corresponding to variations in the depth of the wells will also lead to unequal probabilities for the various directions. This sort of situation is relatively easily envisaged and is most probably very close to that usually found in practice. Multiple multiplicities of wells have been inferred from energy calculations on phase *I* of tertiary butyl chloride.[37] This plastic phase is face-centred cubic with four molecules per unit cell. *t*-butyl chloride has C_{3v} symmetry but sterically corresponds to T_d symmetry. As a result a number of unequivalent spatial orientations can be envisaged for the pseudo T_d molecule. The potential barriers separating the wells, which the computation shows are about equally populated, has a value of 5 ± 0.8 kJ mol^{-1} in agreement with the experimental result from dielectric relaxation. At 238 K the dielectric relaxation time is 7.4×10^{-12} s which corresponds to the first spherical harmonic correlation time of 5.4×10^{-12} s. The second spherical harmonic correlation time measured by spin lattice relaxation[38] is 3×10^{-12} s yielding a ratio of τ_1/τ_2 which is closer to 2 than to 3, indicating that rather large angle jumps are involved in the reorientational motion.

(iii) Potential Surfaces Related by Crystal Operations Involving Reflections. If a molecule possesses a symmetry plane it can jump by a rotation into a new well which is related to the first through an improper rotation of the crystal group. An example of this type of behaviour is that of the hexasubstituted benzenes. At room temperature many of these compounds form rotor phases which are not mechanically plastic. The rate of reorientational motion is usually very slow and typically lies in the range 10^{-2} to 10^6 s^{-1}. Single crystal x-ray studies

of 1,2-dichloro-3,4,5,6-tetramethylbenzene indicates that it has a monoclinic $P2_1/c$ lattice with two molecules per unit cell.[39–42] There is a crystallographic inversion centre at the molecular centres. Each molecule has six possible coplanar equidistant orientations, equally populated at room temperature. Three of these directions are crystallographically independent and generate, through inversion centre, the three others. The molecule performs, C_6, C_3 and C_2 reorientations about an axis which is neither a crystallographic axis nor a molecular symmetry axis but which sterically would be a sixfold axis for the molecule. The dynamics of these systems have been discussed in detail in the chapter on dielectric relaxation. It may be recalled that evidence for some form of cooperative reorientation appears through the slight curvature of the Arrhenius plots and also the breadth of the relaxation curve.

Application of the appropriate form of the above models has met with reasonable success, however more precise comparison has indicated certain limitations of the simple models. For instance, the assumption is usually made that all inertial effects can be neglected leading to an incorrect prediction of the form of the reorientational correlation function at $t = 0$.[43] The inclusion of inertia brings a remedy to this situation and implies that the jump takes a finite time to occur. The inertia term also implies that librational motion may be expected which in turn will be perturbed by phonon scattering and reorientational motion of neighbouring molecules. Such effects might collectively be described as cooperative phenomena and be expected to modify both the far infrared absorption and the neutron inelastic scattering data.

(iv) Composite Models. In an attempt to produce a more flexible approach to this problem composite models have been proposed. These make the assumption that orientation can occur by diffusional motion which is interrupted by orientational jumps. These latter jumps are usually assumed to occur about a molecular axis and bring about indistinguishable orientations of the molecule. This type of approach has met with some success when applied to neopentane[44,45] and cyclohexane[46] but does not lead to a greater understanding of the form of the interaction potential.

In conclusion it is clear that use of a potential surface tied to the molecular and crystallographic structure presents a reasonable formulism of the real situation and appears capable of rationalizing the majority of dynamic observations.

9.8 COOPERATIVE ROTATIONAL MOTION

The question of whether or not the reorientational motions of neighbouring molecules are correlated is intimately connected with the strength of the interaction potential. Only in the strong interaction situation can such cooperative motions be expected to be observed, however it is very likely that such effects might also be present even in the weaker interaction situations. The principal problem of correlated rotational motions is the explicit observation of a property directly related with this phenomena. From the point of view of correlation theory, such effects would appear as cross correlation terms and are virtually indistinguishable from the self terms. One method of at least detecting such correlated motions consists of comparing data of an autocorrelated sensitive technique with one which corresponds to a monomolecular correlation observation; for instance NMR (mono correlation) with depolarized Rayleigh light scattering, or near infrared line profiles with dielectric data. The principal problem associated with this type of comparison is justification of the observation that the reorientational motion is occurring about the same molecular axis. Invariably this is not the case and differences observed may then be a consequence of anisotropy of the diffusion tensor rather than an indication of cooperative motion.

A number of such comparison have been made[47–53] and there would appear to be sufficient evidence to suggest that in a number of systems the reorientational motion is to some extent coupled with the motion of neighbouring molecules. For instance, the dielectric static correlation parameter g in trichlorotrimethylbenzene is smaller than unity, indicating a strong tendency to antiparallel arrangement of molecules. This observation would suggest that the reorientation process retains this antiparallel alignment and is indicative of cooperative reorientation. This type of motion is active in NMR relaxation, while dielectric relaxation 'sees' rather the molecules which are, at a given time, uncorrelated and whose motion is faster. This is a possible explanation of the fact that the dielectric relaxation times are slightly faster than the NMR times.

9.9 CONCLUSIONS

It is very clear that the principle of a high degree of reorientational freedom in the plastic phase is well established.[54] Whether or not such motion is to be considered cooperative is still very much an open question. Recent measurements on succinonitrile[55] and carbon tetrabromide[56,57] indicate that in both cases the reorientational motion occurs *via* a sequence of jumps. Succinonitrile has the additional complication of changes in the distribution of isomeric states with temperature, however evidence from electro-optical Kerr effect studies[55] suggest that both at the plastic–solid and plastic–liquid transitions, marked changes occur in the degree to which correlations of reorientational motion are observed to occur. As the temperature is lowered towards the plastic-solid phase transition, molecular segments become correlated over ever increasing distances. The considerable decrease in the orientational anisotropy observed near the melting point suggests that the motion of the molecules has become virtually isotropic. This observation of a pretransition behaviour can hardly be interpretated as the cause of melting, as this process in succinonitrile is clearly a first order transition.[58] It is most probable that the pretransition is a consequence of the modification in the shape of the potential surface as a result of the gradual increase in volume and increasing defect density as the melt temperature is approached.

The description of reorientational motion in terms of models requiring a specific potential surface with well defined minima is fairly well established for a wide variety of systems. Specific examples have been considered in detail in the chapters on infrared and NMR studies. One conclusion which can be drawn from the wealth of information which is now available is that only when a common model exists for the description of a variety of experiments can any real conclusions be drawn with regard to the form of the potential surface. Recent theoretical calculations have simplified this task[59–63] and are making possible a detailed comparison of potential surfaces, dynamic observation and equilibrium properties.[36,54]

The problems associated with the occurance of dislocations and defects appears to be one which has not been thoroughly investigated and may hold the key to certain of the apparent anomalies which exist between the short and long time observations of the total dynamics. In electro–optical Kerr and Brillouin scattering experiments there appears to exist the possibility of bridging the gap between microscopic and macroscopic observation and elucidation of the differences portrayed by these experiments. It may be hoped that with a better understanding of the factors which control the dynamics of the plastic phase will come a better understanding of the way in which disorder, crystal symmetry and strength of intermolecular interaction influence the properties of molecular solid state.

REFERENCES

1. L. Pauling, *Phys. Rev.*, **36**, 430 (1930).
2. J. Frenkel, *Acta Physiochem, URSS*, **3**, 23 (1935).
3. M. S. Green, *J. Chem. Phys.*, **22**, 398 (1954).
4. R. Kubo, *Statistical Mechanics of Equilibrium and Non-Equilibrium States*, J. Meizner, North Holland, Amsterdam (1965).
5. E. R. Andrews and R. G. Eades, *Proc. Roy. Soc.*, **A 218**, 537 (1953).
6. M. O. Norris, J. H. Strange, J. E. Powles, M. Rhodes, K. Marsden, and K. Krynicki, *J. Phys. C.*, **1**, 422 (1968).
7. E. Helfand, *Phys. Fluids*, **4**, 681 (1961).
8. A. Abragam, *The Principles of Nuclear Magnetism*, Oxford U.P. London (1961).
9. D. A. McQuarrie. *Statistical Mechanics*, Harper and Row, New York, p. 514 (1976).
10. D. Levesque and L. Verlet, *Phys. Rev.*, **A2**, 2514 (1970).
11. R. Zwanzig, *Ann. Rev. Phys. Chem.*, **16**, 67, (1965).
12. L. van Hove, *Phys. Rev.*, **95**, 249 (1954).
13. A. Rahman, **Phys. Rev., 136**, A 405, (1964).
14. M. J. Bird, D. A. Jackson, and J. G. Powles, *Mol. Phys.*, **25**, 1051 (1973).
15. J. N. Sherwood, Personal communication.

16. R. G. Gordon, *Adv. Mag. Res.*, **3**, 1 (1968).
17. B. J. Berne and R. Pecora, *Introduction to the Molecular Theory of Light Scattering*, New York, Wiley (1975).
18. P. Eglestaff, *Thermal Neutron Scattering*, Academic Press New York, 1965.
19. W. Marshall and S. W. Lovesay, *Theory of Thermal Neutron Scattering*, Oxford U.P., London 1971.
20. K. E. Larsson, V. Dahlborg, and K. Skold, *Simple Dense Fluids*, H. L. Frisch and Z. W. Salsburg, New York, Academic Press, (1968).
21. J. P. Hansen and I. R. McDonald *Theory of Simple Liquids*, Academic Press, New York 1976.
22. R. Folland, S. M. Ross, and J. H. Strange, *Mol. Phys.*, **26**, 27 (1973).
23. H. A. Resing, *Mol. Cryst.* **9**, 101 (1969).
24. J. E. Anderson and R. Ullman, *J. Chem. Phys.*, **44**, 1797 (1966).
25. N. I. Lui and J. Jonas, *Chem. Phys. Letters*, **14**, 555 (1972).
26. C. Brot, *Dielectric and Related Molecular Processes*, **2**, 1 Chemical Society, London (1975).
27. S. H. Glarum, *J. Chem. Phys.*, **33**, 1371 (1960).
28. R. G. Gordan, *J. Chem. Phys.*, **44**, 1830 (1966).
29. A. G. St. Pierre and W. A. Steele, *Phys. Rev.*, **184**, 192 (1969).
30. B. Quentrec and C. Brot, *Phys. Rev.*, **12A**, 272 (1975).
31. N. E. Hill, *Proc. Phys. Soc.*, **82**, 723 (1963).
32. G. Wyllie, *J. Phys. C*, **4**, 564 (1971).
33. E. N. Ivanov, *Sov. Phys., J.E.T.P.*, **18**, 1041 (1964).
34. G. B. Guthrie and J. P. McCullough, *J. Phys. Chem. Solids*, **18**, 53 (1961).
35. D. E. Woessner and B. S. Snowder, *J. Phys. Chem.*, **71**, 952 (1967).
36. C. Brot and B. Lassier, Govers, *Ber. Bunsen, Gesell.*, **80**, 31 (1976).
37. B. Lassier and C. Brot, *J. Chim. Phys.*, **65**, 1723 (1968).
38. D. E. O'Reilly, E. M. Paterson, C. E. Scheie and E. Seyfarth, *J. Chem. Phys.*, **59**, 3576 (1973).
39. C. Brot and I. Darmon, *J. Chem. Phys.*, **53**, 2271 (1970).
40. R. Fourme and M. Renaud, *Mol. Cryst. Liq. Cryst.*, **17**, 223 (1972).
41. G. P. Charbonneau and J. Trotter, *J. Chem. Soc. A*, **2032**, (1967).
42. W. Kauzmann, *Rev. Mod. Phys.*, **14**, 12 (1942).
43. B. Lassier and C. Bort, *Chem. Phys. Letters*, **1**, 581 (1968).
44. L. A. De Graaf and J. Scieskinski, *Physica*, **48**, 79 (1970).
45. R. E. Lechner, *Solid State Comm.*, **10**, 1247 (1972).
46. L. A. De Graaf, *Physica*, **40**, 497 (1969).
47. P. A. Eglestaff, *J. Chem. Phys.*, **53**, 2590, (1970).
48. M. J. Bird, D. A. Jackson, and J. C. Powles, *Mol. Phys.*, **25**, 1051 (1973).
49. R. L. Jackson and J. H. Strange, *Mol. Phys.*, **22**, 313 (1971).
50. J. E. Anderson, *J. Chem. Phys.*, **43**, 3575 (1965).
51. A. M. Levelut and M. Lambert, *Mol. Cryst. Liq. Cryst.*, **23**, 111 (1973).
52. J. M. Chezeau, J. H. Strange, and C. Brot, *J. Chem. Phys.*, **56**, 1386 (1964).
53. D. Chapman and S. G. Whittington, *Trans. Faraday Soc.*, **60**, 1369 (1964).
54. J. G. Powles, *Ber. Bunsen Gesells.*, **80**, 261 (1976).
55. T. Bischofberger and E. Courtens. *Phys. Rev.*, **14**, 2278 (1976).
56. G. J. Davies, G. J. Evans, and M. Evans, *J. Chem. Soc.*, Faraday II, 2147 (1976).
57. V. J. Tekippe and L. L. Abels, *Phys. Letters*, **A60**, 129–132, (1977).
58. R. M. MacFarland, E. Courtens, and T. Bischofberger, *Mol. Cryst. Liq. Cryst.*, **35**, 27 (1976).
59. F. Kneubuhl, *Helv. Physica Acta*, **41**, 985 (1968).

60. P. Rigny, *Physica*, **59**, 707 (1972).
61. C. Thibaudier and F. Volino, *Mol. Phys.*, **30**, 1159 (1975).
62. C. Thibaudier and F. Volino, *Mol. Phys.*, **26**, 1281 (1973).
63. C. Brot, *J. Phys. C*, **5**, 223 (1971).
64. H. A. Resing, *Mol. Cryst. Liq. Cryst.*, **9**, 101 (1969).
65. H. Hervet, A. J. Dianoux, R. E. Lechner, and F. Volino, *J. Physique*, **37**, 587 (1976).
66. R. Stockmayer, *Discussions Faraday Society*, **48**, 156 (1969).
67. J. M. Chezeau, J. Dufourcq, and H. H. Strange, *Mol. Phys.*, **20**, 305 (1971).
68. S. Albert, H. S. Gutowsky, and J. A. Ripmeester, *J. Chem. Phys.*, **56**, 1332 (1972).
69. N. Boden, J. Cohen, and P. P. Davis, *Mol. Phys.*, **23**, 819 (1972).
70. E. O. Stejskal, D. E. Woessner, T. C. Farrar, and H. S. Gutowsky, *J. Chem. Phys.*, **31**, 55 (1959).
71. R. Folland, S. M. Ross, and J. H. Strange, *Mol. Phys.*, **26**, 27 (1973).
72. J. E. Anderson and W. P. Slichter, *J. Chem. Phys.*, **41**, 1922 (1964); **44** 1797 (1966).
73. Y. D. Harker and R. M. Brugger, *J. Chem. Phys.*, **46**, 2201 (1967).
74. J. G. Powles, A. Begum, and M. O. Norris, *Mol. Phys.*, **17**, 489 (1969).
75. J. P. Armoureux, Thesis Université des Science et Techniques de Lille, France (1976).
76. W. Longueville, H. Fontaine, and A. Chapoton, *J. Chim. Phys.*, **68**, 439 (1971).
77. N. Boden, S. Hanlon, M. Mortimer, and S. Ross, *private communication.*
78. D. E. Reilly, E. M. Paterson, C. E. Scheie, and E. Seyfarth, *J. Chem. Phys.*, **59**, 3576 (1973).
79. D. E. O'Reilly, E. M. Paterson, C. E. Scheie, and E. Seyfarth, *J. Chem. Phys.*, **59**, 3576 (1973).
80. A. Heidemann, J. C. Lasseques, R. E. Lechner, and M. Schlaak in *Molecular Spectroscopy of Dense Phases-Proc.* 12th European Congress on Mol. Spectroscopy Strasbour, July 1975, Elsevair Sci. Publ. Comp. Amsterdam, (1976).
81. M. Schlaak, J. C. Lassegues, A. Heidemann, and R. E. Lechner, *Mol. Phys.* (1976).
82. N. Boden and R. Folland, *Chem. Phys. Letters*, **10**, 167 (1971).
83. N. Boden and R. Folland, *Chem. Phys. Letters*, **32**, 127 (1975).
84. R. L. Jackson and J. H. Strange, *Mol. Phys.*, **22**, 313 (1971).
85. A. J. Leadbetter and A. Turnball, *J. C. S., Faraday II*, (1977).
86. M. Ito, A. Ozora, N. Niimura, and N. Watanabe, *Chem. Phys. Letters*, **18**, 306 (1973).

10

Theoretical Aspects of Solid Rotator Phases

A. Huller and W. Press

10.1 INTRODUCTION

Crystalline phases with orientational disorder occur in substances containing relatively rigid polyatomic units. These units can be grouped into three categories: (1) Neutral molecules (e.g. N_2 or CH_4) forming molecular crystals with a high degree of rotational freedom. (2) Polyatomic ions (as e.g. CN^- or NH_4^+) that form ionic crystals where the rotational motion generally is more restricted. (3) Side groups, e.g. CH_3-groups, that may exhibit rotational freedom around the bond that ties them to a larger organic unit. Following the notation of Venkataraman and Sahni (1970) all three types of rotating units will be called molecular groups in this article.

The notion of molecular rotation is useful only if the rotation occurs without much distortion of the molecular group, that is for rigid molecules. A molecule is called rigid if the internal vibrations of the molecule have much higher frequencies than the external modes (rotational or translational). For tightly bound small molecules the internal mode frequencies are of the order of several thousand cm^{-1} whereas rotational excitation energies seldom exceed 100 cm^{-1}. The coupling of rotations and internal vibrations therefore may be regarded as a small effect in most cases.

Molecular groups may perform rotational motions with one, two, or three rotational degrees of freedom. (A) examples of rotations with only one rotational degree of freedom, $0 < \alpha \leqq 2\pi$, are provided by long molecules such as hydrocarbon chains. Their long axis is fixed in space by geometrical restrictions and therefore rotation is only possible around this axis. Another example falling into this group are ions like NO_3^- where the atoms are restricted to move in a plane. To these may be added all the side group rotors since these groups may only rotate around the bond which connects them to the rest of the molecule. (B) All linear molecules like N_3^- and CO_2 or the dumb-bells like H_2, N_2, OH^-, CN^-, etc. fall into the group with two rotational degrees of freedom. The appropriate coordinates are the polar angles ϑ and φ of the molecular axis. (C) In all other cases, i.e. for three dimensional molecular groups like CH_4 or NH_4^+ and planar molecules without geometrical restrictions, three angles are needed to define the orientation of the molecule and the rotational motion is three dimensional. Usually the three Euler angles are used as rotational co-ordinates, but other choices are also possible. The set of Euler angles ξ, η, and ζ is usually denoted collectively by ω.

The existence of solid rotator phases depends on the shape and strength of the intermolecular forces. Section 10.2 is devoted to a discussion of the origin of the angle dependent part of the interaction and to a review of phenomenological descriptions of the intermolecular potential. In Section 10.3 the notion of orientational localization is introduced and different mathematical descriptions of orientational density distributions are detailed. The density distribution function is related to the angular potential and the definition of the order parameter at an orientational phase transition is discussed. Theories of order-disorder phase transitions in molecular crystals are critically reviewed in Section 10.4. Section 10.5 finally deals with the rotational excitations in molecular crystals. A distinction is made between single particle excitations like free rotations and tunneling transitions and collective excitations as for instance phonon-like librational modes.

10.2 ANGLE DEPENDENT INTERACTIONS

To calculate from first principles the interaction of two molecular groups one starts from the adiabatic approximation for the motion of the electrons. The interaction potential then is given by the expectation value of the Hamiltonian of nuclei and electrons in the electronic ground state (the ground state energy at infinite separation having been subtracted). Consequently one has to solve the Schrödinger equation for the electronic motion for given configurations of the nuclei and tabulate the energy as a function of distance and orientation of the molecules. Such *ab initio* calculations have only been performed in the case of two interacting hydrogen molecules.[1–3] Similar calculations for molecules with more electrons are not practicable.

The nature of the angular interactions resulting from a quantum mechanical treatment is threefold. There is the Coulomb interaction, i.e. the electrostatic interaction between the permanent multipoles of the molecules. For multipoles of order l and l' its radial dependence is proportional to the factor $I_l I_{l'}/R^{l+l'+1}$ where I_l and $I_{l'}$ are the multipole moments and R is the distance between the molecular centres. In the presence of highly polarizable ions such as O^{--} a further contribution to the Coulomb energy may become appreciable. A molecular group induces dipole moments on its polarizable neighbours and interacts with the induced moments. The radial dependence of this contribution is: $I_l^2\alpha/R^{2l+4}$ (α polarizability).

Secondly there are the dispersion forces due to the quantum fluctuations of the electronic wave functions. The leading contribution is an induced dipole-induced dipole interaction whose strength is determined by $E_0 \alpha^2/R^6$. If the polarizability α is anisotropic, as for example in N_2, the dispersion forces depend on the orientations of the molecules. E_0 is a typical electronic excitation energy (e.g. the ionization energy).

A third contribution to the angle dependent interaction energy stems from the overlap of nonspherical electron distributions. The overlap energy which is also called valence energy dominates at short distances. Its radial dependence is frequently approximated by an inverse power law or an exponential behaviour

$$\epsilon_0(\sigma/r)^n \text{ with } n = 12, \ldots, 18 \text{ or } b \exp(-r_0/r).$$

Since *ab initio* calculations of the potential are impossible for all but the simplest molecules one usually starts from a phenomenological potential with several adjustable parameters. The final aim of such an approach is the explanation of all obtainable data in the gas, liquid and solid phase with just one set of parameters.

Examples of phenomenological potentials are the so-called atom–atom potentials[4] where a potential function $\varphi_{ab}(r_{ab})$ with $r_{ab} = |\mathbf{r}_a - \mathbf{r}_b|$ is assumed between atoms a and b of different molecular groups. Typically one uses

$$\varphi_{\mathrm{ab}}(r_{\mathrm{ab}}) = B_{\mathrm{ab}} \exp(-\sigma_{\mathrm{ab}} r_{\mathrm{ab}}) - A_{\mathrm{ab}} r_{\mathrm{ab}}^{-6} \tag{10.2.1}$$

where B_{ab}, A_{ab}, and σ_{ab} are parameters. In a case like CH_4 with two different atoms per molecule there are three φ_{ab} functions, namely $\varphi_{\mathrm{CC}}, \varphi\hat{C}_{\mathrm{H}}$ and φ_{HH} with 9 parameters altogether. With some success attempts have been made to find a universal potential function for each pair of atoms independent of the chemical structure to which these atoms belong. Inclusion of electrostatic interactions (the dominant term in CH_4) introduces at least one additional parameter (the octopole moment in the case of CH_4). Atom–atom potentials of course lead to central and angle dependent forces between the molecules. Only the latter are of interest in this context.

Another approach is the Kihara core model (see MacRury[5] and references therein) where the interaction energy between two molecules is written as $v(r/\rho)$ where ρ is the closest approach between two molecules. ρ depends on the orientation of both molecules (ω_1 and ω_2) and on their distance R. v is a function like (10.2.1) or an inverse 6–12 power law. The prime difficulty is the calculation of $\rho(R, \omega_1, \omega_2)$. An interesting extension of the Kihara core model is the overlap model[6] where the repulsive part of the potential is approximated by the overlap volume of two spheroids. The explicit form of the potential is $\epsilon v(r/\rho)$ where ϵ and ρ depend on the relative orientation of the molecules.

A third approach towards a phenomenological description of the interaction potential of two molecular groups is a two centre expansion of the interaction.[7,8]

The potential energy is expanded in a systematic way into a series of multipole–multipole terms. Symmetry tells which moments are zero. The non-vanishing coefficients are phenomenological constants. Usually the series is terminated after one or two nonvanishing terms.

The two centre expansion has been developed for the Coulomb interaction of two charge distributions:

$$V = \int \rho(\mathbf{r})\,\rho(\mathbf{r}')\,v(|\mathbf{r} - \mathbf{r}'|)\,d\mathbf{r}\,d\mathbf{r}' \tag{10.2.2}$$

where $v(x) = 1/x$. $\rho(\mathbf{r})$ and $\rho(\mathbf{r}')$ are expanded into a multiple series and the addition theorem for surface harmonics is used to expand $1/|\mathbf{r} - \mathbf{r}'|$ into a double series. The two centre expansion is however not restricted to the Coulomb case. If $\rho(\mathbf{r})$ and $\rho(\mathbf{r}')$ are replaced by the densities of the outer electron charges and $v(x)$ by their interaction, e.g. dispersion forces or Pauli repulsion, V may be expanded into a multipole–multipole series in the same way. Atom–atom potentials also can be transformed into a multipole–multipole series.[9] In this case $\rho(\mathbf{r})$ and $\rho(\mathbf{r}')$ are δ-functions at the nuclear position and $v(x)$ the respective potential between the two atoms. For the following discussion the multipole–multipole series is split up into the monopole–multipole contribution and all other multipole–multipole terms.

The combined potential of all monopoles in the neighbourhood of one molecule is the crystal field. The crystal field does not depend on the orientation of these neighbours and therefore the rotational potential energy of a molecule in the crystal field $V(\omega)$ is a single particle potential which depends only on the Euler angles ω of the particle under consideration. The multipole–monopole contribution is especially important for ionic crystals where it is governed by the electrostatic monopoles. In the ammonium halides, this is the overwhelming part of the rotational potential and has a strength of a few thousand cm^{-1}.

The multipole–monopole term is not restricted to ionic substances. The expansion of the short range overlap repulsion and of dispersion energies in terms of a multipole–multipole series also leads to a multipole–monopole term. Consider the interaction of two neighbouring methane molecules as an example: Averaging over all angles of one of the two molecules it looses its tetrahedral structure and resembles a rare gas atom. Nevertheless it exerts an orienting field on its nonspherical neighbour.

The crystal field can be expanded into a set of symmetry adapted surface harmonics. If we use cubic harmonics[10] as an example such an expansion reads:

$$\phi(\mathbf{r}) = \sum_{l=0}^{\infty} \sum_{\mu=1}^{2l+1} c_{l\mu}(r)\, K_{l\mu}(\Omega) \tag{10.2.3}$$

where Ω denotes polar coordinates ϑ and φ, the $K_{l\mu}(\Omega)$ are cubic harmonics. They are linear combinations of spherical harmonics $Y_{lm}(\Omega)$ and form a basis for an irreducible representation of the cubic symmetry group. The main advantage of an expansion into symmetry adapted harmonics in comparison with other surface harmonics is the rather small number of terms in the series. In a cubic crystal expansion (10.2.3) reduces to:

$$\varphi(\mathbf{r}) = c_{01}(r)K_{01}(\Omega) + c_{41}(r)K_{41}(\Omega) + c_{61}(r)K_{61}(\Omega) + \ldots \qquad (10.2.4)$$

Up to order $l = 6$ there are only 3 nonvanishing coefficients. Cubic harmonics are most efficiently expressed in terms of cartesian coordinates x, y, and z.

$$K_{01} = 1/\sqrt{4\pi},\ K_{41} = 5\sqrt{21}\,(x^4 + y^4 + z^4 - 3r^4/5)/4\sqrt{4\pi}r^4, \text{ etc.,}$$
$$\text{with } r^2 = x^2 + y^2 + x^2.$$

As an example let us calculate the coefficient $c_{41}(r)$ from atom–atom potentials for the situation in Figure 10.1, where an atom at position $\mathbf{r}$ (the origin is at the centre of the cube) interacts with its 8 neighbours on the corners of a cube with an inverse power law potential $v(\mathbf{R}_i - \mathbf{r}) = b\,|\mathbf{R}_i - \mathbf{r}|^{-n}$.

$$c_{41}(r) = -\sqrt{4\pi}\,8b(135\sqrt{21})^{-1}n(n+2)(n+4)(n+6)(r/R)^4 \times \left[1 + \right.$$
$$\left. + \frac{(n-1)(n+8)}{2\cdot 11}(r/R)^2 + \frac{(n-1)(n+1)(n+8)(n+10)}{2\cdot 4\cdot 11\cdot 13}(r/R)^4 + \ldots\right] \qquad (10.2.5)$$

To describe the short range overlap repulsion between electron shells, n usually is chosen equal to 12. For dispersion forces $n = 6$. The expansion becomes especially simple in the Coulomb case where the coefficient of order l is proportional to $(r/R)^l$. Here $n = 1$ and $b = Qq$ where Q is the charge of an ion at $\mathbf{R}_i$

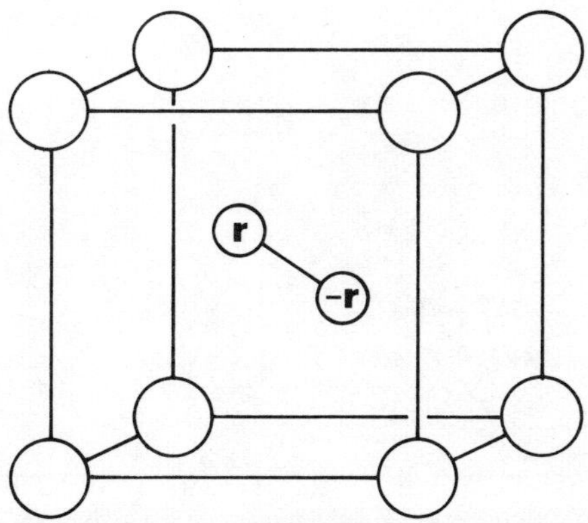

Figure 10.1 A dumb-bell molecule at a lattice site with cubic symmetry. An expansion of the potential into cubic harmonics (equation (10.2.4)) is favourable

and q is the charge of the probing particle at $\mathbf{r}$. In the Coulomb case the summation clearly has to be extended over more neighbouring atoms. For a dumb-bell molecule with one atom at $\mathbf{r}$ and the other at $-\mathbf{r}$ equation (10.2.5) gives directly the rotational potential (apart from a factor 2) since $\phi(\mathbf{r})$ is symmetric for $\mathbf{r} \to -\mathbf{r}$. For a more complicated rotating molecule the particle density (or charge density in the Coulomb case) is also expanded into a series of symmetry adapted surface harmonics. The expansion is especially simple if we use a (primed) coordinate system that is fixed in the molecule. In Figure 10.2 we show the case of a tetrahedron with the appropriate coordinate system. Here the expansion reads:

$$\rho'(\mathbf{r}') = b'_{01}(r)K_{01}(\Omega') + b'_{31}(r)K_{31}(\Omega') + \\ + b'_{41}(r)K_{41}(\Omega') + b'_{61}(1)K_{61}(\Omega') + \ldots \tag{10.2.6}$$

Tetrahedral symmetry allows a nonzero coefficient $b'_{31}(r)$ with $K_{31} = \sqrt{105}\, xyz/\sqrt{4\pi}r^3$. The potential energy of the tetrahedron in the crystal field $\phi(\mathbf{r})$ is:

$$V_c = \int \rho(\mathbf{r})\, \phi(\mathbf{r})\, d\mathbf{r}, \tag{10.2.7}$$

where $\rho(\mathbf{r}) \equiv \rho'(\mathbf{r}')$.
Using expansions (10.2.6) and (10.2.4) for $\rho'(\mathbf{r}')$ and $\phi(\mathbf{r})$ the angular integrations in (10.2.7) are easily performed when use is made of the definition of cubic rotator functions $U^{(l)}_{\mu\mu'}(\omega)$[11]

$$\int K_{l\mu}(\Omega)\, K_{l'\mu'}(\Omega')\, d\Omega = \delta_{ll'}\, U^{(l)}_{\mu\mu'}(\omega). \tag{10.2.8}$$

Here ω denotes the set of Euler angles ξ, η, ζ that describe the rotation from

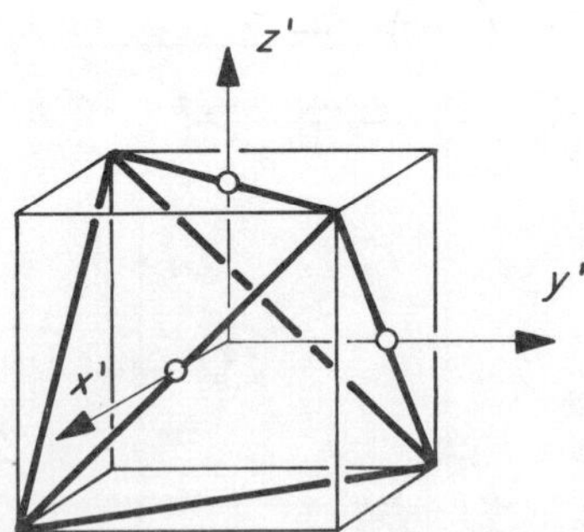

Figure 10.2 The primed coordinate system fixed in the molecule. The expansion of the particle density into symmetry adapted surface harmonics in this coordinate-system has a particularly simple form

the primed to the unprimed coordinate system. With the definition (10.2.8) the potential of a tetrahedron in a cubic crystal field becomes

$$V_c(\omega) = A_{11}^{(0)} + A_{11}^{(4)} U_{11}^{(4)}(\omega) + A_{11}^{(6)} U_{11}^{(6)}(\omega) + \dots \tag{10.2.9}$$

where use has been made of the fact that $U_{11}^{(0)}(\omega) = 1/8\pi^2$. The coefficients $A_{\mu\mu'}^{(l)}$ are obtained from the radial integration

$$A_{\mu\mu'}^{(l)} = \int c_{l\mu}(r)\, b'_{l\mu'}(r)\, r^2\, \mathrm{d}r. \tag{10.2.10}$$

A description of the rotation from the primed to the unprimed coordinate system in terms of quaternions (τ_1, τ_2, τ_3, τ_4 with $\sum_{i=1}^{4} \tau_i^2 = 1$[12–15] is in less common use than Euler angles, but it is of some advantage. Expressed in terms of quaternions the cubic rotator functions have a particularly simple form they are polynomials in τ_i of order $2l$.

So far we have restricted ourselves to monopole–multipole interactions which depend only on the orientation of one of the partners. In the general case of multipole–multipole interactions the potential energy of two molecules depends on both sets of Euler angles. A prominent example is the octopole–octopole interaction of two methane tetrahedra in the Hamiltonian of James and Keenan.[11]

$$W(\omega_i, \omega j) = \frac{I_3^2}{R_{ij}^7} C_{\mu\nu}(\mathbf{R}_{ij})\, U_{'\mu}^{(3)}(\omega_i)\, U_{'\nu}^{(3)}(\omega_j) \tag{10.2.11}$$

Here i and j denote the interacting molecules, I_3 is the octopole moment and $C_{\mu\nu}(\mathbf{R}_{ij})$ is a numerical matrix which depends on the distance $\mathbf{R}_{ij}$ of the two molecules. The appearance of cubic rotator functions of order $l = 3$ only, is due to the restriction to octopole terms.

From the interaction $W(\omega_i, \omega_j)$ we proceed to the molecular field $V_m(\omega_i)$ by averaging over the angular probability distribution functios $f(\omega_j)$ of all neighbours of one molecule and summing over the neighbours:

$$V_m(\omega_i) = \sum_{j \neq i} \int W(\omega_i, \omega_j)\, f(\omega_j)\, \mathrm{d}\omega_j \tag{10.2.12}$$

The molecular field approximation does not hold close to a continuous phase transition where correlations between the molecules become important.

10.3 ORIENTATIONAL LOCALIZATION

Before starting a discussion of orientational order–disorder transitions one has to develop some kind of description for the orientational localization of molecules. In particular we want to know how the nuclei forming a molecule are distributed orientationally in the disordered phase.

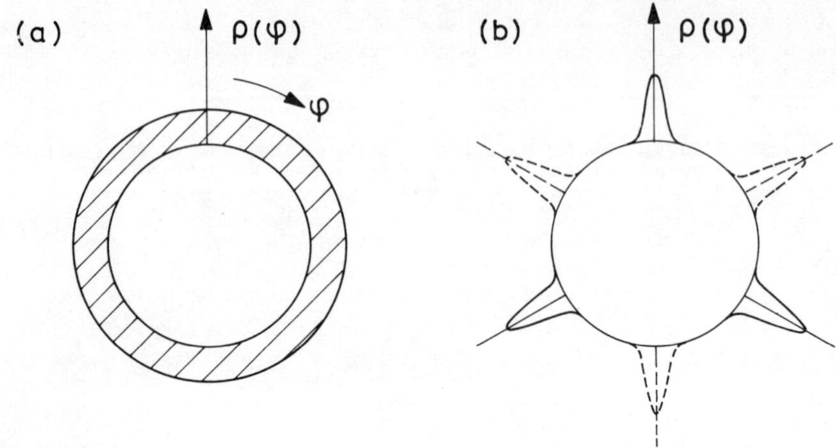

Figure 10.3 (a) The density distribution ρ (φ) in the case of complete orientational disorder; (b) The density distribution in the presence of a strong crystal field. The figure illustrates two alternative orientations (solid and dashed lines respectively) in a disordered phase

10.3.1 Model Systems

First we shall examine two examples of orientationally disordered phases which represent two extremes and almost ideally illustrate the alternatives:

(i) the high-temperature phase of solid ortho-hydrogen ($o\mathrm{H}_2$(I) or $p\mathrm{D}_2$(I).
(ii) NH_4Cl II

The first example, ortho-hydrogen[16] in its high temperature phase is characterized by complete orientational disorder. $o\mathrm{H}_2$ molecules have a rotational ground state with the rotational quantum number $J = 1$ and in this state have an electrostatic quadrupole moment Q. The high temperature phase of solid $o\mathrm{H}_2$ is orientationally disordered: There is no long range orientational order in the system of electrostatic quadrupoles. The angle-dependent density distribution $\rho(\Omega)$ of the nuclei at a molecular site is almost isotropic, that is $\rho(\Omega) =$ const (Figure 10.3). Apart from a molecular field, which is absent due to the lack of long-range orientational order, there might be a (hexagonal) crystal field effect on the orientational distribution in $o\mathrm{H}_2$. This crystal field, however, has been estimated to be extremely weak[17] in solid hydrogen and does not lead to a significant modulation of $\rho(\Omega)$ = constant.

For ammonium chloride, the second example, the situation looks quite different. The reason for this is a strong ionic contribution to the (cubic) crystal field (equation (10.2.9): The NH_4^+ tetrahedra are surrounded by a cage of eight negatively charged Cl^- -ions. These cause a strong orientation dependent potential and a pronounced orientational localization of the molecules within the potential well (Figure 10.4). Two distinguishable sets of 12 orientations exist (described by 24 different sets of Euler angles). In a shorthand version, this situation very often is referred to as two alternative orientations of NH_4^+ tetrahedra (symmetry $\bar{4}3m$) in an environment with full cubic symmetry ($m3m$) Figure 10.4). The resultant density distribution can be approximated by a sum of 8 δ-functions at the corners of a cube

$$\rho(\Omega) = \frac{1}{2} \sum_{i=1}^{8} \delta(\Omega - \Omega_i)/\sin \vartheta$$

where the Ω_i denote the polar angles of the corners.

Concerning the phase transition in NH_4Cl there is an apparent analogy to the Ising case.[18] Whilst two alternative orientations exist in phase II, in the low temperature phase a preference for one of them develops: the system becomes 'ferro-ordered'. This has led to the formulation of a pseudospin model for the ammonium halides.[19] It involves the approximation that narrow peaks in the density distribution (the librational amplitudes in NH_4Cl are about 7 degrees) are replaced by δ-functions.

The other extreme, oH_2, bears more resemblance with a Heisenberg antiferromagnet.[20] It should be noted, however, that the quadrupole–quadrupole interaction, which is the dominant orientation-dependent interaction in solid H_2, is very anisotropic.

10.3.2 General Systems

In general the orientational localization in disordered phases is somewhere between the two limiting cases.[21] All plastic crystals which mostly have either

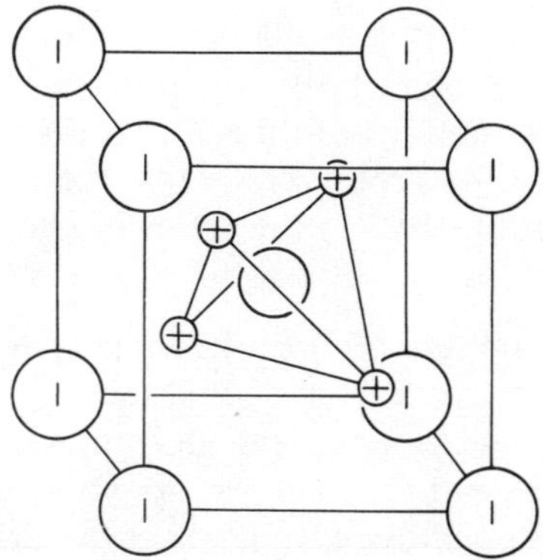

Figure 10.4 The orientation of an ammonium ion in the low temperature phase of NH_4Cl. ($\oplus$ = hydrogen, $\ominus$ = chlorine). The alternative orientation in the disordered phase is obtained by inversion

cubic (f.c.c., b.c.c.) or hexagonal (h.c.p.) structures, display a more or less pronounced crystal field giving rise to an asphericity of the density distribution. This is a consequence of the hard core repulsion of neighbouring molecules, which in some cases leads to an obvious 'steric hindrance'.

An adequate description of the particular situation may be attempted by generalizing the models used for the aforementioned examples. Let us start with the Ising case. A generalization of the Ising picture to more than two equivalent orientations (omitting the symmetry operations which transform a molecule into itself) has been suggested by Guthrie and McCullough.[22] In their approach to highly disordered systems they assume that the molecules take on orientations with less than the full molecular symmetry. As an example, the possible orientations of a tetrahedron ($\bar{4}3m$) on a lattice site with cubic symmetry ($m3m$) are discussed. If the symmetry elements of the subgroup $mm2$ of molecular and site symmetry coincide, this leads to 12 distinguishable molecular orientations. In this way one can generate the whole catalogue of subgroups common to $\bar{4}3m$ and $m3m$ and the number of equivalent orientations (given in parenthesis): $\bar{4}3m(2)$, $\bar{4}2m(6)$, $3m(8)$, $mm2(12)$, ... In order to obtain full cubic symmetry a random distribution of the equivalent orientations in the crystal is assumed. The approach can be extended readily to different molecular and site symmetries. Its merit is that it provides a method of how to simulate disorder in a plastic crystal when two discrete orientations are not sufficient. As such it has been used very frequently. However in most instances it provides a rather artificial description of the disorder. This becomes particularly true when entropy changes ΔS_t at an orientational order disorder transition (as obtained from calorometric measurements) are used for estimating the number of distinguishable molecular orientations in a plastic crystal with the above model.[22]

$$\Delta S_t = -k \ln N_1/N_2. \tag{10.3.1}$$

N_1 and N_2 are the numbers of possible orientations of the molecule above and below the phase transition respectively. Equation (10.3.1) is valid only if a number of idealizing assumptions are made and therefore its applicability to any real system (even to the prototype NH_4Cl) can be heavily questioned.[23] For the validity of equation (10.3.1) the following conditions have to be fulfilled:

(1) The distribution function $f(\omega)$ of the molecules over Euler angles must consist of narrow peaks above and below the phase transition. This is the case if there is a strong crystal field such that the molecules lock in at the minimum of the potential energy well and that neither quantum mechanical fluctuations nor thermal motion can appreciably smear out the orientational localization. The crystal field equation (10.2.9) has to be large in comparison with the rotational constant $B = \hbar^2/2\theta$ and also to the thermal energy $k_B T$.

(2) If so, only one set of equivalent positions may be used to calculate

N_1. Nonequivalent positions differ in $V_c(\omega)$ and therefore have different occupation.

(3) The shape of the peaks of $f(\omega)$ must not change at the transition. This third condition is almost inevitably violated. Below the transition the potential wells become deeper at the preferred orientations thus leading to narrower peaks.

(4) The transition must be clearly discontinuous. For almost continuous or continuous transitions the entropy change is distributed over a wide temperature range which necessitates a rather tedious integration.

(5) The transition must be purely orientational. Most of the orientational phase transitions are coupled to a change of symmetry of the centre of mass structure. ΔS_t then is not entirely due to the orientational ordering.

Obviously, continuous density distributions describe reality better than models using discrete peaks. One starts from an isotropic (angle-independent) density distribution and then modulates this constant distribution with appropriate symmetry adapted surface harmonics. We shall assume that correlation between the centre-of-mass motion and orientational motions of a molecule can be neglected and that the molecules are essentially rigid. Then the translational and orientational part of the density distribution can be treated separately and we can expand the nuclear density distribution ρ at a crystal site.[24–26] (For example D-density of a CD_4 molecule[27])

$$\rho(\Omega) = \sum_{l=0}^{\infty} \sum_{m=1}^{2l+1} a_{lm} K_{lm}(\Omega) \qquad (10.3.2)$$

where $K_{lm}(\Omega)$ denotes symmetry adapted surface harmonics, (e.g. cubic harmonics). In order to illustrate this expansion, let us consider a site with tetrahedral symmetry (43m). Only very few terms of low order are non-zero at a site of high symmetry.

$$\rho_T = \frac{1}{4\pi}(1 + a_{31}K_{31}(\Omega) + a_{41}K_{41}(\Omega) + \ldots) \qquad (10.3.3)$$

where explicit forms of $K_{31}(\Omega)$ and $K_{41}(\Omega)$ have been given in Section 10.2. Figure 10.5 illustrates equation (10.3.3) for the simple case of a planar rotator. If full cubic symmetry is taken, additional terms are dropped. For example all terms with l odd disappear, as they do not have inversion symmetry. Where is the expansion into symmetry adapted surface harmonics appropriate? It obviously is appropriate, if only a few terms in the expansion are significant, which is the case in most plastic crystals. If, on the other hand, $\rho(\Omega)$ is strongly localized, many terms have to be retained in the expansion and other descriptions will be better.

Another advantage concerns the rather direct connection between rotational potentials in the solid and $\rho(\Omega)$.[11,28] As an example we consider a three-dimen-

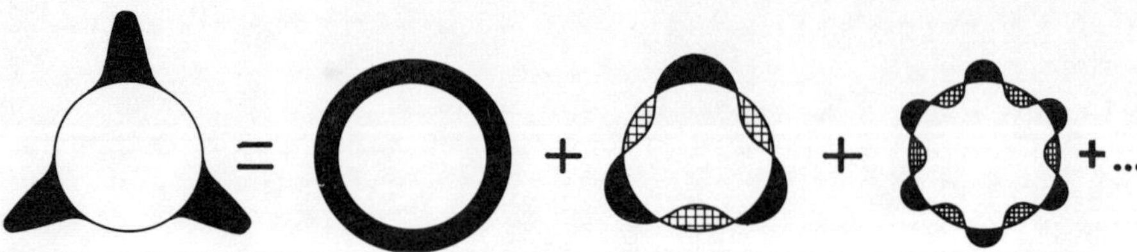

Figure 10.5 Expansion of the angle dependent density distribution into symmetry adapted surface harmonics. $\rho(\varphi) = 1/2\pi\,(1 + a_3 \cos 3\varphi + a_6 \cos 6\varphi + \ldots)$. Negative contributions are cross-hatched

sional molecule at a site with potential $V(\omega)$. The potential depends on Euler angles ω (or quaternions) as does the orientation of the molecule. The probability of finding a molecule in a particular orientation is $f(\omega)$ and for a classical rotator in an orientation dependent potential $V(\omega)$ one assumes a Boltzmann distribution

$$f(\omega) = G \exp(-\beta\, V(\omega)), \tag{10.3.4}$$

where G is the normalization constant and $\beta = 1/kT$. Both $f(\omega)$ and $V(\omega)$ can be expanded in an orthogonal set of functions. In the cubic case these are the cubic rotator functions $U^{(l)}_{mm'}(\omega)$ with $m, m' \leqq 2l + 1$.[11]

$$V(\omega) = \sum_{lmm'} A^{(l)}_{mm'}\, U^{(l)}_{mm'}(\omega) \tag{10.3.5a}$$

$$f(\omega) = \sum_{lmm'} B^{(l)}_{mm'}\, U^{(l)}_{mm'}(\omega) \tag{10.3.5b}$$

Again only very few of these coefficients are non-zero at a site of high symmetry. At a site with full cubic symmetry for example

$$V(\omega) = U^{(0)}_{11}(\omega) + A^{(4)}_{11}\, U^{(4)}_{11}(\omega) + A^{(6)}_{11}\, U^{(6)}_{11}(\omega) + \ldots \tag{10.3.6}$$

$$f(\omega) = U^{(0)}_{11}(\omega) + B^{(4)}_{11}\, U^{(4)}_{11}(\omega) + B^{(6)}_{11}\, U^{(6)}_{11}(\omega) + \ldots \tag{10.3.7}$$

The expansion coefficients $B^{(l)}_{mm'}$ are determined by the potential $V(\omega)$ (equation (10.3.4)) and are temperature dependent. An expression of the entropy in terms of $f(\omega)$ is given in Section 10.4. Entropy changes at orientational order–disorder transitions certainly can be calculated provided $f(\omega)$ is known in both phases. The connection between $V(\omega)$ and $\rho(\Omega)$ can be formulated, by first expanding the nuclear density distribution $\rho'(\Omega')$ within a molecular frame and then transforming to the crystal frame.[26]

$$\rho'(\Omega') = \sum_{l'm'} b^0_{l'm'}\, K_{l'm'}(\Omega') \tag{10.3.8}$$

$$\tilde{\rho}(\Omega, \omega) = \sum_{l'mm''} b^0_{l'm'}\, U^{(l')}_{m''m'}(\omega)\, K_{l'm''}(\omega). \tag{10.3.9}$$

$\rho'(\Omega')$ consists of δ-peaks in the coordinate system of the molecule (see Figure

10.3b) as does $\tilde{\rho}(\Omega, \omega)$ which is the nuclear density in the crystal frame for a definite orientation ω of the molecular frame. To obtain the density distribution of the nuclei we average over the probability function for the Euler angles.

$$\rho(\Omega) = \int \mathrm{d}\omega\, f(\omega)\, \tilde{\rho}(\Omega, \omega). \tag{10.3.10}$$

With the orthogonality relations for the cubic rotator functions one obtains

$$\rho(\Omega) = \sum_{lm} a_{lm} K_{lm}(\Omega) \tag{10.3.11}$$

where

$$a_{lm} = \frac{1}{2l+1} \sum_{m'} \mathrm{B}^{(l)}_{mm'}\, b^0_{lm'} \tag{10.3.12}$$

The coefficients a_{lm} can be determined in a neutron scattering experiment. The $b^0_{lm'}$ only depend on the arrangement of the nuclei within the molecule and not on the rotational distribution. The above relation not only helps to determine the crystalline field in a plastic crystal[28] but also may be used to determine rotational potentials at low temperatures, where rotational tunneling is observed.[15]

10.3.3 Orientational Order Parameter

The question remains how to define an order parameter at an orientational order–disorder transition. Concepts applicable to both approaches which have been described above, are rather straightforward. In the Ising case the two possible orientations are denoted by + and −, and the number of spins in the + and − state by n_+ and n_-. The obvious choice for the orientational order parameter is

$$\eta = (n_+ - n_-)/(n_+ + n_-) \tag{10.3.13}$$

In the high temperature phase $n_+ = n_-$ and hence $\eta = 0$.

In the other (Heisenberg) case the disordered phase was characterized by a spherical distribution—maybe somewhat modulated by symmetry-allowed surface harmonics. At the phase transition a breaking of symmetry occurs. The site symmetry is reduced and additional symmetry allowed surface harmonics contribute. In some cases one (low order) harmonic which becomes symmetry allowed is dominant (in $o\mathrm{H}_2$ just one additional harmonic contributes) and its expectation value or the one of the corresponding rotator function obviously has to be identified with the order parameter. Questions related with the latter description have been discussed in more detail[20,21,26] in connection with $o\mathrm{H}_2$, $p\mathrm{D}_2$ and CD_4, respectively.

10.4 ORIENTATIONAL ORDER–DISORDER TRANSITIONS

An interesting aspect in the field of molecular crystals is the occurrence of orientational phase transitions with a spontaneous decrease of the molecular site symmetry as the temperature is lowered. Consequently the angular distribution function $f(\omega)$ of the molecules contains more symmetry elements above the transition temperature than below.

In general, phase transitions may occur discontinuously with some hysteresis or continuously at the critical temperature T_c. Well known examples for continuous transitions are the liquid-gas transition at critical conditions, many magnetic phase transitions, the λ-transition in liquid helium and the transition from the normal to the superconducting state in metals. Singularities of physical quantities $A_i(T)$ like the correlation length, the order parameter, or the susceptibility at T_c are expressed in terms of power laws

$$A_i(T)\alpha \left|\frac{T - T_c}{T_c}\right|^{x_i} \tag{10.4.1}$$

with critical exponents x_i. Close to T_c (in the critical region) long-range fluctuations dominate the behaviour of the system and determine the values of the critical exponents. The common cause for all singularities at T_c is the increase of the correlation length to infinity. Different systems become similar when the correlation length grows beyond the range of the interactions. This implies relations between different exponents for one transition (scaling laws) and equality of corresponding exponents for different phase transitions within one universality class.

For some systems with longe range interactions (e.g. superconductors) the critical region is extremely small. Outside this region the molecular field theory (MFA) which neglects all correlations provides an adequate description. Deviations of critical exponents from MFA-values are usually attributed to the importance of critical fluctuations.

The MFA, which is also called Bragg–Williams approximation in lattice gas models or Weiss approximation in magnetic systems, is the common approach towards a theoretical description of orientational phase transitions. The breakdown of MFA in the critical region is not crucial for orientational phase transitions which with almost no exceptions occur discontinuously. At a discontinuous transition (or first order transition) with a small discontinuity, the system jumps into the new phase before T_c has been approached. The region close to T_c where fluctuations dominate cannot be reached and MFA should be sufficient in most cases. Theoretical considerations that take into account fluctuations at first order transitions with a small discontinuity have only recently become possible, but they have not been applied to molecular crystals.[29]

One should be careful with attempts to fit the order parameter $\eta(T)$ below a discontinuous phase transition by a power law with a critical exponent β

$$\eta(T) = A\left(\frac{T_c - T}{T_c}\right)^{\beta} \tag{10.4.2}$$

From the many examples in the literature where equation (10.4.2) has been applied to first order phase transitions, reference is made only to the determination of the orientational order parameter of solid CD_4.[30] At a discontinuous transition a critical temperature does not exist, therefore equation (10.4.2) is meaningless and the introduction of β is misleading. A simple consideration[31–33] shows that the spurious exponent 'β' at a discontinous phase transition will be smaller than 1/4. If we start from a Landau-type free energy

$$F = \tfrac{1}{2}a'(T - T_0)\eta^2 + \tfrac{1}{4}b\eta^4 + \tfrac{1}{6}c\eta^6 + \ldots \tag{10.4.3}$$

all fluctuations are obviously neglected as F depends only on the mean value η of the order parameter. In Figure 10.6 F has been plotted for negative b. Figure 10.7 shows the order parameter which in this approximation is given by

$$\eta_L(T) = \left\{\frac{-b}{2c}\left[1 + \left(1 + \frac{4a'c}{b^2}(T_0 - T)\right]^{1/2}\right\}^{1/2} \tag{10.4.4}$$

for $T < T_0 + 3b^2/16a'c$ and $\eta = 0$ otherwise. If one tries to fit $\eta_L(T)$ from equation (10.4.4) by the expression (10.4.2) the 'critical exponent β' depends on the temperature range over which the fit is extended. If we form $d/dT[\eta/(d\eta/dT)]$ from equation (10.4.2) we obtain $1/\beta$. Application of the same operation on

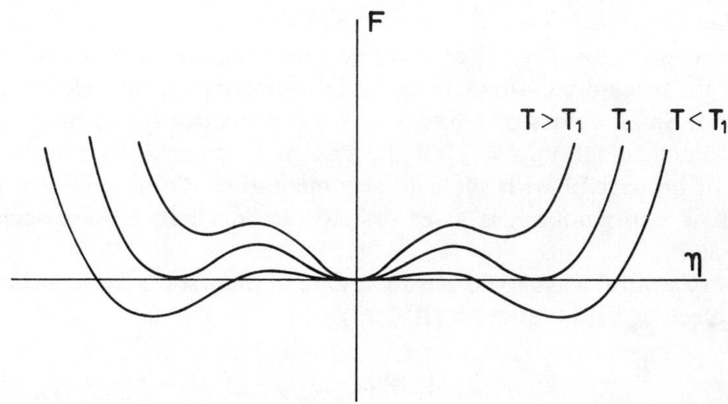

Figure 10.6 The free energy as a function of the order parameter in the vicinity of a first order phase transition. Convex parts of F belong to thermodynamically unstable states. The free energy is minimized by $\eta = 0$ above $T_1 = T_0 + 3b^2/16a'c$ and by a finite η below T_1

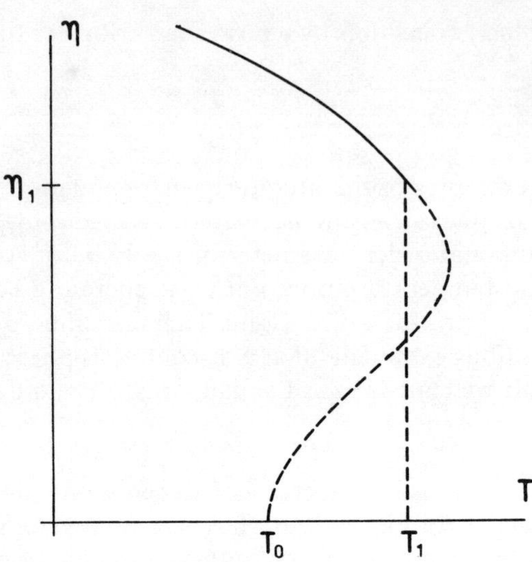

Figure 10.7 The temperature depencence of the order parameter corresponding to the free energy in Figure 10.6. The η values along the dotted parts of the curve do not lead into the absolute minimum of F

η_L yields

$$\mathrm{d}/\mathrm{d}T\left[\eta_L/(\mathrm{d}\eta_L/\mathrm{d}T)\right] = 4 + 2(1 - 4a'c(T - T_0)/b^2)^{-1/2}. \qquad (10.4.5)$$

That is, 'β' = 1/4 for very small T which is diminishing to 'β' = 1/6 at $T = T_0$ and further to 'β' = 1/8 for $T = T_0 + 3b^2/16a'c$.

The MFA result for β is 1/2 at a continuous transition, whereas the correct result for the three-dimensional Ising model as obtained from series expansions is $\beta = 0.31$. Small values of β like $\beta = 1/8$ are frequently attributed to a low effective dimensionality ($d = 2$) of the system. Our consideration shows that one has to be careful with such an argumentation. To be sure that a given transition is continuous it is safer first to check whether the susceptibility diverges at T_c.

There are several ways to derive MFA. One of them starts from the definition of the molecular field (equation (10.2.12)).

$$V_m = \sum_{j \neq i} \int W(\omega_i, \omega_j) f(\omega_j)\, \mathrm{d}\omega_j$$

To the molecular field we add the crystal field and an external field

$$V(\omega_i) = V_m(\omega_i) + V_c(\omega_i) + V_e(\omega_i) \qquad (10.4.6)$$

The distribution $f(\omega_i)$ then is calculated from a Boltzmann distribution

$$f(\omega_i) = \exp(-\beta V(\omega_i)) / \int \exp(-\beta V(\omega_i))\, d\omega_i \qquad (10.4.7)$$

The solution of these equations for $f(\omega)$ is the molecular field approximation. As mentioned before, in MFA the distribution functions of different particles are taken to be statistically independent, thus all correlations are neglected. The order parameter is defined from the expansion of $f(\omega)$ into symmetry adapted rotator functions (10.3.5b). If one or more of the coefficients $A^{(l)}_{mm'}$ are nonzero at low and zero at high temperatures (for vanishing $V_e(\omega)$) there is a phase transition and the lowest order $A^{(l)}_{mm'} = \eta(T)$ serves as an order parameter. The susceptibility is defined by $\lim_{V_e \to 0}(\partial \eta / \partial V_e)$. Within MFA the free energy is given by

$$F = -\frac{1}{\beta} \ln \int e^{-\beta V(\omega)}\, d\omega \qquad (10.4.8)$$

which permits a calculation of all thermodynamic quantities for example the entropy:

$$S = -k \int f(\omega) \ln f(\omega)\, d\omega \qquad (10.4.9)$$

MFA equations have been formulated for the case of ferro-ordering where all molecular groups order parallel. In the case of antiordering with n differently ordered sublattices one obtains n sets of equations which determine the value of the distribution functions $f_a(\omega_i)$ ($a = 1, 2, \ldots n$) on the n sublattices. A very elegant solution of such a problem is the James and Keenan theory of solid methane.[11]

So far the kinetic part of the rotational energy has been neglected. It becomes important for molecules with a small moment of inertia Θ, in which cases the quantum mechanical single particle problem has to be solved for the Hartree Hamiltonian:

$$\mathscr{H}_H = \frac{\hbar^2}{2\Theta} K + V(\omega_i). \qquad (10.4.10)$$

where $(\hbar^2/2\Theta)\, K$ is the kinetic energy. The molecular field now is defined by:

$$V_m = \sum_{j \neq i} \int W(\omega_i, \omega_j)\, \rho(\omega_j, \omega_j)\, d\omega_j \qquad (10.4.11)$$

where $\rho(\omega_j, \omega_j)$ is the diagonal element of the single particle density matrix $\rho(\omega_j, \omega_j')$.[34]

$$\rho(\omega_j, \omega_j) = \exp[-\beta\, \mathscr{H}_H(\omega_j)] / \mathrm{Tr}\{\exp[-\mathscr{H}_H(\omega_j)]\}. \qquad (10.4.12)$$

This quantum mechanical version of the MFA is also called Hartree approximation. It simultaneously averages over the thermal motions and the quantum fluctuations of the molecules. The importance of quantum effects for the light molecules can be estimated from the effect of deuteration on phase transition temperatures (36% in H_2–D_2, 22% in CH_4–CD_4).

A possible simplification of the MFA-equations has been discussed in Section 10.3. If $V_c(\omega) \gg kT$ and $V_c(\omega) \gg \hbar^2/2\Theta$, then the distribution function $f(\omega)$ may be replaced by δ-functions at the minima of the potential

$$f(\omega) = \frac{1}{N_\alpha} \sum_{m=1}^{N_\alpha} \delta(\omega - \omega_m) \tag{10.4.13}$$

where N_1 and N_2 are the numbers of minima above and below the transition. If (10.4.13) is substituted into (10.4.9) the transition entropy simplifies to the expression (10.3.1). A thorough discussion of the conditions for which (10.3.1) in meaningful has been given in Section 10.3.

Angle dependent forces in an assembly of molecules work towards orientational order, whereas central forces promote translational order, i.e. the crystalline state. Spherical molecules with small angle dependent interactions thus undergo the liquid–solid phase transition before orientational order sets in. For long chain molecules the reverse is true and one finds the phenomenon of the liquid crystalline state.[35] Pople and Karasz[36] (see also Chandrasekhar *et al.*[37]) have combined the Lennard–Jones and Devonshire[38] theory of melting with Ising-type models of orientational phase transitions. The L.–J. and D.-theory is based on a lattice model with two different types of lattice sites. The ordered state where all sites of one type are occupied is identified with the solid, the disordered state with the liquid. Depending on the relative strength of site dependent and spin dependent interactions the 'liquid–solid' phase transition may be above or below the 'orientational' transition. The model has only been solved in MFA, furthermore it is remote from the original Hamiltonian with respect to both the angle dependent and the central forces. A direct connection between the model parameters and the interaction constants is not possible and therefore quantitative conclusions should not be drawn. This is especially true for extensions of the Pople and Karasz theory for systems where several sets of inequivalent equilibrium orientations have to be invoked.[39] Problems with the interpretation of transition entropies in terms of such models have been discussed in Section 10.3.

10.5 ROTATIONAL EXCITATIONS

One of the most fascinating aspects of molecular crystals is the rotational dynamics and one easily could write a review on rotational motions alone. As, however, many questions with respect to this topic have been raised in papers reviewing experimental work (in this book), only a relatively brief description of this matter is given here. First, we shall discuss the single particle rotation in molecular solids. The experimental techniques most frequently applied to investigate single particle rotations in solids are incoherent inelastic neutron scattering and nuclear magnetic resonance. On the other hand IR- and Raman scattering as well as coherent neutron scattering yield information on collective excitations. Problems connected with the librational motion in orientationally ordered phases as well as theories on the critical dynamics at a phase transition will be discussed in the second part.

10.5.1 Single Particle Rotation

In order to characterize single particle rotation of molecules or molecular groups, we need to define a number of characteristic quantities:

(i) the rotational constant $B = \hbar^2/2\Theta$ of a molecule.
(ii) the static part of the angle dependent potential $V_{ST} = \overline{V(t)}$; the bar means a time average.
(iii) the fluctuating part of the potential $V_{FL}(t) = V(t) - V_{ST}$, the magnitude of which may be defined as $|V_{FL}| = [\overline{V_{FL}^2(t)}]^{1/2}$. $|V_{FL}|$ vanishes at $T = 0$ and increases with rising temperature.

Only the relative magnitude of the above quantities is important. We therefore introduce reduced quantities $V'_{ST} = V_{ST}/B$ and $V'_{FL} = |V_{FL}|/B$. Four different situations may be distinguished, depending on whether V'_{ST} or V'_{FL} are weak or strong, respectively.

Table 10.1 summarizes the various possibilities. Let us start by considering the situation at low temperatures (ordered phases). Usually it is characterized by strong potentials V_{ST} and only weak fluctuations. The excitations to be observed are librational modes, which in general are collective rotational excitations. Only to the extent to which the potential V_{ST} is a single particle potential (in the sense discussed in Section 10.2) librational excitations have the character of local excitations. An example for a system in which single particle librations have been observed is ND_4Cl:[40] Inelastic neutron scattering does not reveal a q-dependence of the mode energies, thus demonstrating the single particle character.

Table 10.1.

V'_{ST} \ V'_{FL}	Strong	Weak
Strong	(a) Rotational jumps	(b) Librations + rotational tunneling
Weak	(c) Rotational diffusion	(d) Quantum mechanical free rotation

Rotators in a periodic potential may show another interesting phenomenon which often is referred to as 'rotational tunneling'. Molecular wavefunction in adjacent potential wells may overlap considerably, which gives rise to a splitting of the librational ground state. In a number of recent NMR and neutron scattering experiments ground state splittings for protonated molecules or molecular groups in various solids have been observed: Some examples in neutron scattering are CH_4[41] (Figure 10.8) NH_4^+–ions in NH_4ClO_4[42] and $(NH_4)_2$ $SnCl_6$[43], CH_3-groups in γ-piccolene[44], DMA [45] and NH_3^+ groups in Ni $(NH_3)_6I_2$.[46] Observed energy splittings range between 1 μeV and 1 meV. It should be noted that transitions in the rotational ground state not only involve a change of symmetry of the rotational wave function, but also do change the symmetry of the nuclear spin function. This is due to the symmetry character of the total wave function under the exchange operator for two identical particles. As the spin species is unambigously labelled by the total nuclear spin of the molecule for H_2, XH_3, XH_4, transitions are accompanied by a spin flip of the molecule. Therefore, these transitions are 'forbidden' in the crystal. They may, however, be induced by a scattered neutron. Level schemes measured at low temperatures yield information on the single particle potential $V(\omega_i)$ of a molecular site. Theoretically the situation is well-described

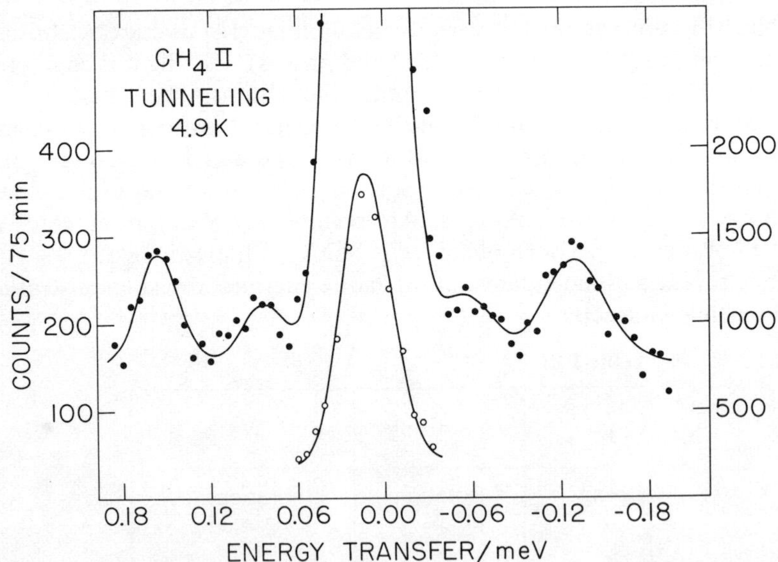

Figure 10.8 Inelastic spectrum of solid CH_4(II) (T = 4.9 K) measured by incoherent neutron scattering:[41] $Q = 1.4\ \mathring{A}^{-1}$; incoming energy 3.8 meV; energy resolution 42 μeV. The solid lines represent a computer fit; the right scale refers to the elastic peak

by stationary Hamiltonians. Exact solutions may be obtained in the case of one-dimensional rotators e.g. $-CH_3$ groups which are restricted to rotate around a single axis. If only a single Fourier component of the potential needs to be considered, the eigenvalue problem leads to Mathieu's equation (e.g. $V = V_3/2\ \cos 3\varphi$ for CH_3–groups). For two and three dimensional rotators the eigenvalue problem is somewhat more complex and numerical techniques must be applied (Figure 10.9). This is particularly true in an intermediate regime (between (b) and (d) in Table 10.1) where neither free rotator functions nor harmonic oscillator functions are suitable functions. The latter are the

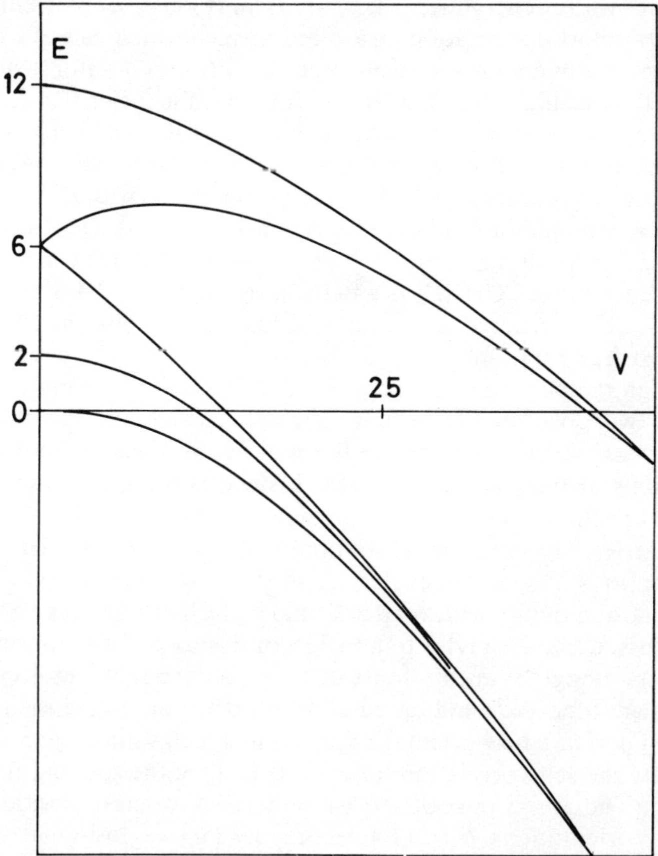

Figure 10.9 Low lying rotational energy levels of a tetrahedron in a tetrahedral field: $V_c(\omega) = V\, U^3_{11}(\omega)$. The potential strength V and the energy levels are given in units of the rotational constant $B = \hbar^2/2\theta$. $U^3_{11}(\omega)$ is a cubic rotator function which was defined in equation (10.2.8). The splitting of a tunnelling multiplet decreases exponentially with increasing V

appropriate wavefunctions in the limit of weak and strong potentials, respectively. Instead, in the intermediate regime, trial wavefunctions may be constructed (pocket state wavefunctions) the width of which is a variational parameter.[15] Such a variational approach is possible for one, two, and three-dimensional rotators and also produces the correct results in the limit of weak and strong potentials.

An alternative approach uses free rotator functions as basis functions.[47] This procedure is best-suited in the region of weak potentials where only few basis functions have to be included (regime labelled (b) in Table 10.1) and it breaks down for large potentials.

Whereas many examples exist in the range of intermediate or high potentials (for XH_4: activation energies ≤ 1 kcal/mol), only very few molecular crystals close to the limit of free rotation (in a quantum mechanical sense) are presently known. Free rotator energy eigenvalues are $(\hbar^2/2\Theta)J(J+1)$ for two-and three-dimensional molecules and $(\hbar^2/2\Theta)\ J^2$ for one-dimensional rotation. Solid hydrogen represents the most prominent example displaying almost free rotation: for the $J = 1 \rightarrow 0$ transition (ortho–para conversion) an energy $E_{1-0} = 14.2$ meV is observed, which has to be compared with $E^{\text{free}}_{1-0} = 14.8$ meV. Other known examples are solid CH_4(II), where 1/4 of the molecules are disordered and rotate almost freely[41,48] and γ-piccolene[44] with little hindrance to CH_3-group rotation. CH_4(II) is a peculiar example, as 3/4 of the molecules in the unit cell are ordered and rotational tunneling is observed[41] in addition to the almost free rotation.

In the high temperature regime, potential fluctuations dominate. We may distinguish two possibilities, namely the high barrier and low barrier limit. In the high barrier limit despite the thermal fluctuations in the crystal, librational motions of the molecules persist. Instead of rotational tunneling, however, there are thermally activated angular jumps of the molecules across the potential barrier. The jumps are reorientational motions between indistinguishable orientations of the molecules. A number of models dealing with the stochastic motion of molecules can be found in the literature. As not all of these can be discussed here, we refer to a review of this topic[49] and only mention the very simplest model[50,51] in this context. It is assumed that a molecule librates for an average time τ around an equilibrium orientation and then performes an angular jump to a new orientation in a time much shorter than τ. One then can calculate the self-correlation function $G_s(\mathbf{r}, t)$ of the rotating atoms within a molecule. The atoms pass through a number of discrete position $\mathbf{r}_i$ during successive reorientations. $G_s(\mathbf{r}, t)$ is decomposed into a positional part, (which still may contain the librational motion) and a conditional probability $p_i(t)$ of finding an atom at position i at time t, given that it was at position i_0 at $t = 0$. Assuming, that successive jumps are uncorrelated, $p_i(t)$ (e.g. the time evolution of the proton in a NH_4^+ group) can be described by a set of rate equations.

The incoherent neutron scattering law is now obtained by Fourier transformation of $G_s(\mathbf{r}, t)$. $S_{\text{ine}}(Q, \omega) = \int\int G_s(\mathbf{r}, t) \exp(\mathrm{i}\mathbf{Q}\mathbf{r} - i\omega t)\ \mathrm{d}\mathbf{Q}\mathrm{d}\omega$.

It represents the dominant part of the scattered intensity if one is dealing with protonated molecules. $S_{inc}(\mathbf{Q}, \omega)$ may be decomposed into a strictly elastic part, which originates from $p_i(t) \to$ const. $\neq 0$ for $t \to \infty$ and an inelastic part, which contains a sum of Lorentzians each weighted with a **Q**-dependent form factor. The energy width Γ of a Lorentzian only depends on the average time at which rotational jumps occur. It does not depend on **Q** as the width Γ for translational diffusion.

In the low barrier limit, that is in the absence of a time-independent potential, the molecules perform rotational diffusion. Expressions for the incoherent neutron scattering cross section have been derived by Sears.[52–54] It can be expressed in terms of an orientational self-correlation function $G_s(\omega_0, \omega; t)$ which for $V_{ST} = 0$ only depends on the difference of Euler angles $\omega - \omega_0$ (for one dimensional rotation: Janik *et al.*[55]) After a partial wave expansion of $G_s(\omega - \omega_0; t)$ into a series of rotator functions a scattering law

$$S(q, \omega) = \sum_{l=0}^{\infty} (2l + 1) j_l^2(\mathbf{Q}\rho)\, S_l(\omega)$$

is obtained. $S_l(\omega)$ is the Fourier transform of a relaxation function $F_l(t)$, with the initial condition $F_l(t = 0) = 1$ for all l. Under various model assumptions particularly simple results are obtained for spherical molecules. If use is made of a macroscopic diffusion equation

$$D_R \Delta\, G_s(\omega - \omega_0; t) = \partial G_s(\omega - \omega_0; t)/\partial t$$

for the orientational self-correlation function $G_s(\omega - \omega_0; t)$ one obtains

$$F_l(t) = \exp[-l(l + 1)D_R t]$$

where D_R is a diffusion constant. An exponential decay for small t is unphysical. This defect can be overcome by including an internal friction coefficient ζ which leads to a modified expression for 'rotational Brownian motion'

$$F_l(t) = \exp\{-l(l + 1)D_R[t - (1 - \exp(\zeta(t))/\zeta]\}$$

While there is no selection rule which singles out a particular order l in incoherent neutron scattering, in optics and NMR only orders $l = 1$ and $l = 2$, respectively, contribute.

The above treatment originally was performed in view of rotational motions in liquids, but later was applied to plastic crystals as well. A problem results from the fact that in most plastic crystals a sizeable crystalline field is present, which means that one is somewhere in between the high barrier and low barrier limit. What is needed is a theory which treats diffusion in a rotational potential. Recently one-dimensional translational Brownian motion in a periodic potential has been discussed.[56] Perhaps the rotational Brownian motion in a potential can be treated along similar lines.

Obviously one can imagine all possible intermediate situations with regard to the four possibilities tabulated in Table 10.1. We have already been going along the rows of Table 10.1. Similarly one can ask for the vertical lines connecting the limiting cases in this diagram. This is a particularly fascinating aspect of rotator solids since the transition from the quantum mechanical regime (low temperatures) to the corresponding diffusion limit can be studied in one system merely by changing temperature. This transition is continuous in systems with no 1st order phase transition at temperatures below about 100 K and at present is actively investigated both experimentally (Clough and Hill[57]) and theoretically (Allen,[58] Clough[59]). In systems without phase transition (Figure 10.10) the ground state splitting observed with neutrons strongly reduces in a narrow temperature range and at higher temperatures merges into quasi-elastic scattering, characteristic of diffusive motion. Attempts have been made to adapt a method developed by Anderson[60] and later reviewed by Abragam[61] to the temperature dependence of the rotational tunneling.[58] It consists of the calculation of response functions for a system with different characteristic frequencies and random (Markoffic type) transitions between them. Recently a theory has been put forward which attributes the reduction of the overlap to a coherent mixing of ground and excited state wave functions.[59]

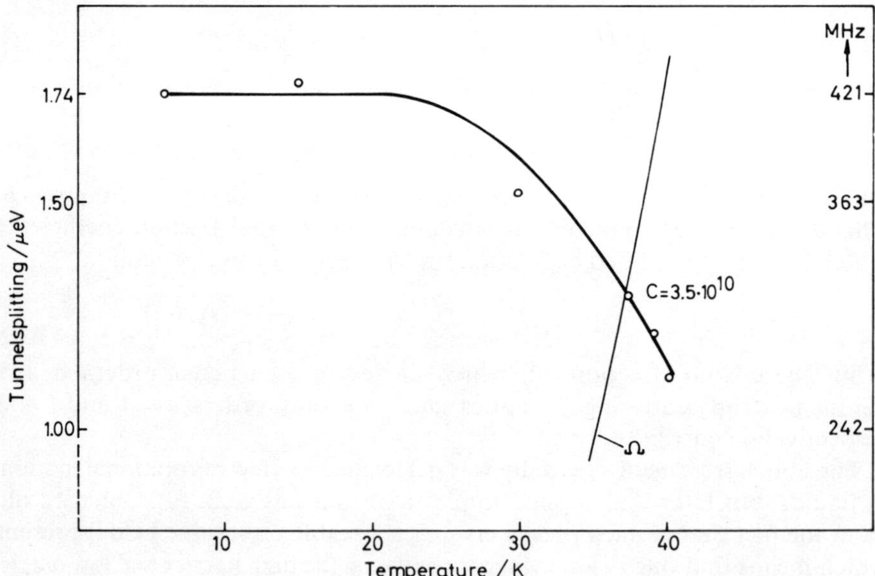

Figure 10.10 Alefeld and Kollmar[45] have measured the temperature dependence of the tunnel splitting in dimethylacetylene. The solid line is a fit to Allen's theory[58] which also predicts a width $\Omega(T)$ of the transition which does not agree with experiment

10.5.2 Collective Rotations

10.5.2(a) Orientationally-ordered Solids

There are much less experimental data on collective rotational excitations than there are on the single particle problem. This is also reflected in the number of theoretical works in this field. The development of these theories followed lines very similar to the ones for phonons in atomic solids. The starting point is the harmonic approximation and only recently theories including anharmonicities and large librational amplitudes of the molecules have been developed. A very helpful concept is the external mode approach,[62] where the degrees of freedom of molecular groups in solids are approximated by rigid-body motions. This means an important simplification and leads to a considerable reduction of the number of modes in a system with crystal periodicity. Only six external modes remain for each group in the primitive cell: 3 translational and 3 rotational modes. In the case of a XH_4 group, for example, this has to be compared with 15 modes in theories that start from displacements of single atoms. The external mode approach has been extensively reviewed by Venkataraman and Sahni.[62]

The authors follow the lines of a phenomenological Born-von-Karman theory. The dispersion of librational as well as of translational modes is calculated within the harmonic approximation. An expansion of the potential energy of the displacements (angular and positional) is truncated after the quadratic terms. It should be noted that translational displacements are represented by polar vectors and angular displacements by axial vectors. The Born–von Karman force constants of the external modes can be calculated from interatomic force constants extracted from phenomenological atom–atom potentials. Even if no set of potentials is available for a specific system, group theoretical methods can be used to reduce the number of independent elements in the dynamical matrix merely on the basis of symmetry. This aspect is discussed in great detail by Venkataraman and Sahni. It has found its application in a computer programme in which degeneracies of external modes at different points in the Brillouin zone can be calculated for a variety of structures.[63]

The external mode approach is rather general and the effect of non-rigidity of the molecules may be treated as a perturbation.[64] The harmonic approximation, however, is satisfactory for small displacements only. Angle-dependent periodic potentials in ordered molecular crystals quite naturally give rise to anharmonicities, particularly if there are large librational amplitudes. This is the situation encountered in systems with weakly interacting small molecules, but not only there. More adequate theories are required in this case.

One attempt is the time-dependent Hartree approximation or random phase approximation.[65,66] Here one calculates the linear response of the lattice to an external field $A(\omega, q)$ and then looks for singularities of the response function. Obviously there are excited states (librational modes) of frequency ω and wave-vector q in the systems if it can dissipate energy of an external disturbance $A(\omega, q)$. The excited state wavefunctions are obtained as linear combinations of single-particle excited state wavefunctions. The starting point of the time-dependent Hartree approximation clearly is a single particle picture. It does include anharmonicities in the sense that no expansion of the Hartree potential in the (angular) displacements with consequent truncation after the quadratic term is made. The theory does not, however, include a damping of librational excitations, due to libron–libron interactions. The time-dependent Hartree approximation has been successfully applied to a number of molecular crystals.[67–70] Another approach consists in the adaption of the self-consistent phonon theory[71,72] to the situation in molecular crystals.[73] In contrast to the time-dependent Hartree approximation one immediately concentrates on collective aspects. Anharmonicities (quartic terms) are included in the self-consistent phonon theory by modifying the force constants. These are no longer calculated from derivatives at the equilibrium positions or orientations, but are averaged over the displacements of the molecules (allowing for the thermal motion). By doing so displacement–displacement correlation functions are introduced and self-consistently constructed from eigenfrequencies and eigenvectors of the system.

10.5.2(b) Orientationally-disordered Solids

Recently a new approach towards the rotational dynamic in molecular crystals has been reported.[74] Its aim is a unified description of ordered and disordered phases with proper inclusion of anharmonicities. The secular variables are symmetry-adapted functions and angular momenta as conjugate quantities, standing for the orientation and the rotation of molecules, respectively. Zwanzig–Mori's[75,76] projection operator technique is used to derive response functions expressed in the secular variables and leads to a coupled set of dynamical matrix equations. Its application is not restricted to systems with partial order, but it also has been used to describe the critical dynamics in the vicinity of an orientational order–disorder phase transition. As a specific example it has been used to analyse the critical slowing down of orientational correlations in CD_4(I)[74] which was observed by coherent neutron scattering.[77]

NOTE ADDED IN PROOF

Since the material for this section was collected, considerable progress has been made in several fields of interest. We mention particularly the neutron incoherent scattering law for uniaxial rotational diffusion in a potential,[78] the treatment of coupled rotational and translational motion[79] of molecules, and the application of the method of weak graphs on the problem of steric hinderance.[80] There is no space for an adequate appreciation of the progress in these fields.

REFERENCES

1. H. Margenau, *Phys. Rev.*, **64**, 131 (1943).
2. A. E. Evett and H. Margenau, *Phys. Rev.*, **90**, 1021 (1953).
3. E. A. Mason and J. O. Hirschfelder, *J. Chem. Phys.*, **26**, 756 (1957).
4. A. I. Kitaigorodsky, *Molecular Crystals and Molecules*, Academic Press, New York, 1973.
5. T. B. MacRury and W.A. Steele, *J. Chem. Phys.*, **64**, 1288 (1976).
6. B. J. Berne and P. Pechukas, *J. Chem. Phys.*, **56**, 4213 (1972).
7. B. C. Carlson and G. S. Rushbrooke, *Proc. Cambridge Phil. Soc.*, **46**, 626 (1950).
8. J. O. Hirschfelder, C. F. Curtiss, and R. B. Bird, *Molecular Theory of Gases and Liquids*, Wiley, New York, 1954.
9. H. Yasuda, *Prog. Theor. Phys.*, **45**, 1361 (1971).
10. F. C. von der Lage and H. A. Bethe, *Phys. Rev.*, **71**, 612 (1947).
11. H. M. James and T. A. Keenan, *J. Chem. Phys.*, **31**, 12 (1959).
12. F. Hund, *Z. Phys.*, **51**, 1 (1928).
13. J. W. Foreman, Jr., *Ph.D. thesis, Purdue University* (1964).
14. A. Huller and J. W. Kane, *J. Chem. Phys.*, **61**, 3599 (1974).
15. A. Huller and D. M. Kroll, *J. Chem. Phys.*, **63**, 4495 (1975).
16. I. F. Silvera, *Low Temperature Physics, LT 14*, North-Holland, Amsterdam, 1975, Vol. 5, p. 123.
17. J. H. Constable and J. R. Gaines, *Solid State Comm.*, **9**, 155 (1971).
18. C. W. Garland and R. Renard, *J. Chem. Phys.*, **44**, 1120 (1966).
19. Y. Yamada, M. Mori, and Y. Noda, *J. Phys. Soc. Japan*, **32**, 1565 (1972).
20. A. B. Harris, J. Appl. Phys., **42**, 1574 (1971).
21. W. Press and A. Huller, *Anharmonic Lattices, Structural Transitions and Melting,* edited T. Riste, Noordhoff, Leiden, 1974.
22. G. B. Guthrie and J. P. McCulloch, *J. Phys. Chem. Solids*, **18**, 53 (1961).
23. T. Clark, M. A. McKervey, H. Mackle, and J. J. Rooney, *J. C. S. Faraday I,* **70**, 1279 (1974).
24. K. Kurki-Suonio, *Ann. Acad. Sci. Fenn.,* **A VI 241**, 1 (1967).
25. R. S. Seymour and A. W. Pryor, *Acta Cryst.*, **B26**, 1487 (1970).
26. W. Press and A. Huller, *Acta Cryst.*, **A29**, 257 (1973).
27. W. Press, *Acta Cryst.*, **A29**, 252 (1973) erratum *Acta Cryst.*, **A32**, 170 (1976).
28. E. Filter and W. Biem, unpublished.
29. W. Klein, D. J. Wallace, and R. K. P. Zia, to be published (1976).
30. W. Press and A. Huller, *Phys. Rev. Letters.*, **30**, 1207 (1973).
31. B. Dorner, J. D. Axe, and G. Shirane, *Phys. Rev.*, **B6**, 1950 (1972).
32. E. Banda, R. A. Craven, R. D. Parks, P. M. Horn, and M. Blume, *Sol. State Comm.*, **17**, 11 (1975).
33. J. P. Bacheimer and G. Dolino, *Phys. Rev.*, **B11**, 3195 (1975).
34. L. D. Landau and E. M. Lifshitz, *Quantum Mechanics*, Pergamon, London (1958).
35. G. W. Smith, *Adv. in Liquid Cryst.*, **1**, 189 (1975).
36. J. A. Pople and F. E. Karasz, *J. Phys. Chem. Solids*, **18**, 28 (1961).
37. S. R. Chandrasekhar, R. Shashidar, and N. Tara, *Mol. Cryst. Liq. Cryst.*, **10**, 337 (1970) and **12**, 245 (1971).

38. J. E. Lennard-Jones and A. F. Devonshire, *Proc. Roy. Soc.*, **170**, 464 (1939).
39. L. M. Amzel and L. N. Becka, *J. Phys. Chem. Solids*, **30**, 521 (1969).
40. H. C. Teh and B. N. Brockhouse, *Phys. Rev.*, **B3**, 2733 (1971).
41. W. Press and A. Kollmar, *Solid State Comm.*, **17**, 405 (1975).
42. B. Alefeld, and M. Prager, *J. Chem. Phys.*, **65**, 4927 (1976).
43. M. Prager, W. Press, B. Alefeld, and A. Huller, *J. Chem. Phys.*, **67**, 5126 (1977).
44. B. Alefeld, A. Kollmar, and B. Dasannacharia, *J. Chem. Phys.*, **63**, 4415 (1975).
45. B. Alefeld and A. Kollmar, *Proc. Gatlinburg Conf. on Neutron Research* (1976).
46. J. Eckert and W. Press, to be published (1976).
47. Y. Kataoka, K. Okeda, and T. Yamamoto, *Chem. Phys. Letters*, **19**, 365 (1973).
48. H. Kapulla and W. Glaser, *Inelastic Scattering of Neutrons in Solids and Liquids*, IAEA, Vienna (1973), p. 841.
49. T. Springer, *Springer Tracts in Modern Physics*, Springer–Verlag, Berlin, 1972, Vol. 64.
50. R. Stockmeyer and H. Stiller, *Phys. Stat. Sol.*, **27**, 269 (1968).
51. K. Skold, *J. Chem. Phys.*, **49**, 2443 (1968).
52. V. F. Sears, *Can. J. Phys.*, **44**, 1279 (1 966).
53. V. F. Sears, *Can. J. Phys.*, **44**, 1299 (1966).
54. V. F. Sears, *Can. J. Phys.*, **45**, 237 (1967).
55. J. A. Janik, J. M. Janik, K. Otnes, and K. Rosciszewski, *to be published* (1976).
56. P. L. Fulde, L. Pietronero, W. R. Schneider, and S. Strassler, *Phys. Rev. Letters*, **35**, 1776 (1975).
57. S. Clough ane J. R. Hill. *J. Phys. C*, **7**, L20 (1974).
58. P. S. Allen, *J. Phys. C*, **7**, L22 (1974).
59. S. Clough, to be published (1976).
60. P. W. Anderson, *J. Phys. Soc. Japan*, **9**, 316 (1954).
61. A. Abragam, *Principles of Nuclear Magnetism*, Oxford U.P. London 1961.
62. G. Venkataraman and V.C. Sahni, *Rev. Mod. Phys.*, **42**, 409 (1970).
63. T. G. Worlton and J. L. Warren, *Computer Physics Comm.*, **3**, 88 (1972).
64. G. Dolling, G. S. Pawley, and B. M. Powell, *Proc. Roy. Soc.*, **A333**, 363 (1973).
65. W. Brenig, *Z. Phys.*, **171**, 60 (1963).
66. D. R. Fredkin and N. R. Werthamer, *Phys. Rev.*, **138**, 1527 (1965).
67. A. Huller, *Phys. Rev.*, **B10**, 4403 (1974).
68. O. Schnepp and N. Jacobi, *Adv. Chem. Phys.*, **22**, 205, (1972).
69. N. Jacobi, *J. Chem. Phys.*, **57**, 2505 (1972).
70. P. V. Dunmore and D. A. Goodings, Can. J. Phys., **55**, 554 and 573 (1977).
71. N. R. Werthamer, *Am. J. Phys.*, **37**, 763 (1969).
72. N. R. Werthamer, *Phys. Rev.*, **B1**, 572 (1970).
73. J. C. Raich, N. S. Gillis, and A. B. Anderson, *J. Chem. Phys.*, **61**, 1399 (1974).
74. K. H. Michel and H. De Raedt, *J. Chem. Phys.*, **65**, 977 (1976).
75. R. W. Zwanzig, *J. Chem. Phys.*, **33**, 1388 (1960).
76. H. Mori, *Progr. Theoret. Phys.*, **23**, 423 (1965).
77. W. Press, A. Huller, H. Stiller, W. G. Stirling, and R. Currat, *Phys. Rev. Letters,* **32**, 1354 (1974).
78. A. J. Dianoux and F. Volino, *Molecular Physics*, **34**, 1263 (1977).
79. K. H. Michel and J. Naudts, *J. Chem. Phys.*, **67**, 547 (1977).
80. M. Descamps and G. Coulon, *Chem. Physics*, **19**, 347 (1977) and references therein.

Compound Index

Subject Index